Advancements in Non-Conventional Cooling and Thermal Storage Strategies

Advancements in Non-Conventional Cooling and Thermal Storage Strategies

Edited by

Bidyut Baran Saha
Kyushu University
Fukuoka, Japan

Dibakar Rakshit
Indian Institute of Technology Delhi
Delhi, India

WILEY

Published by John Wiley & Sons, Inc., Hoboken, New Jersey.
Published simultaneously in Canada.

Library of Congress Cataloging-in-Publication Data:

Names: Saha, Bidyut Baran, editor. | Rakshit, Dibakar, editor.
Title: Advancements in non-conventional cooling and thermal storage strategies / edited by Bidyut Baran Saha, Dibakar Rakshit.
Description: Hoboken, New Jersey : Wiley, [2024] | Includes bibliographical references and index.
Identifiers: LCCN 2024010515 (print) | LCCN 2024010516 (ebook) | ISBN 9781394189922 (hardback) | ISBN 9781394189939 (adobe pdf) | ISBN 9781394189946 (epub)
Subjects: LCSH: Cooling. | Heat storage devices. | Buildings–Energy conservation.
Classification: LCC TH7687 .A28 2024 (print) | LCC TH7687 (ebook) | DDC 697–dc23/eng/20240402
LC record available at https://lccn.loc.gov/2024010515
LC ebook record available at https://lccn.loc.gov/2024010516

Cover Design: Wiley
Cover Image: © Auttawit Jindaloung/Shutterstock

Set in 9.5/12.5pt STIXTwoText by Straive, Chennai, India

Contents

List of Contributors

E. Anil Kumar
Department of Mechanical Engineering
Indian Institute of Technology Tirupati
Tirupati
India

Hector Bastida
School of Engineering
Cardiff University
Cardiff
Wales
UK

Jorge Duarte Benther
School of Engineering
University of Tasmania
Hobart
Tasmania
Australia

Anutosh Chakraborty
School of Mechanical and Aerospace Engineering
Nanyang Technological University
Singapore

Akshay Chate
Department of Mechanical Engineering
Indian Institute of Science
Bangalore
India

Ivan De la Cruz-Loredo
School of Engineering
Cardiff University
Cardiff
Wales
UK

Susmita Dash
Department of Mechanical Engineering
Indian Institute of Science
Bangalore
India

Pradip Dutta
Department of Mechanical Engineering
Indian Institute of Science
Bangalore
India

and

Interdisciplinary Centre for Energy Research
Indian Institute of Science
Bangalore
India

Sana Fatima Ali
Department of Energy Science and Engineering
Indian Institute of Technology Delhi
New Delhi
India

Evan Franklin
School of Engineering
University of Tasmania
Hobart
Tasmania
Australia

Bo Han
School of Mechanical and Aerospace Engineering
Nanyang Technological University
Singapore

Md. Amirul Islam
International Institute for Carbon-Neutral Energy Research (WPI-I^{2}CNER)
Multiscale Science and Engineering for Energy and the Environment Thrust
Kyushu University
Fukuoka
Japan

and

Department of Electrical and Electronic Engineering
Bangabandhu Sheikh Mujibur Rahman Science & Technology University
Gopalganj
Bangladesh

Kartik Jain
Department of Mechanical Engineering
Indian Institute of Science
Bangalore
India

Daniel Morales
School of Engineering
Cardiff University
Cardiff
Wales
UK

Animesh Pal
International Institute for
Carbon-Neutral Energy Research
(WPI-I^2CNER)
Kyushu University
Fukuoka
Japan

and

Department of Nuclear Engineering
University of Dhaka
Dhaka
Bangladesh

Dibakar Rakshit
Department of Energy Science and
Engineering
Indian Institute of Technology Delhi
New Delhi
India

Bidyut Baran Saha
International Institute for
Carbon-Neutral Energy Research
(WPI-I^2CNER)
Kyushu University
Fukuoka
Japan

and

Department of Mechanical
Engineering
Kyushu University
Fukuoka
Japan

Pranaynil Saikia
School of Engineering
Cardiff University
Cardiff
Wales
UK

Arslan Saleem
School of Engineering
Cardiff University
Cardiff
Wales
UK

Rahul Kumar Sharma
Department of Energy Science and
Engineering
Indian Institute of Technology Delhi
New Delhi
India

Rakesh Sharma
Department of Mechanical
Engineering
National Institute of Technology
Raipur
Raipur
India

Carlos E. Ugalde-Loo
School of Engineering
Cardiff University
Cardiff
Wales
UK

Rahul Verma
Department of Energy Science and Engineering
Indian Institute of Technology Delhi
New Delhi
India

Xiaolin Wang
School of Engineering
University of Tasmania
Hobart
Tasmania
Australia

Sai Yagnamurthy
School of Engineering
University of Warwick
Coventry
UK

Preface

Amid an ever-expanding world where buildings stand as symbols of human progress and achievement, we must address the critical issues of energy consumption, sustainability, and environmental responsibility. The nexus of energy efficiency, environmental responsibility, and cutting-edge technology has led us to explore new horizons in cooling and thermal storage strategies. This book, *Advancements in Non-Conventional Cooling and Thermal Storage Strategies*, emerges as an essential compendium for those who seek to comprehend and engage with the transformative developments in this dynamic field.

The urgency of our current global climate crisis necessitates a shift away from conventional cooling and thermal storage methods toward innovative and eco-conscious alternatives. As we face growing energy demands, shifting climate patterns, and the realization that traditional air conditioning systems are not only resource-intensive but also detrimental to our environment, the exploration of non-conventional solutions becomes paramount.

The chapters that follow delve into various facets of non-conventional cooling, thermal storage, and space conditioning, covering a broad spectrum of topics, including passive cooling strategies, renewable energy integration, and thermal storage strategies. The authors, experts in their respective fields, share their expertise and experiences to provide a holistic understanding of these emerging technologies. These innovative approaches aim to mitigate the environmental impact of space conditioning, reduce energy consumption, and address the challenges posed by climate change. We explore the importance, fundamental concepts, categories, and application areas of these strategies, with a particular focus on absorption and adsorption technologies, as well as thermal energy storage systems. Each chapter is designed to offer not only theoretical insights but also practical applications, making this book a valuable resource for both novices and seasoned practitioners.

The chapters within this book present a comprehensive exploration of various aspects of non-conventional cooling and thermal storage, ranging from the utilization of adsorbents to the development of metal–organic frameworks and the synthesis of phase-change materials. Each chapter represents a unique contribution to the field, offering fresh perspectives and innovative solutions to the challenges posed by traditional HVAC systems.

Furthermore, thermal energy storage (TES) systems have the potential to reshape the future of space conditioning by storing excess thermal energy for later use. These systems enhance the reliability of renewable energy sources, reduce peak grid loads, and help balance energy supply and demand. TES systems encompass sensible heat storage, latent heat storage, and thermochemical energy storage, each offering distinct benefits and applications.

In the pages that follow, readers will embark on a journey through the exciting world of non-conventional cooling and thermal storage strategies, unlocking the potential for a brighter and more sustainable future. We invite you to explore, learn, and participate in the transformation of how we condition our spaces and store thermal energy in a way that is not only technologically advanced but also environmentally responsible. The book you hold in your hands is your gateway to the next generation of space conditioning.

This book is more than just a collection of chapters; it is a collaborative effort that reflects the dedication and passion of researchers, engineers, and the scientific community who have recognized the urgency of addressing the energy crisis and environmental concerns. The research trends depicted in this book highlight the growing interest in non-conventional cooling and thermal energy storage, demonstrating its potential to reshape the future of space conditioning.

The journey from conventional to non-conventional cooling and thermal storage is challenging but essential. It is a testament to human ingenuity, resilience, and the spirit of collaboration. We are profoundly grateful to the Scheme for Promotion of Academic and Research Collaboration (SPARC) for their generous support, which has empowered us to embark on this journey, explore uncharted territories, and share our findings with the world.

We thank all the authors who have contributed to this book. We wish to express our sincere thanks to the reviewers' services for their constructive comments, which further augmented the quality of the chapters chosen for this edited book.

We extend an invitation to you to embark on this journey of exploration and discovery into the realm of non-conventional cooling and thermal storage. We trust that the insights presented in the ensuing chapters will inspire a new era of sustainable and energy-efficient space conditioning solutions, thereby contributing to the well-being of our planet and its inhabitants.

13 June 2024

Bidyut Baran Saha
Fukuoka, Japan

Dibakar Rakshit
New Delhi, India

1

Introduction to Advancements in Non-Conventional Cooling and Thermal Storage Strategies: Technologies for More Sustainable Space Conditioning

Animesh Pal[1,2], Bidyut Baran Saha[1,3], and Dibakar Rakshit[4]

[1] *International Institute for Carbon-Neutral Energy Research (WPI-I2CNER), Kyushu University, Fukuoka, Japan*
[2] *Department of Nuclear Engineering, University of Dhaka, Dhaka, Bangladesh*
[3] *Department of Mechanical Engineering, Kyushu University, Fukuoka, Japan*
[4] *Department of Energy Science and Engineering, Indian Institute of Technology Delhi, New Delhi, India*

1.1 Background

Buildings, encompassing residential, commercial, and industrial structures, play a significant role in global energy consumption patterns. As our world becomes increasingly urbanized, understanding and optimizing the energy use of these buildings become crucial for sustainability, environmental responsibility, and economic efficiency. Space cooling is a notable driver of energy consumption, accounting for about one gigaton of CO_2 emissions and approximately 5% of global energy usage in 2020. The worldwide building floor area is projected to surge by 75% from 2020 to 2050, with emerging and developing economies contributing to 80% of this growth [1]. This growth highlights a pivotal moment in implementing strategies that reduce cooling energy consumption through bio-climatic architecture and adopting passive building design approaches. Furthermore, it is observed that energy consumption for space cooling has seen a twofold increase since 2000, escalating from 1000 terawatt-hours to 1945 terawatt-hours [2]. This surge can be attributed to rising temperatures, swift urban expansion, increased accessibility and ownership of air-conditioning units, and the prevalence of inefficient cooling systems. In a scenario aiming for net-zero energy consumption, the energy required for space cooling has the potential to be cut by half by 2050 when contrasted with the existing policy trajectory.

The global temperatures are rising, and it can be observed from Figure 1.1 that the average temperature is higher for the countries in between the tropics. Thus, it is evident from Figure 1.2 that in 2020, the maximum number of air conditioners

Advancements in Non-Conventional Cooling and Thermal Storage Strategies, First Edition.
Edited by Bidyut Baran Saha and Dibakar Rakshit.

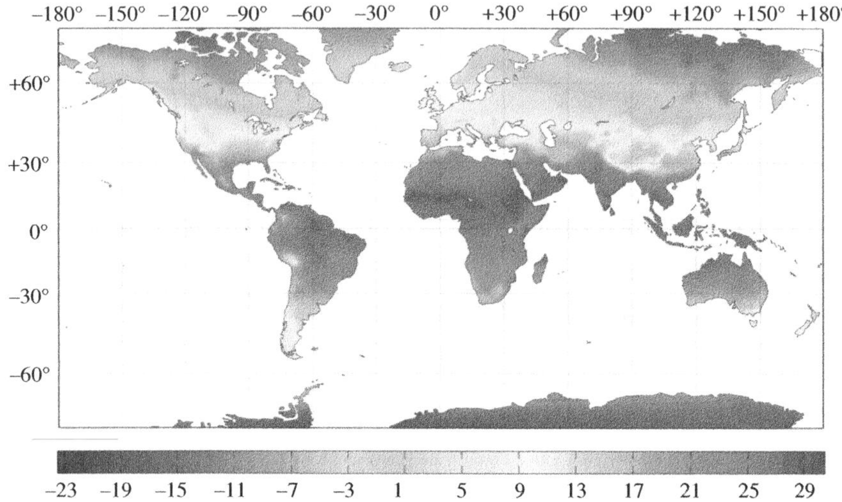

Figure 1.1 Global distribution of annual mean temperature. Source: Mourshed [3]/with permission of Elsevier.

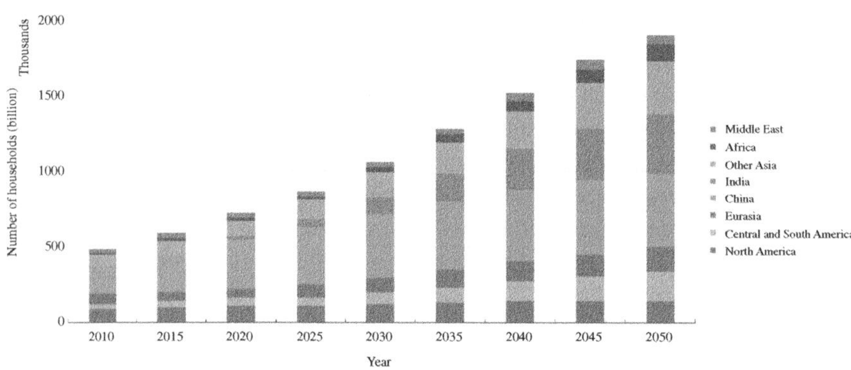

Figure 1.2 Number of homes fitted with air-conditioning across regions from 2010 to 2050. Source: Adapted from IEA [4].

was installed in developing nations. With the rising income levels and increased affordability, the penetration of air conditioners will be more significant in under-developed and developing nations, and as per IEA, the number of households will double by 2050 [4]. As the world grapples with the challenges of climate change and increasing energy demands, the need for sustainable and energy-efficient space-conditioning solutions has become paramount. Conventional HVAC (Heating, Ventilation, and Air-Conditioning) systems contribute significantly to energy consumption and greenhouse gas emissions.

These emissions and energy demands are expected to increase further, as with the increasing income levels of developing nations, the number of air conditioners per household is expected to increase to 2.1 by 2030 compared to the current status of 0.6 [5]. With the growing concerns about climate change and the need for energy conservation, there is a pressing need to explore non-conventional cooling and thermal storage strategies.

The non-conventional (sorption) cooling technologies can be driven by thermal energy. The sorption cooling system can be simply classified into absorption (liquid–gas) and adsorption (solid–gas), which are shown in Figure 1.3. Adsorption (solid–gas) is a surface phenomenon where gas molecules, atoms, or ions adhere to the surface of a solid material. In adsorption, the adsorbate (substance being adsorbed) accumulates or concentrates on the surface of the adsorbent (the material on which adsorption occurs) rather than entering the bulk of the adsorbent. Adsorption processes can be classified into physical adsorption (physisorption) or chemical adsorption (chemisorption). Physical adsorption is relatively weak intermolecular forces that are caused by van der Waals forces between the adsorbate and the adsorbent's surface. It is a reversible process. The adsorbate can be easily desorbed by changing temperature or pressure. However, chemical adsorption involves the formation of chemical bonds between the adsorbate and adsorbent. Chemisorption is a more potent and specific form of adsorption. These chemical bonds are typically covalent or ionic in nature. Chemisorption is usually not a reversible process and requires the breaking of chemical bonds to desorb the adsorbate. Generally, the main difference between adsorption and absorption is the sorbent pair's nature and the cycle duration time. It is stated that the performance of the absorption cycle is higher than that of the adsorption cycle. However, the absorption cycle needs a high-temperature heat source to operate. Absorption cycles are always closed cycles. Adsorption cycles, however, can be split into open and closed cycles. Both desiccant cooling and dehumidification applications use the open cycle. Silica gel, calcium chloride, and lithium bromide are generally used as desiccants. Depending on the kind of adsorbent–adsorbate (refrigerant) used in the system and the application, closed adsorption systems can be divided into high-pressure (above atmospheric pressure) and low-pressure (subatmospheric pressure) categories. It should be noted that practically all solid adsorbents employed in adsorption cooling systems (ACSs) exhibit physical adsorption. The most commonly employed adsorbent + adsorbate pair include activated carbon (AC) + ethanol/CO_2/methanol/ammonia [6–17], silica gel + water [18–21], metal–organic frameworks (MOFs) + water/ethanol/CO_2 [22–26], zeolite + water/CO_2 [27–29], composite + ethanol/CO_2 [30–40], and AC fiber + ethanol/methanol [41–43].

Thermal energy storage (TES) systems have the potential to revolutionize space conditioning by enabling the storage of excess thermal energy for later use. These

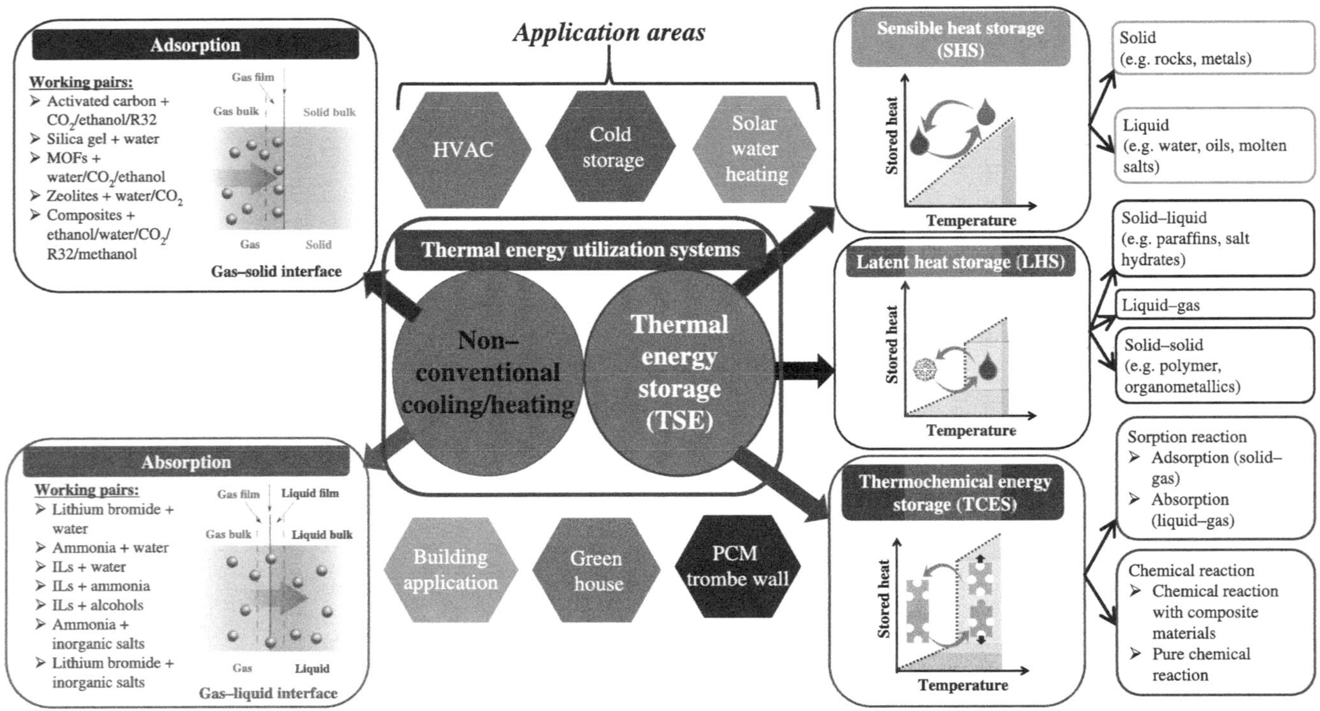

Figure 1.3 Schematic diagram of the classifications of sorption systems, TES systems, and their application ranges.

systems can help balance energy demand and supply, reduce peak load on the grid, and enhance the reliability of renewable energy sources. TES systems have been used for district heating and cooling purposes. Studies have shown that integrating a TES with a building can store thermal energy up to 93.2 MW, thereby reducing the heat requirements by 14.8% [44]. Furthermore, incorporating cold TES reduces energy imprint by reducing electricity consumption along with reducing the size of the cooling system [45]. Depending on the energy storage mechanism, the TES system can be classified into three main categories: sensible heat storage (SHS), latent heat storage (LHS), and thermochemical energy storage (TCES), as shown in Figure 1.3. SHS systems make use of specific heat capacity of materials (solid or liquid) to store heat energy without inducing a change in the material's phase. Depending on the temperature of the heat source, commonly employed materials for these systems include rocks, metals, water, thermal oils, molten salts, and liquid metals. The heat storage capacity of these systems is directly proportional to the specific heat, temperature change, and amount of the chosen material [46, 47]. In contrast, LHS entails the process of heating a material until it reaches a critical temperature, causing it to undergo a phase transition. This transition can involve changing from a solid to a solid or a solid to a liquid, or from a liquid to a gas. When the material reaches this phase change temperature, it absorbs a significant quantity of heat energy to facilitate the transition. This absorption of heat is referred to as the latent heat of fusion or vaporization depending on the specific case, and it serves as a means to store energy. Phase change materials (PCMs) have gained attention as an innovative way to efficiently store and release thermal energy. These materials undergo a phase transition, from solid to liquid, liquid to vapor, or vice versa, at a specific temperature range. This property allows them to absorb and release heat energy during the phase change. Furthermore, the magnitude of stored energy density in latent heat is much higher than in sensible heat [48]. For solid–liquid heat storage systems, a solid material like paraffin wax or salt hydrates is employed for heat storage. In solid–solid systems, polymers or organometallics are used to store heat. In the case of liquid–gas systems, a liquid that can undergo vaporization at relatively low temperatures and pressures, such as propane, alcohols, paraffins, esters, ammonia, or water, is utilized for heat storage [46, 47]. TCES systems comprise storing and releasing heat through chemical reactions, including both exothermic (heat-releasing) and endothermic (heat-absorbing) processes. These systems offer the advantage of storing a substantial amount of energy within a relatively compact space, which can be beneficial in various applications. Moreover, their ability to retain energy over extended durations makes them valuable for load management and peak demand reduction tasks. To minimize environmental impact, these systems can employ nontoxic, noncorrosive, and nonflammable materials, and they do not produce greenhouse gas emissions [46, 47].

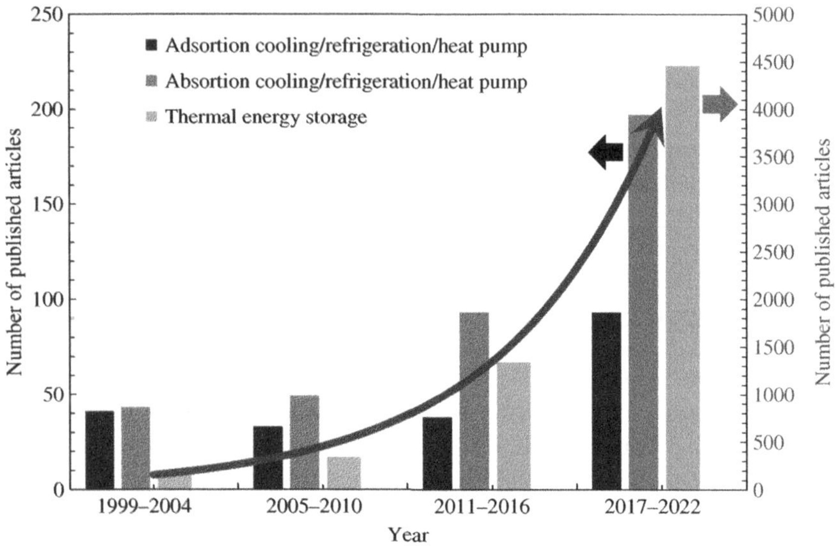

Figure 1.4 Research trends in adsorption cooling/refrigeration/heat pump, absorption cooling/refrigeration/heat pump, and thermal energy storage fields. Source: SCOPUS.

Due to the energy crisis and environmental concerns, there has been a growing interest among engineers, researchers, and the scientific community in exploring alternative methods for cooling, heating, and TES over the past few decades. Figure 1.4 illustrates the research trends in the areas of absorption-based cooling, refrigeration, heat pumps, and TES. This data was extracted from the SCOPUS database through a search using "article title" and specific keywords, including "absorption AND cooling OR absorption AND refrigeration OR absorption AND heat pump," "adsorption AND cooling OR adsorption AND refrigeration OR adsorption AND heat pump," and "thermal AND heat OR energy AND storage." It is evident from the graph that the number of published articles in these fields began to surge after 2010, with TES research experiencing remarkably rapid growth compared to the other two areas.

1.2 Key Contribution of Each Chapter

This book covers topics such as fundamentals of non-conventional heating and cooling technologies, a thorough review of the research conducted on such strategies, energy storage technologies for space-conditioning applications through both active and passive means, latest developments in cooling technologies with more focus on the adsorption cooling, exploration of refrigerant and adsorbent materials for adsorption-based space-conditioning applications, study of their adsorption characteristics and performance, design and optimization of heat exchangers, and related experimentation studies. The key contribution of each chapter is summarized as follows.

Chapter 2 explores the potential of hydrofluoro refrigerants of R134a, R32, R410A, and R1234ze(E) with Maxsorb III AC as an adsorbent for adsorption cooling applications. The system's performance has been studied for the hottest month of June for Delhi's, the capital city of India, and local weather conditions. Furthermore, the performances of stand-alone and vapor compression refrigeration (VCR) cascaded various configurations have been studied. The authors observed that the working pair of Maxsorb III + R1234ze(E) is seen to yield the best cooling performance among the considered pairs in terms of coefficient of performance (COP), specific cooling effect (SCE), and lower regeneration temperatures. Furthermore, a COP enhancement of over 60% is estimated for the VCR system with cascaded configuration B, where the compressor follows the ACS in the refrigerant path, with the working pair of Maxsorb III + R1234ze(E). The authors claim that the findings of this study are anticipated to pave the way for further design and optimization of the hydrofluoro refrigerant-based ACSs to contribute toward the climate goals.

Chapter 3 presents the synthesis technique of highly microporous with extremely high surface area and pore volume containing ACs from waste mangroves and waste palm trunks. The authors also assess the adsorption characteristics of R32 refrigerant, its thermodynamic properties, and the environmental impact of these systems. The authors claim that the experimental data of the best sample exhibits an extremely high uptake (2.677 kg/kg) of R32, which is a benchmark for any HFC/AC pairs ever reported. Experimentally obtained isotherm data are correlated with well-known Dubinin–Astakhov (DA) and Tóth models with an acceptable error margin. The findings of this chapter show their (M-AC + R32 and P-AC + R32) great potential to build an energy-efficient and environmentally benign ACS.

Chapter 4 focuses on the fabrication and functionalization of UiO-66 (Zr) and MIL-125 (Ti) MOFs with various functional groups. Later, this chapter investigates the porous properties and water adsorption capacity of these synthesized MOFs under static and dynamic conditions. According to the findings, the functionalized MOFs outperform the original MOF in terms of enhanced adsorption cooling performance, more water transfer per operating cycle, faster water adsorption/desorption kinetics, and higher thermal energy storage density (TESD). The chiller designed with functionalized MOFs provides relatively higher specific cooling power (SCP) and the COP. Furthermore, the modified MOFs exhibit higher TESD up to 1.06 MJ/l.

Chapter 5 presents the development of ammonia and metal halide-based various thermochemical devices for cooling, heating, power generation, and energy storage. It is reported that the availability of a wide variety of halide salts interacting with ammonia as a natural refrigerant at different operating conditions encourages their usability with solar thermal energy and other sources of waste heat like industry or automobile. The ammoniated-salt sorption system can be developed by coupling halide salt with an ammonia tank, termed an adsorption process, or by linking two halide salts between which ammonia is exchanged, called the resorption process.

Chapter 6 focuses on various aspects of the gas–solid pair-based TCES systems. It includes various classifications of available thermochemical storage materials (TSMs), inherent characteristics, and significant challenges and potential applications. This chapter also highlights the critical aspects affecting the performance of respective chemical reactions of TSMs, which encompass concepts such as solid–gas reaction equilibrium, reaction kinetics, reaction rate expressions, etc. The thermodynamic analysis methodology is discussed with the help of a case study of a TESS designed for performing numerous heating applications in cold climates. Advantages and challenges related to different reactor configurations used in the practical TESS are also presented.

Chapter 7 presents an innovative methodology for the quantification of the state-of-charge (SoC) of TES based on sensible heat and latent heat. The method consists of three main steps, leading to a SoC estimation where the need for disrupting instrumentation is reduced. The performance of the methodology is demonstrated with three examples, where simulation results are compared with experimental data.

Chapter 8 outlines adsorption cooling qualitatively, covering its basic concepts, mathematical modeling, performance enhancements, applications, and a discussion and potential prospects. The fundamental concepts of adsorption cooling have been covered, including one-bed, two-bed, multi-bed, and multistage configurations, providing a solid foundation for understanding these systems' operational principles and design considerations. Mathematical models encompass the fundamental equations governing mass and heat transfer, including adsorption isotherms, providing a foundation for system analysis and performance prediction. Moreover, performance enhancement strategies are discussed, including mass and heat recovery and combined mass–heat methods. Furthermore, various practical applications of adsorption cooling are explored, including air-conditioning, refrigeration, waste heat utilization, and solar cooling. This chapter also mentions several intriguing possibilities for future research in adsorption cooling. Creating advanced adsorbent materials with improved adsorption properties and thermal stability holds promise for improving system performance. Adsorption cooling, integration with renewable energy sources and direct air capture (DAC), and advanced heat exchanger designs may improve efficiency and sustainability.

Chapter 9 comprehensively reviews existing studies regarding passive thermal conditioning through PCMs. It presents a PCM identification approach for passive air-conditioning based on its thermophysical characteristics and meteorological parameters suitable for specific climatic conditions. The utilization of a PCM air heat exchanger is being investigated as a viable strategy for enhancing the energy efficiency of buildings by lowering their internal temperatures within the framework of active air-conditioning. It discusses what happens when a PCM air heat exchanger is installed inside an air-conditioning unit to cool the fraction of fresh air. This chapter mentions all the possible effects of PCM for building applications, such as the position of the PCM layer, location and its weather, thickness of PCM, thermal comfort conditions, and regions with composite climates.

The integration of non-conventional cooling and thermal storage strategies into smart building technologies is essential for optimizing energy use and providing occupants with a comfortable indoor environment. While non-conventional cooling and thermal storage technologies offer immense promise, they also face challenges and barriers to widespread adoption. These challenges include high initial costs, regulatory hurdles, and the need for public awareness and education.

References

1 IEA(c) (2022). Global Energy and Climate Model. 129 p.

2 IEA (2021). India Energy Outlook 2021. IEA, Paris. https://www.iea.org/reports/india-energy-outlook-2021, Licence: CC BY 4.0.

3 Mourshed, M. (2016). Climatic parameters for building energy applications: a temporal-geospatial assessment of temperature indicators. *Renewable Energy* 94: 55–71. https://doi.org/10.1016/j.renene.2016.03.021.

4 IEA (2020). Is cooling the future of heating? IEA, Paris. https://www.iea.org/commentaries/is-cooling-the-future-of-heating.

5 IEA (2021). World Energy Outlook 2021. IEA, Paris. https://www.iea.org/reports/world-energy-outlook-2021.

6 Singh, V.K. and Kumar, E.A. (2017). Experimental investigation and thermodynamic analysis of CO_2 adsorption on activated carbons for cooling system. *Journal of CO_2 Utilization* 17: 290–304. https://doi.org/10.1016/j.jcou.2016.12.004.

7 El-Sharkawy, I.I., Uddin, K., Miyazaki, T. et al. (2014). Adsorption of ethanol onto parent and surface treated activated carbon powders. *International Journal of Heat and Mass Transfer* 73: 445–455. https://doi.org/10.1016/j.ijheatmasstransfer.2014.02.046.

8 Jribi, S., Miyazaki, T., Saha, B.B. et al. (2017). Equilibrium and kinetics of CO_2 adsorption onto activated carbon. *International Journal of Heat and Mass*

Transfer 108: 1941–1946. https://doi.org/10.1016/j.ijheatmasstransfer.2016.12.114.

9 Pal, A., Thu, K., Mitra, S. et al. (2017). Study on biomass derived activated carbons for adsorptive heat pump application. *International Journal of Heat and Mass Transfer* 110: 7–19. https://doi.org/10.1016/j.ijheatmasstransfer.2017.02.081.

10 Pal, A., Uddin, K., Saha, B.B. et al. (2020). A benchmark for CO_2 uptake onto newly synthesized biomass-derived activated carbons. *Applied Energy* 264: 114720. https://doi.org/10.1016/j.apenergy.2020.114720.

11 Uddin, K., Pal, A., and Saha, B.B. (2020). Improved CO_2 adsorption onto chemically activated spherical phenol resin. *Journal of CO_2 Utilization* 41: 101255. https://doi.org/10.1016/j.jcou.2020.101255.

12 Jribi, S., Saha, B.B., Koyama, S., and Bentaher, H. (2014). Modeling and simulation of an activated carbon-CO_2 four bed based adsorption cooling system. *Energy Conversion and Management* 78: 985–991. https://doi.org/10.1016/j.enconman.2013.06.061.

13 El-Sharkawy, I.I., Saha, B.B., Koyama, S. et al. (2008). Experimental investigation on activated carbon-ethanol pair for solar powered adsorption cooling applications. *International Journal of Refrigeration* 31: 1407–1413. https://doi.org/10.1016/j.ijrefrig.2008.03.012.

14 Pal, A., Rocky, K.A., and Saha, B.B. (2021). Thermodynamic analysis of promising biomass-derived activated carbons/CO_2 based adsorption cooling systems. *Journal of CO_2 Utilization* 46: 101457. https://doi.org/10.1016/j.jcou.2021.101457.

15 Gordeeva, L. and Aristov, Y. (2014). Dynamic study of methanol adsorption on activated carbon ACM-35.4 for enhancing the specific cooling power of adsorptive chillers. *Applied Energy* 117: 127–133. https://doi.org/10.1016/j.apenergy.2013.11.073.

16 El-Sharkawy, I.I., Hassan, M., Saha, B.B. et al. (2009). Study on adsorption of methanol onto carbon based adsorbents. *International Journal of Refrigeration* 32: 1579–1586. https://doi.org/10.1016/j.ijrefrig.2009.06.011.

17 Wang, L.W., Wu, J.Y., Wang, R.Z. et al. (2003). Study of the performance of activated carbon–methanol adsorption systems concerning heat and mass transfer. *Applied Thermal Engineering* 23: 1605–1617. https://doi.org/10.1016/S1359-4311(03)00104-2.

18 Ng, K.C., Chua, H.T., Chung, C.Y. et al. (2001). Experimental investigation of the silica gel-water adsorption isotherm characteristics. *Applied Thermal Engineering* 21: 1631–1642. https://doi.org/10.1016/S1359-4311(01)00039-4.

19 Mitra, S., Kumar, P., Srinivasan, K., and Dutta, P. (2015). Performance evaluation of a two-stage silica gel + water adsorption based cooling-cum-desalination

system. *International Journal of Refrigeration* 58: 186–198. https://doi.org/10.1016/j.ijrefrig.2015.06.018.

20 Wang, D., Zhang, J., Yang, Q. et al. (2014). Study of adsorption characteristics in silica gel–water adsorption refrigeration. *Applied Energy* 113: 734–741. https://doi.org/10.1016/j.apenergy.2013.08.011.

21 Saha, B.B., Akisawa, A., and Kashiwagi, T. (1997). Silica gel water advanced adsorption refrigeration cycle. *Energy* 437–447. https://doi.org/10.1016/S0360-5442(96)00102-8.

22 Palash, M.L., Jahan, I., Rupam, T.H. et al. (2020). Novel technique for improving the water adsorption isotherms of metal-organic frameworks for performance enhancement of adsorption driven chillers. *Inorganica Chimica Acta* 501: 119313. https://doi.org/10.1016/j.ica.2019.119313.

23 Rezk, A., Al-Dadah, R., Mahmoud, S., and Elsayed, A. (2013). Investigation of Ethanol/metal organic frameworks for low temperature adsorption cooling applications. *Applied Energy* 112: 1025–1031. https://doi.org/10.1016/j.apenergy.2013.06.041.

24 Kayal, S., Chakraborty, A., and Teo, H.W.B. (2018). Green synthesis and characterization of aluminium fumarate metal-organic framework for heat transformation applications. *Materials Letters* 221: 165–167. https://doi.org/10.1016/j.matlet.2018.03.099.

25 Henninger, S.K., Habib, H.A., and Janiak, C. (2009). MOFs as adsorbents for low temperature heating and cooling applications. *Journal of the American Chemical Society* 131: 2776–2777. https://doi.org/10.1021/ja808444z.

26 Solovyeva, M.V., Gordeeva, L.G., Krieger, T.A., and Aristov, Y.I. (2018). MOF-801 as a promising material for adsorption cooling: equilibrium and dynamics of water adsorption. *Energy Conversion and Management* 174: 356–363. https://doi.org/10.1016/j.enconman.2018.08.032.

27 Kayal, S., Baichuan, S., and Saha, B.B. (2016). Adsorption characteristics of AQSOA zeolites and water for adsorption chillers. *International Journal of Heat and Mass Transfer* 92: 1120–1127. https://doi.org/10.1016/j.ijheatmasstransfer.2015.09.060.

28 Solmuş, I., Yamali, C., Kaftanoğlu, B. et al. (2010). Adsorption properties of a natural zeolite-water pair for use in adsorption cooling cycles. *Applied Energy* 87: 2062–2067. https://doi.org/10.1016/j.apenergy.2009.11.027.

29 Goldsworthy, M.J. (2014). Measurements of water vapour sorption isotherms for RD silica gel, AQSOA-Z01, AQSOA-Z02, AQSOA-Z05 and CECA zeolite 3A. *Microporous and Mesoporous Materials* 196: 59–67. https://doi.org/10.1016/j.micromeso.2014.04.046.

30 Pal, A., Uddin, K., Thu, K., and Saha, B.B. (2019). Activated carbon and graphene nanoplatelets based novel composite for performance enhancement of adsorption cooling cycle. *Energy Conversion and Management* 180: 134–148. https://doi.org/10.1016/j.enconman.2018.10.092.

31 Pal, A., Shahrom, M.S.R., Moniruzzaman, M. et al. (2017). Ionic liquid as a new binder for activated carbon based consolidated composite adsorbents. *Chemical Engineering Journal* 326: 980–986. https://doi.org/10.1016/j.cej.2017.06.031.

32 Rupa, M.J., Pal, A., and Saha, B.B. (2020). Activated carbon-graphene nanoplatelets based green cooling system: adsorption kinetics, heat of adsorption, and thermodynamic performance. *Energy* 193: https://doi.org/10.1016/j.energy.2019.116774.

33 Pal, A., El-Sharkawy, I.I., Saha, B.B. et al. (2016). Experimental investigation of CO_2 adsorption onto a carbon based consolidated composite adsorbent for adsorption cooling application. *Applied Thermal Engineering* 109: 304–311. https://doi.org/10.1016/j.applthermaleng.2016.08.031.

34 Wang, L.W., Tamainot-Telto, Z., Thorpe, R. et al. (2011). Study of thermal conductivity, permeability, and adsorption performance of consolidated composite activated carbon adsorbent for refrigeration. *Renewable Energy* 36: 2062–2066. https://doi.org/10.1016/j.renene.2011.01.005.

35 Younes, M.M., El-sharkawy, I.I., Kabeel, A.E. et al. (2019). Synthesis and characterization of silica gel composite with polymer binders for adsorption cooling applications. *International Journal of Refrigeration* 98: 161–170. https://doi.org/10.1016/j.ijrefrig.2018.09.003.

36 Brancato, V., Frazzica, A., Sapienza, A. et al. (2015). Ethanol adsorption onto carbonaceous and composite adsorbents for adsorptive cooling system. *Energy* 84: 177–185. https://doi.org/10.1016/j.energy.2015.02.077.

37 Pino, L., Aristov, Y.I.U., Cacciola, G., and Restuccia, G. (1997). Composite materials based on zeolite 4A for adsorption heat pumps. *Adsorption* 3: 33–40. https://doi.org/10.1007/BF01133005.

38 Pal, A., Uddin, K., Rocky, K.A. et al. (2019). CO2 adsorption onto activated carbon–graphene composite for cooling applications. *International Journal of Refrigeration* 106: 558–569. https://doi.org/10.1016/j.ijrefrig.2019.04.022.

39 Pal, A. (2018). *Study on novel functional activated carbon and composites for adsorption heat pump systems – Collections* I. Kyushu University Library, Kyushu University https://catalog.lib.kyushu-u.ac.jp/opac_download_md/1959153/tj1065.pdf (accessed 14 May 2020).

40 El-Sharkawy, I.I., Pal, A., Miyazaki, T. et al. (2016). A study on consolidated composite adsorbents for cooling application. *Applied Thermal Engineering* 98: 1214–1220. https://doi.org/10.1016/j.applthermaleng.2015.12.105.

41 El-Sharkawy, I.I., Kuwahara, K., Saha, B.B. et al. (2006). Experimental investigation of activated carbon fibers/ethanol pairs for adsorption cooling system application. *Applied Thermal Engineering* 26: 859–865. https://doi.org/10.1016/j.applthermaleng.2005.10.010.

42 Saha, B.B., El-Sharkawy, I.I., Chakraborty, A. et al. (2006). Adsorption rate of ethanol on activated carbon fiber. *Journal of Chemical & Engineering Data* 51: 1587–1592. https://doi.org/10.1021/je060071z.

43 Saha, B.B., El-Sharkawy, I.I., Chakraborty, A., and Koyama, S. (2007). Study on an activated carbon fiber-ethanol adsorption chiller: part I – system description and modelling. *International Journal of Refrigeration* 30: 86–95. https://doi.org/10.1016/j.ijrefrig.2006.08.004.

44 Turski, M. and Sekret, R. (2018). Buildings and a district heating network as thermal energy storages in the district heating system. *Energy and Buildings* 179: 49–56. https://doi.org/10.1016/J.ENBUILD.2018.09.015.

45 Mazzoni, S., Sze, J.Y., Nastasi, B. et al. (2021). A techno-economic assessment on the adoption of latent heat thermal energy storage systems for district cooling optimal dispatch & operations. *Applied Energy* 289: 116646. https://doi.org/10.1016/J.APENERGY.2021.116646.

46 Saha, B.B. and Rupam, T.H. (2023). Specialty grand challenge: thermal energy storage and conversion. *Frontiers in Thermal Engineering* 3: 1157794. https://doi.org/10.3389/fther.2023.1157794.

47 Cárdenas, B. and León, N. (2013). High temperature latent heat thermal energy storage: phase change materials, design considerations and performance enhancement techniques. *Renewable and Sustainable Energy Reviews* 27: 724–737. https://doi.org/10.1016/j.rser.2013.07.028.

48 Kumar Sharma, R., Kumar, A., Rakshit, D. (2022). Building energy efficiency enhancement through load reduction in the air conditioning system. *EuroSun2022 Proceedings*. International Solar Energy Society, https://doi.org/10.18086/eurosun.2022.01.12.

2

Potential Assessment of Hydrofluoro Refrigerant-Based Adsorption Cooling Systems

Sai Yagnamurthy[1], Md. Amirul Islam[2], Bidyut Baran Saha[2,3], and Dibakar Rakshit[4]

[1] *School of Engineering, University of Warwick, Coventry, UK*
[2] *International Institute for Carbon-Neutral Energy Research (WPI-I²CNER), Kyushu University, Fukuoka, Japan*
[3] *Department of Mechanical Engineering, Kyushu University, Fukuoka, Japan*
[4] *Department of Energy Science and Engineering, Indian Institute of Technology Delhi, New Delhi, India*

2.1 Introduction

In the attempts toward decarbonizing the energy sector, the heating and cooling applications assume the primary focus due to them occupying around 50% of the energy demand. The demand for space cooling has seen the highest growth than for any other application in buildings [1]. Space cooling is also seen to occupy the major share of the peak electricity demand, reaching over 70% in the United States. Catering to the peak energy demands poses a challenge to the grid. Furthermore, the cooling systems employ refrigerants with high Global Warming Potentials (GWPs) ranging up to 23,000 times of CO_2 [2]. The rising cooling energy demand along with the ongoing usage of high GWP refrigerants is raising concerns toward the rising average global temperatures. In order to achieve the goal set by the Kigali amendment [3] to limit the rise in global average temperature to below 0.1 °C by 2100, the phasing down of the widely used hydrofluorocarbon (HFC) refrigerants has been recommended. There is also a need for the adoption of alternate energy-powered thermal technologies for meeting the cooling demand. In this regard, adsorption cooling technologies play an important role in decarbonizing the environment through the usage of waste or renewable heat for meeting the cooling applications. Besides, the adsorption cooling technology is capable of employing a wide variety of environmentally benign refrigerants.

Activated carbon adsorbent has been a popular choice in adsorption cooling systems due to its affinity for many refrigerants, owing to its high surface area and

Advancements in Non-Conventional Cooling and Thermal Storage Strategies, First Edition.
Edited by Bidyut Baran Saha and Dibakar Rakshit.

porosity characteristics [4–7]. It has also been found to be effective for the HFC refrigerant recovery and separation applications [2, 8]. The potential assessment of activated carbon adsorbents with the commercial refrigerants assumes significance for cooling applications, to serve as a replacement, or for load reduction of the compressor in vapor compression refrigeration (VCR) systems. Several authors studied the adsorption and cooling performance characteristics of activated carbon-based adsorption cooling systems in the literature. Banker et al. [9] studied the experimental performance of a four-bed adsorption cooling system based on an activated carbon–R134a pair. A cooling capacity of up to 5 W was observed with a low waste heat temperature of 90 °C. Loh et al. [10] studied the adsorption isotherms of the HFC refrigerants of R134a, R290, R410A, and R507A with activated carbon and fitted suitable equilibrium uptake correlations for each pair. It was observed that the maximum uptake value depends on the refrigerant's density. The following table provides a brief summary of various activated carbon and commercial hydrofluoro refrigerant-based working pairs in the literature (Table 2.1).

The hydrofluoro refrigerant-based systems are seen to offer lower cooling performance with activated carbon adsorbents, over those with the natural refrigerants of methanol, ammonia, etc. The lower latent heat capacities of refrigeration for the hydrofluoro refrigerants compared to that of the natural refrigerants are the primary reason for their lower performance. In this regard, combined or cascaded cycles are seen to enhance the performance of the adsorption cooling systems by reducing the temperature lift between the evaporator and condenser [18]. The temperature lift between the condenser and evaporator has been lowered in cascaded configurations typically through an intermediate heat exchanger between the condenser of an adsorption cooling system with the evaporator of another cooling system, as shown in Figure 2.1. Dakkama et al. [19] analyzed the cooling performance of the cascaded configurations of Maxsorb activated carbon with the refrigerants of R507A, R134a, propane, ethanol, and methanol. The combination of Maxsorb/R134a + Maxsorb/propane yielded the highest cooling performance resulting from the high latent heat values of propane and the high operating pressure of R134a, resulting in better uptake capacities for the Maxsorb carbon. Palomba et al. [20] studied the performance of a cascaded silica gel–water-based adsorption cooling system with a R410A-based commercial VCR system. The electric COP of the compressor of the VCR system is seen to double and reach values of up to 8 in the cascaded configuration.

A working pair with a commercial hydrofluoro refrigerant offers a better scope of cascading with a commercial VCR system employing the same refrigerant, through avoiding additional condenser and evaporator components and/or additional intermediate heat exchangers. Though several researchers studied the adsorption and cooling performance characteristics of the commercial hydrofluoro refrigerants for adsorption cooling, their distinct potential in the

Table 2.1 Overview of activated carbon-based adsorption characteristics of hydrofluoro refrigerants in the literature.

Adsorbent	Refrigerant	Key findings
Pitch-based ACFs OG-A7 ($S_{BET} = 860\,m^2/g$) OG-A15 ($S_{BET} = 1574\,m^2/g$) OG-A20 ($S_{BET} = 2062\,m^2/g$) B1300 ($S_{BET} = 1084\,m^2/g$) [11]	R134a	• OG-A15 showed highest R134a adsorption than the other ACFs due to the slightly larger pore width. • Repeated adsorption and desorption of R32 onto the ACFs revealed very similar weight change patterns over four cycles. • The adsorption capacity did not deteriorate despite repeated adsorption at −20 °C and desorption at 30 °C until the 5th cycle.
EcoSorb ($S_{BET} = 1100\,m^2/g$) BAC ($S_{BET} = 1250\,m^2/g$) AC1 ($S_{BET} = 1695\,m^2/g$) AC2 ($S_{BET} = 1280\,m^2/g$) [2]	R-32 R-125 R-134a	• AC1 and AC2, which are coconut shell and coal-based activated carbons, have the highest adsorption capacities for the three pure F-gases. • R-32 shows the highest adsorption for all the ACs. • Selectivity test of refrigerant mixtures (R-410A and R-407F) shows that all ACs are more selective to R-125 and R-134a.
Maxsorb III [12]	HFC404A	• A maximum uptake of 2.2 kg/kg of R404A is found at a temperature of 25 °C. • Optimum performance can be obtained for hot- and cold-water temperatures of 85 and 25 °C, respectively. • COP and SCP were found to be 0.23 and 275 W/kg, respectively, at a half cycle time of 200 s.
Maxsorb III [13]	HFO-1234ze(E)	• Performance evaluation of a four-bed adsorption chiller was analyzed. • The system is capable of producing 2 kW of cooling power by using 80 kg of Maxsorb III with a driving heat source temperature of 85 °C. The cooling water temperature is 30 °C. • The performance is slightly higher than that of the Maxsorb III/R134a-based adsorption cooling cycle.

(Continued)

Table 2.1 (Continued)

Adsorbent	Refrigerant	Key findings
Maxsorb III ($S_{BET} = 3150\ m^2/g$) ACF-A20 [10]	R134a R290 R410A R507A	• Maximum equilibrium uptake from the D-A equation is found highest for the Maxsorb III/R134a pair, and the value is 2.22 kg/kg. • For ACF-A20/R134a, the uptake is 1.29 kg/kg.
Maxsorb III [14]	HFO-1234ze(E)	• The maximum experimental adsorption uptake is 1.73 kg/kg at 0.607 relative pressure. • Saturated uptake from the D-A equation is 2.53 kg/kg. • Theoretical maximum coefficient of performance is computed as 0.9.
Maxsorb III [15]	HFC-152a	• The equilibrium uptake is about 1.3 kg/kg at 0.6 relative pressure. • Saturation uptake from the D-A equation is 3.438 cm^3/g. • The isosteric heat of adsorption values has been found to vary from 406 to 975 kJ/kg depending on adsorbate loading.
Maxsorb III [16]	R134a R507A	• The adsorption uptake of R134a is 1.6 kg/kg at 25 °C after 1200 s. • When the adsorption temperature is increased to 60 °C, the equilibrium uptake of R134a falls to 1.0 kg/kg. • The adsorption uptake of R507A is 1.3 kg/kg at 20 °C after 1100 s.
Carbosieve G ($S_{BET} = 1130\ m^2/g$) [17]	R114 R114a R115 R124 R125 R134 R134a	• Saturation uptakes of Carbosieve G and the mentioned refrigerants are 625, 581, 477, 505, 496, 505, and 478 mg/g, respectively.

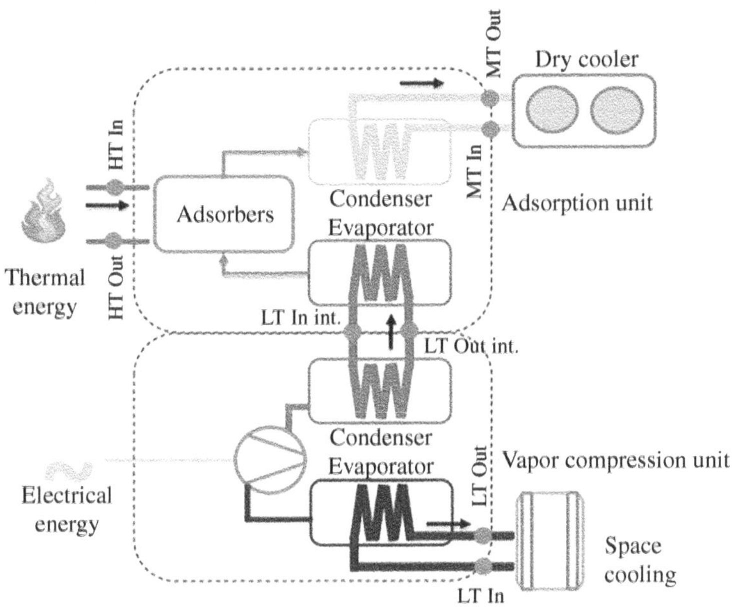

Figure 2.1 Layout of the cascaded configuration of VCR and adsorption cooling units. Source: Ref. [19])/Stichting HPC.

VCR cascaded configurations has not been explored as far as the knowledge of the authors is concerned. The present study attempts to address this gap through the study of an activated carbon-based adsorption system with the commercial hydrofluoro refrigerants. The performance of the adsorption cooling system is analyzed in stand-alone and VCR cascaded configurations for an air-conditioning application. The findings of the present study aim to facilitate the selection of the appropriate working pair as well as the suitable configuration of the adsorption cooling system to maximize the system performance.

2.2 System Description

The present study attempts to evaluate the cooling potential of some of the widely used and upcoming commercial refrigerants for VCR applications. In this regard, the refrigerants of R134a, R410A, R32, and R1234ze(E) have been considered for a typical air-conditioning application for the local weather conditions of Delhi. The refrigerants have been chosen in accordance with the goals of the latest Kigali agreement to gradually move toward the low GWP HFOs. Maxsorb III activated carbon has been selected as the adsorbent based on the literature review

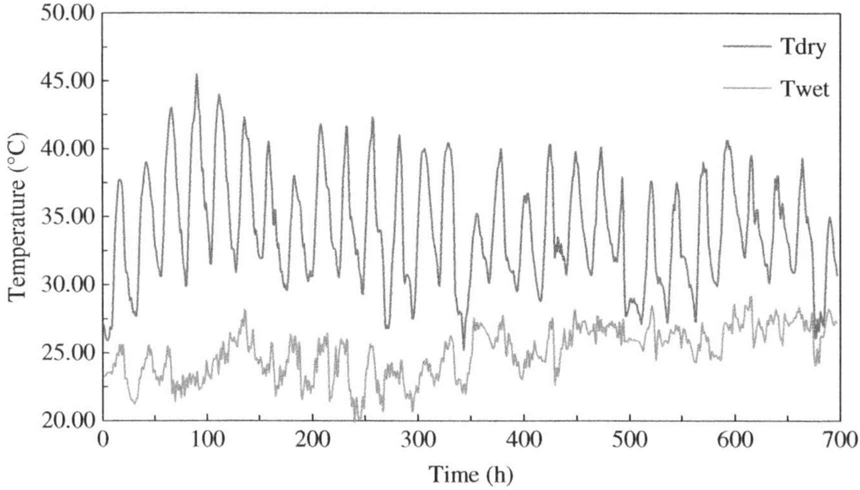

Figure 2.2 Ambient temperature variation of Delhi in the month of June.

presented in Section 2.1. The typical ambient temperature profile for the hottest month of June in Delhi is shown in Figure 2.2. It can be seen that the average dry-bulb temperature is around 35 °C, while the wet-bulb temperature is around 25 °C. The evaporator temperature is assumed to be constant at around 7 °C.

The schematic of a typical VCR-based air-conditioning system and a typical two-bed adsorption cooling system are shown in Figure 2.3. As shown in the figure, the adsorption cooling system requires a thermal energy input and a cooling source for the desorption and adsorption processes, respectively. A temperature difference of 5 °C is assumed across the heat exchangers of various components of the adsorption cooling system. Thus, the average temperature of the condenser is considered to be 40 °C, while that of the adsorbing bed is considered to be 30 °C for the simulation study in accordance with the average dry- and wet-bulb temperatures shown in Figure 2.2.

As mentioned in the previous section, the distinct advantage of employing an adsorption system compatible with HFCs is the ability to effectively integrate with conventional VCR systems without the need for additional evaporator and/or condenser components. The study further explores integration strategies for the HFC-based two-bed adsorption system with a conventional VCR system for load reduction and performance enhancement of both the VCR and adsorption cooling systems. Two configurations are considered as shown in Figure 2.4. Figure 2.4a shows the schematic for configuration A, where the adsorption system follows the compressor in the refrigerant flow path. The refrigerated vapor from the evaporator enters the compressor in this configuration from where it is compressed to a pre-desired pressure before getting adsorbed by the adsorbent

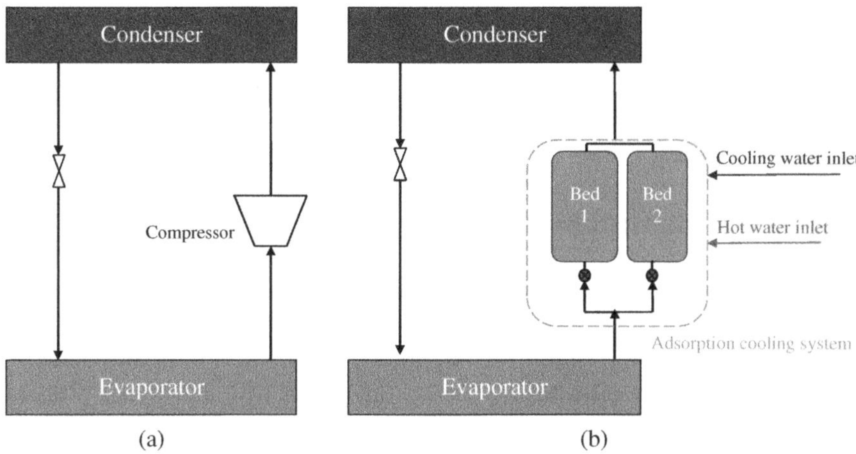

Figure 2.3 Stand-alone configurations of (a) VCR system and (b) adsorption cooling system.

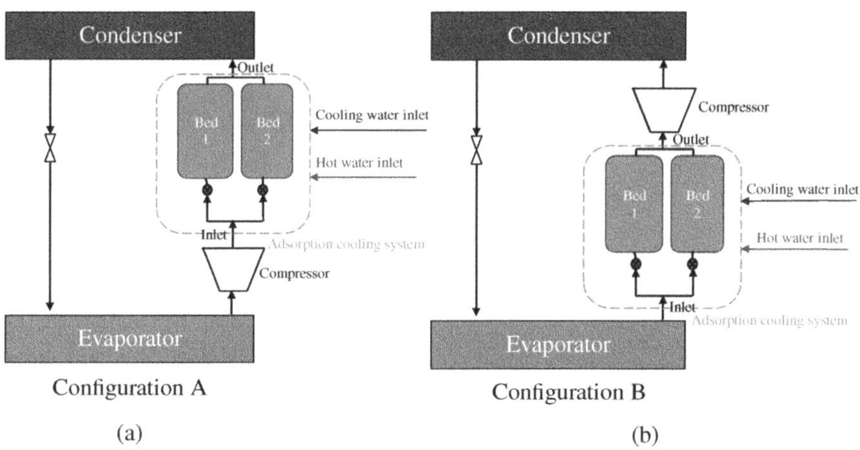

Figure 2.4 Cascaded configurations of VCR–adsorption cooling systems: (a) configuration A and (b) configuration B.

bed. The desorbing bed is connected to the condenser. Thus, configuration A offers the flexibility of lowering the compressor outlet pressure to enhance the COP of the cooling system along with the supplying the adsorbate vapor at an elevated pressure for the COP enhancement of the adsorption cooling system. Figure 2.4b shows the schematic for configuration B, where the compressor follows the adsorption cooling system in the refrigerant path. In this case, the refrigerant vapor from the evaporator is adsorbed by the adsorbing bed, while the desorbed vapor from the desorber enters the compressor from where it is elevated to the condenser pressure. This configuration offers the flexibility of having a lower desorption pressure for the desorber as well as a reduction in the pressure difference for the compressor, thus enhancing the COP of both the systems.

2.3 Mathematical Modeling

The mathematical model attempts to evaluate the performance of the adsorption cooling system in terms of the specific cooling energy (SCE) and Coefficient of Performance (COP). The impact of uptake efficiency is also considered in the present evaluation as it becomes more prominent with high vapor pressure refrigerants. The uptake efficiency is evaluated using the correlation shown in Eq. (2.1) [21]. Equation (2.2) shows the mass of the adsorbate vapor trapped in the adsorbent's pores,

$$\eta = \frac{W_{max} - W_{min} - \Delta W}{W_{max} - W_{min}} \tag{2.1}$$

where

$$\Delta W = (\rho_{outlet} - \rho_{inlet}) \times \left(\frac{1}{\rho_{bulk}} - \frac{1}{\rho_{solid}} \right) \tag{2.2}$$

The inlet density of the refrigerant vapor ρ_{inlet} are considered at the adsorbent temperature and the inlet pressure, while the outlet density ρ_{outlet} is considered at the desorbent temperature and outlet pressure, as shown in Eqs. (2.3)–(2.5) for various configurations. The respective refrigerant properties are taken from the work of Bell et al. [22],

$$\rho_{inlet} = \rho(T_{ads}, P_{eva}) \text{ and } \rho_{outlet} = \rho(T_{des}, P_{cond}) \text{ for stand} - \text{alone configuration} \tag{2.3}$$

$$\rho_{inlet} = \rho(T_{ads}, P_{cascade}) \text{ and } \rho_{outlet} = \rho(T_{des}, P_{cond}) \text{ for configuration A} \tag{2.4}$$

$$\rho_{inlet} = \rho(T_{ads}, P_{eva}) \text{ and } \rho_{outlet} = \rho(T_{des}, P_{cascade}) \text{ for configuration B} \tag{2.5}$$

The bulk and solid densities of Maxsorb III are 310 and 2200 kg/m^3, respectively [23].

The cooling energy output per unit mass is evaluated through Eq. (2.3), and the thermal energy input per unit mass of the adsorbent is evaluated through Eq. (2.4). A half cycle time is considered such that 80% of the equilibrium uptake occurs so as to strike a balance between the COP and cooling power [24]. The system considers an aluminum-based finned flat-tube heat exchanger offering a high surface

Table 2.2 Adsorption parameters of activated carbon-based working pairs considered for the present study.

Working pair	Equilibrium equation	Parameter values			
		W_o (kg/kg)	E (J/mol)	n	Q_{st} (kJ/kg)
Maxsorb III–R134a [10]	D–A $W = W_o \exp\left(-\left[\left(\frac{RT}{E}\ln\left(\frac{P_s}{P}\right)\right)\right]^n\right)$	2.22	7332.69	1.29	290
Maxsorb III–R410A [10]		2.07	5254.38	1.36	237
Maxsorb III–R32 [26]		1.51	4794.68	1.32	420.7
Maxsorb III–R1234ze(E) [14]	Tóth $W = W_o\left(\frac{bP}{(1+(bP)^t)^{1/t}}\right)$, Where $b = b_o \exp\left(\frac{Q_{st}}{RT}\right)$	W_o (kg/kg) 2.53	b_o (kPa) 9.84×10^{-8}	t 0.48	300.3

area per unit adsorbent mass ratio of around 4 m^2/kg [25]. The heat exchanger also offers a low metal mass per unit mass of the adsorbent (~1.8 kg/kg). The thermal energy input considers both the sensible heat component accounting for the temperature rise of the heat exchanger and adsorber and the adsorbed refrigerant along with the heat of adsorption. The uptake efficiency is accounted for in the evaluation of both the cooling and thermal energy inputs, as shown in Eqs. (2.6) and (2.7). These quantities are further used to estimate the key performance index of COP for the adsorption cooling system, as shown in Eq. (2.8). The adsorption uptake limits W_{max} and W_{min} along with the heat of adsorption are evaluated from the data given in the literature, as shown in Table 2.2. For temperatures beyond the critical temperature of the refrigerants, a pseudo-saturation pressure correlation given by Islam [27] has been used for the equilibrium concentration evaluation,

$$SCE = 0.8 \times (W_{max} - W_{min}) \times h_{fg} \times \eta \tag{2.6}$$

$$Q_{thermal} = \frac{m_{hex}}{m_{ads}} \times C_{p,hex} \times (T_{reg} - T_{ads}) + (C_{p,ads} + \int_{T_{ads}}^{T_{reg}} WC_{p,ref}\, dT$$
$$+ 0.8 \times (W_{max} - W_{min}) \times Q_{st} \times \eta \tag{2.7}$$

$$COP = \frac{SCE}{Q_{thermal}} \tag{2.8}$$

The COP of the VCR system is estimated through an empirical correlation shown in Eq. (2.10) for a 5 kW heat pump system proposed by Renaldi et al. [28],

$$COP_{VCR} = -0.066 \times T_{lift} + 4.7 \tag{2.9}$$

where $T_{lift} = T_{cascade} - T_{eva}$ for configuration A
and

$$T_{lift} = T_{cond} - T_{cascade} \text{ for configuration B} \tag{2.10}$$

The equations are computed using the Matlab® 2022 platform for the given working pairs under various operating conditions.

2.4 Results and Discussion

2.4.1 Performance Evaluation of Stand-alone Cooling Systems

The performance of various configurations of the adsorption cooling system is studied for the condenser temperature of 40 °C, adsorption cooling temperature of 30 °C, and varying regeneration temperatures. Figure 2.5 shows the SCE and COP variations of the adsorption cooling system shown in Figure 2.3b for various regeneration temperatures and refrigerants. It can be seen from Figure 2.5a that the SCE of all the working pairs increases with increasing regeneration temperatures. Figure 2.5b shows that the COP of the working pairs follows a similar trend as that of the SCE curves for all the regeneration temperatures. The COP curve of the working pair of Maxsorb III–R32 is shown to have a steep rise immediately after the supercritical temperature of 78.1 °C due to the steep rise of the specific heat capacity of the adsorbed vapor around the supercritical temperature, resulting in the increase of the sensible heat energy input and correspondingly lower COP values. To further analyze the performance of the various working pairs, the cyclic uptake limits and uptake efficiencies are evaluated, as shown in Figure 2.7.

From Figure 2.6a, it can be observed that the cyclic uptake limits increase with increasing regeneration temperatures for all the refrigerants with Maxsorb III adsorbent. A minimal regeneration temperature of 55 °C is seen to be necessary for R1234ze(E) adsorbate, while a regeneration temperature of at least 65 °C is seen to be required for other working pairs to have a positive uptake limit. Figure 2.6b shows the uptake efficiencies of the working pairs, where it is seen that higher regeneration temperatures are required for the working pairs to have nonzero values over those required for the positive cyclic uptake limit values. This implies that for a condenser temperature of 40 °C and an evaporator temperature of 7 °C, higher regeneration temperatures are required to cause the pressure lift within the adsorbent to cause it to desorb the vapor to the condenser. The SCE and

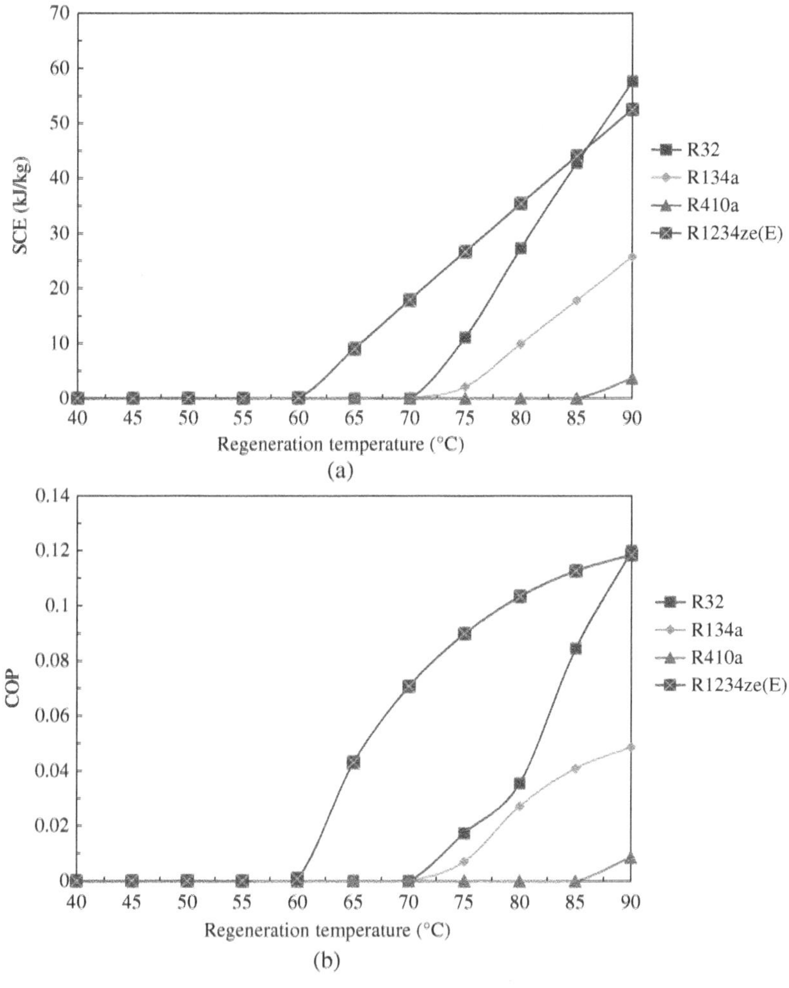

Figure 2.5 Performance variation of the stand-alone configuration of Maxsorb III-based adsorption cooling system with regeneration temperatures: (a) SCE and (b) COP.

COP trends are in line with the uptake efficiency trends for the refrigerants, as achieving the pressure lift becomes easier with higher regeneration temperatures. It can be observed that the Maxsorb III–R1234ze(E) pair yields the highest performance and uptake efficiency values followed by the Maxsorb III–R32 pair. Maxsorb III–R410A yields the lowest performance owing to its low cyclic uptake limits as well as the high density of R410A, resulting in low uptake efficiencies.

The COP of the VCR system operating for the similar condenser and evaporator temperature limits is evaluated to be 2.52 from Eq. (2.6). To enhance the operating

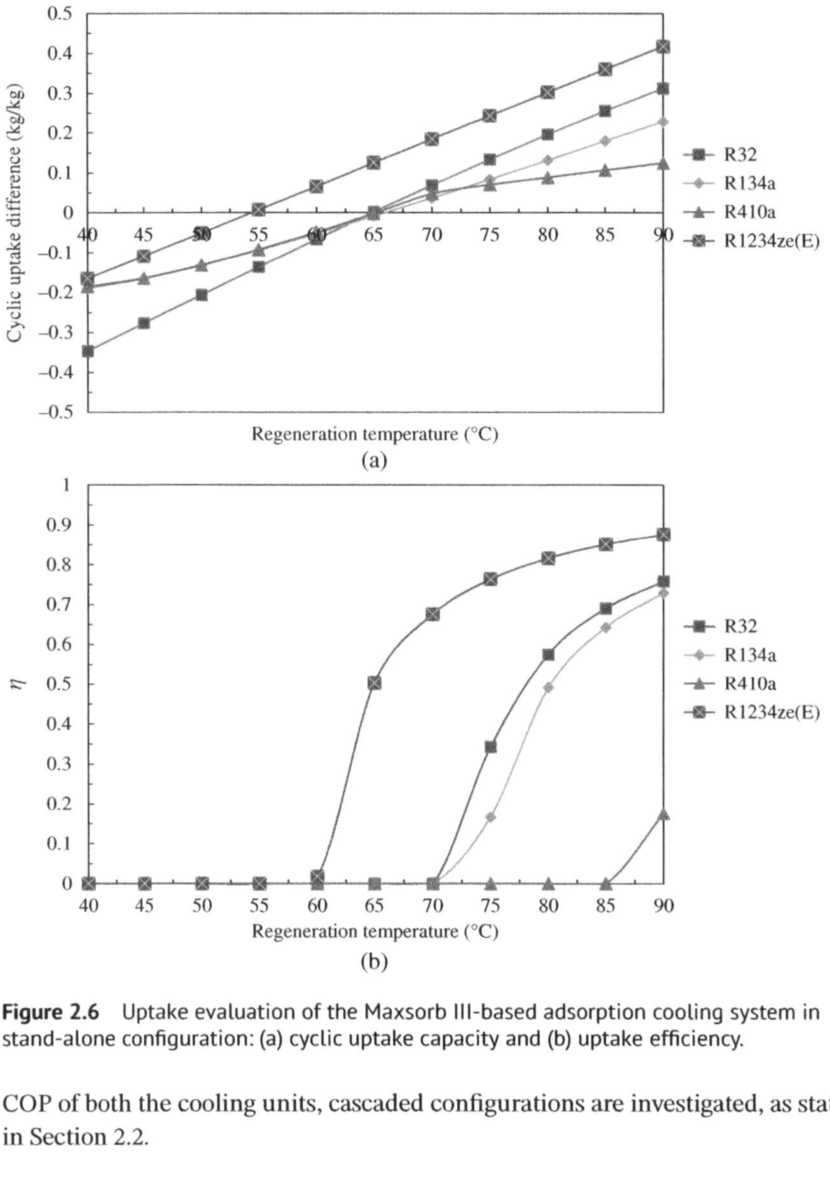

Figure 2.6 Uptake evaluation of the Maxsorb III-based adsorption cooling system in stand-alone configuration: (a) cyclic uptake capacity and (b) uptake efficiency.

COP of both the cooling units, cascaded configurations are investigated, as stated in Section 2.2.

2.4.2 Performance Evaluation of Cascaded Configurations

The performance of the cascaded configurations is investigated for various cascading temperatures between the VCR system and the adsorption cooling system, along with the regeneration temperatures. The cascading temperature is another

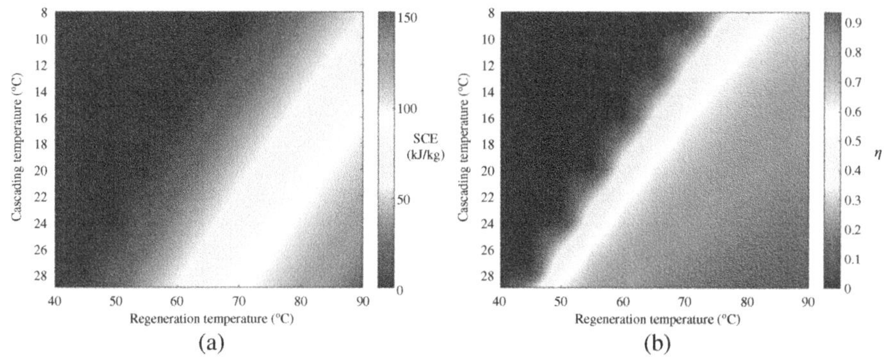

Figure 2.7 (a) SCE and (b) η contours of the Maxsorb III–R32 pair in configuration A at various regeneration and cascading temperatures.

variable parameter that can be controlled by the VCR system to extract the optimal cooling performance from the combined system.

2.4.2.1 Performance Evaluation of Configuration A

The condenser, evaporator, and cooling water inlet temperatures are considered to be 40, 7, and 30 °C, similar to that of the stand-alone system. The permissible range for the cascading temperature is seen to be between 8 and 29°C so as to avoid condensation of the vapor inside the adsorption chamber. The regeneration temperature range is considered to be between 40 and 90 °C. The SCE, COP, and η are evaluated using Eqs. (2.1)–(2.5) and are plotted for the considered operating temperature ranges for performance analysis. Figure 2.7 shows the SCE and uptake efficiency contours of the pair Maxsorb III–R32 for configuration A. Similar to the trend shown in Figure 2.6, the SCE tends to increase toward higher regeneration temperature values owing to the increase of the cyclic uptake limits, as shown in Figure 2.7a. Furthermore, the SCE values are seen to increase with cascading temperatures owing to the same reason. A higher cascading temperature increases the adsorption pressure and thereby the maximum adsorption limit, while a higher regeneration temperature results in a lower minimal adsorption limit. Figure 2.7b shows the uptake efficiency contours of the pair under the same range of operating conditions. Similar to the SCE trend shown in Figure 2.7a, the uptake efficiency values tend to increase toward higher regeneration and cascading temperatures. While a higher regeneration temperature lowers the vapor density for the given condensation pressure, a higher cascading temperature increases the vapor density for the given adsorption temperature. Both of these factors positively contribute toward the uptake efficiency, as can be understood from Eq. (2.4).

Figure 2.8 shows the COP contour of the Maxsorb III–R32 pair for the same set of operating conditions. Similar to the SCE and η trends shown in Figure 2.7, the

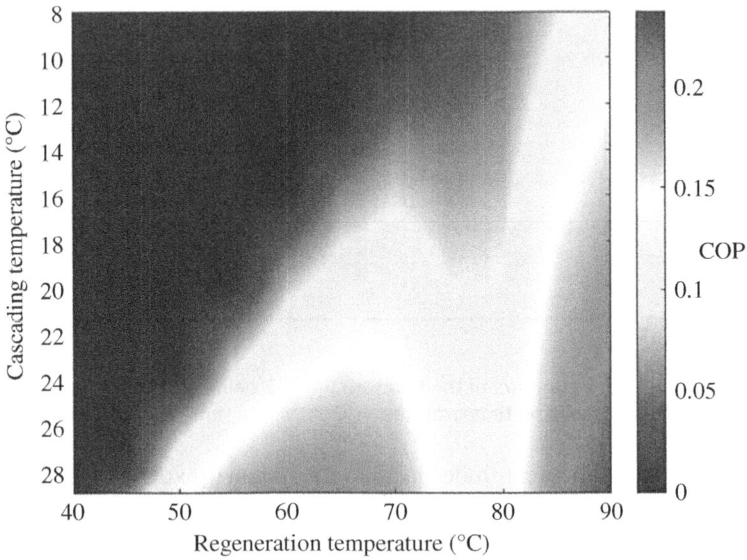

Figure 2.8 COP contour of Maxsorb III–R32 at varying regeneration and cascading temperatures.

COP values increase toward the higher regeneration and cascading temperatures. However, a drop in the COP values can be observed between the regeneration temperatures of 73–83 °C. This is due to the rise in specific heat capacity of the adsorbed vapor near the supercritical temperature, as observed in the stand-alone configuration shown in Figure 2.5b.

Similarly, the performance of the other working pairs has been evaluated for configuration A and the respective COP contours are shown in Figure 2.9. The COP contours of the three pairs can be seen to increase toward higher regeneration and cascading temperatures similar to that of the Maxsorb III–R32 pair due to higher cyclic uptake and uptake efficiency values. The Maxsorb III–R410A pair is seen to have a COP drop at around 65–75 °C. This can be attributed to the rise of the adsorbed vapor's specific heat capacity around the supercritical temperature of 72.1 °C, increasing the thermal energy consumption as it has been with the case of the Maxsorb III–R32 pair.

Figure 2.10 compares the COP variation of various refrigerant-based working pairs along with that of the VCR system at various cascading temperatures and a regeneration temperature of 65 °C. Similar to the stand-alone configuration, the working pairs of Maxsorb III–R32 and Maxsorb III–R1234ze(E) are seen to offer among the highestt COP values, while the working pair of Maxsorb III–R410A offers the lowest COP. The high latent heat capacity of the R32 refrigerant and the highest cyclic uptake capacity and efficiencies with the R1234ze(E) refrigerant

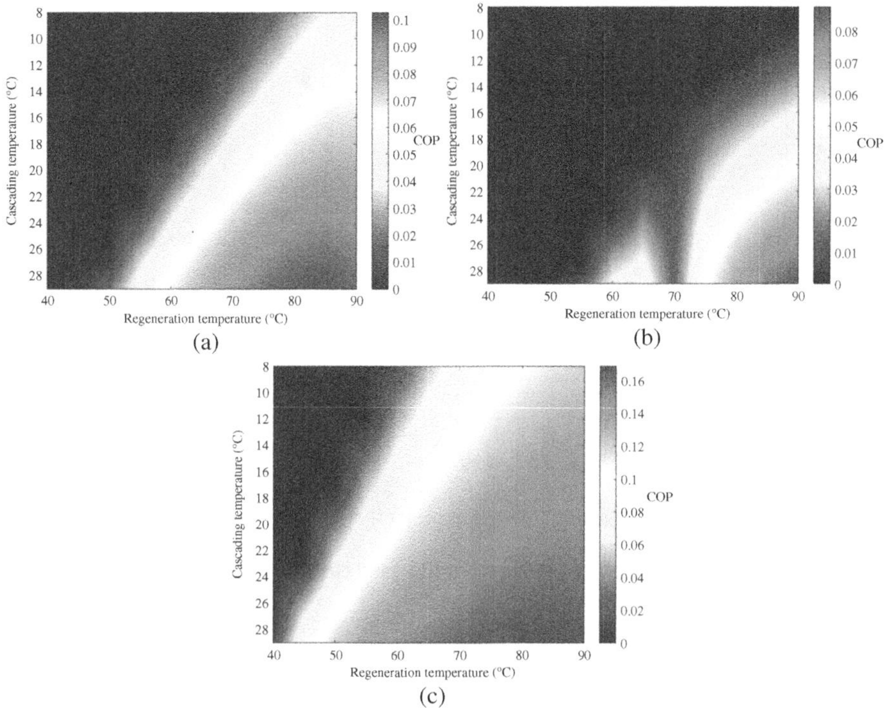

Figure 2.9 COP contours of the Maxsorb III adsorbent with (a) R134a, (b) R410A, and (c) R1234ze(E) refrigerants in configuration A.

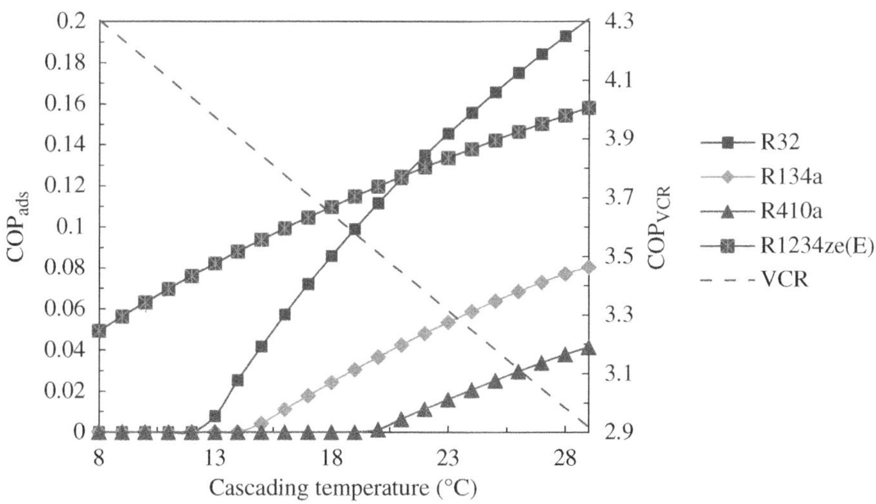

Figure 2.10 COP variation of the Maxsorb III-based cascaded adsorption system (configuration A).

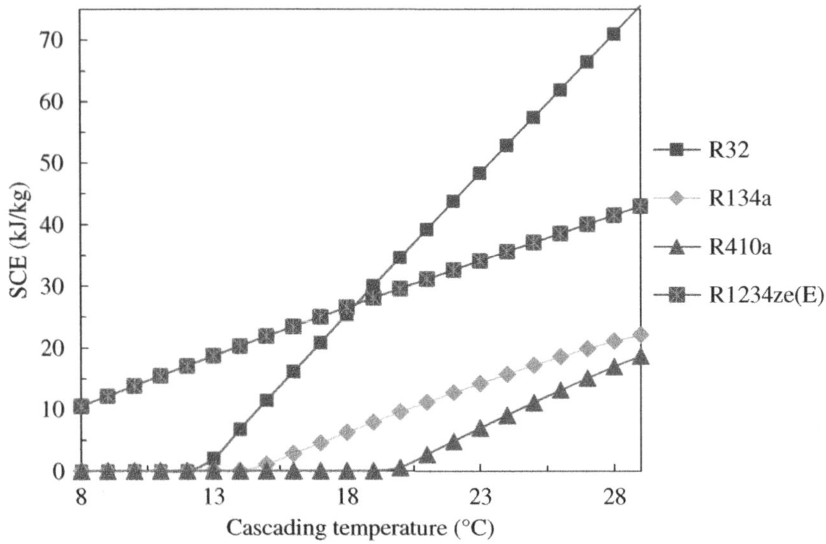

Figure 2.11 SCE variation of the Maxsorb III-based cascaded adsorption system with cascading temperature.

contribute to their high performance. The COP of the VCR system is seen to reduce with increasing cascading temperatures, owing to the increase of temperature lift between the evaporator and adsorber. The lowest COP is seen to be 2.9 at a cascading temperature of 29 °C which is around 16.7% higher than that of the VCR system in a stand-alone configuration. Figure 2.11 compares the SCE values of the working pairs for the regeneration temperature of 65 °C. The working pair with the R32 refrigerant is seen to offer the highest SCE values, primarily owing to the highest latent heat capacity of the R32 refrigerant. The R410A refrigerant despite having the second highest latent heat capacity offers the lowest performance due to the low cyclic uptake capacity and efficiency values, as discussed in Section 2.4.1.

2.4.2.2 Performance Evaluation of Configuration B

The performance of the working pairs is evaluated for various regeneration and cascading temperatures in configuration B. This configuration has a wider cascading temperature range (8–39 °C) compared to that of A due to the compressor being on the desorption side. The SCE and η values computed for the pair Maxsorb III–R32 are shown in Figure 2.12. Both the parameters are seen to have higher values toward higher regeneration temperatures due to higher cyclic uptake limits and lower density of the refrigerant at higher temperature for a given cascading pressure. A lower cascading temperature is seen to be favored for higher SCE and η values. At a lower temperature lift, there is a higher uptake

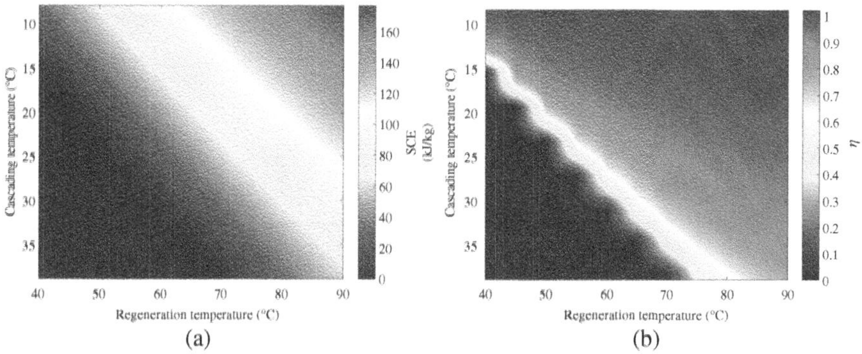

Figure 2.12 SCE and η contours of the Maxsorb III–R32 pair in configuration B for various regeneration and cascading temperatures.

along with a reduced density difference between the adsorbed and desorbed states. This favors the η increase as shown by Eq. (2.2).

Figure 2.13 shows the COP contours of the working pairs. Similar to the SCE and η contours, lower cascading temperatures yield higher COP values due to higher cyclic outputs. The COP is also seen to largely increase with increasing regeneration temperatures. For the working pairs of Maxsorb III–R32 and Maxsorb III–R410A, the COP is seen to increase with regeneration temperatures until their respective supercritical temperature zones followed by a sudden dip, as observed in the stand-alone and A configurations. Overall, it can be deduced that similar to the stand-alone and A configurations, the working pairs of Maxsorb III–R32 and Maxsorb III–R1234ze(E) yield higher performance over the other working pairs.

Figure 2.14 compares the COP values of the working pairs and the VCR system at various cascading temperatures for a regeneration temperature of 65 °C. The COP of the VCR system is seen to increase with cascading temperature due to the lower temperature lift requirement of the compressor. The working pair of Maxsorb III–R1234ze(E) is seen to yield the highest COP values for various cascading temperatures. The SCE values of the working pairs are shown in Figure 2.15, where it is seen that the pair Maxsorb III–R32 offers the highest SCE values until a cascading temperature of 26 °C. It is followed by the pair Maxsorb III–R1234ze(E) similar to that of Figure 2.11. To provide a common basis for the performance comparison of the adsorption cooling system in configurations A and B, Figure 2.16 has been plotted to compare the COP and SCE values of the adsorption cooling system with the working pair of Maxsorb III–R1234ze(E). It can be seen that compared to configuration A, configuration B yields higher values of COP and SCE for identical values of the VCR system's COP. It can also be observed that configuration B enables operation at lower regeneration temperatures over that of configuration A.

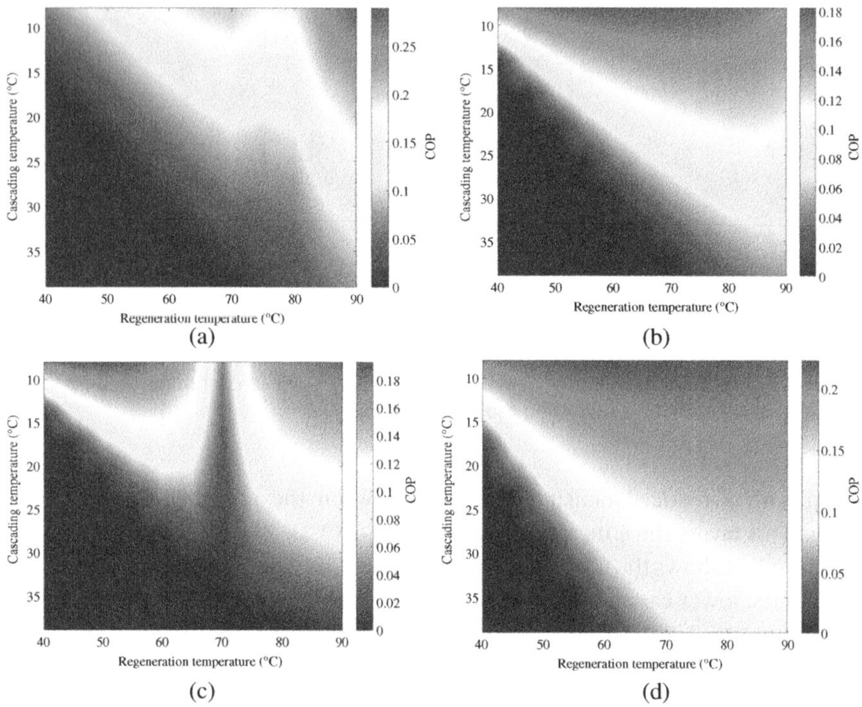

Figure 2.13 COP contours of working pairs in configuration B: (a) Maxsorb III–R32, (b) Maxsorb III–R134a, (c) Maxsorb III–R410A, and (d) Maxsorb III–R1234ze(E).

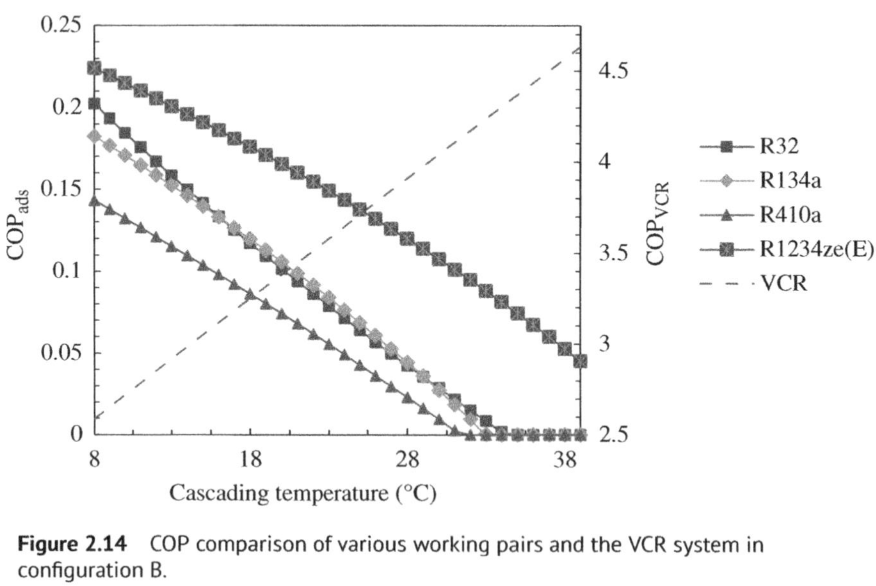

Figure 2.14 COP comparison of various working pairs and the VCR system in configuration B.

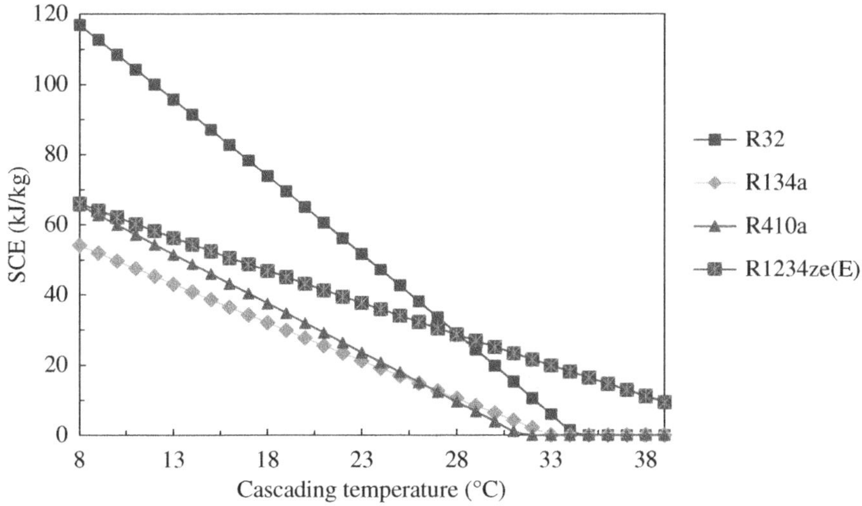

Figure 2.15 SCE comparison of various working pairs in configuration B.

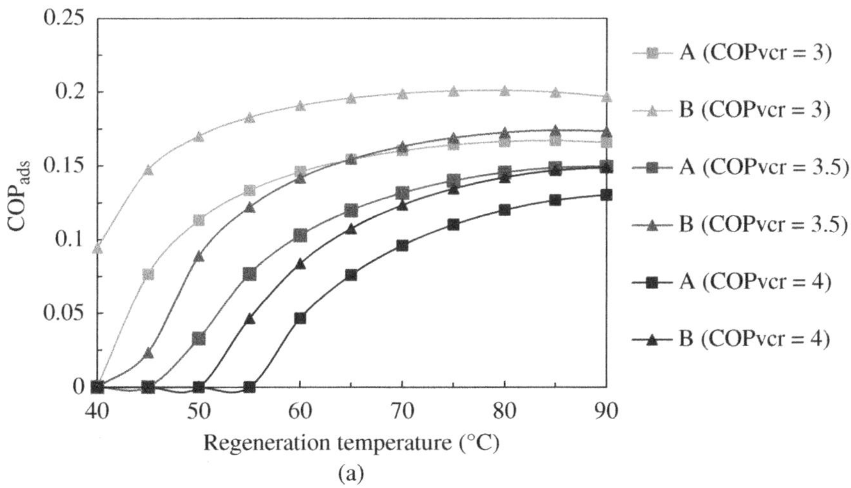

Figure 2.16 (a) COP and (b) SCE comparison of the Maxsorb III–R1234ze(E) pair in configurations A and B.

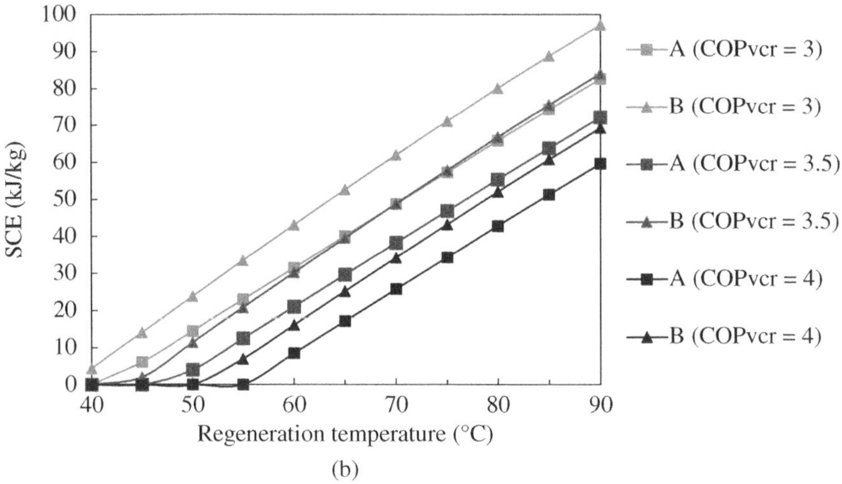

Figure 2.16 (*Continued*)

This implies that the adsorption cooling system has a higher sensitivity for the condenser temperature, and, therefore, configuration B is seen to be better for the local weather conditions of Delhi. While a higher COP reduces the thermal energy load due to the adsorption cooling system, a higher SCE can help to realize compact systems.

2.5 Conclusions

The present work attempted to estimate the cooling potential of HFC-based adsorption systems for domestic space cooling applications. The refrigerants of R32, R134a, and R410A along with the HFO R1234ze(E) have been considered with Maxsorb III as the adsorbent. The performance of the system has been studied for the hottest month of June for Delhi local weather conditions. The performance of stand-alone and VCR cascaded configurations has been studied, and the key observations from the analysis are as follows:

- The uptake efficiency is seen to be a critical parameter affecting the performance of the adsorption cooling systems, particularly at low regeneration temperatures and for refrigerants with high vapor densities.
- The working pair of Maxsorb III–R1234ze(E) is seen to be the most suitable pair for the stand-alone and cascaded configurations, yielding the highest COP values and among the highest SCE values of other working pairs.

- In comparison to the stand-alone configuration, the cascaded configurations are seen to enable operation at lower regeneration temperatures and yield significantly higher SCE and COP values.
- Among the cascaded configurations, configuration B is seen to be the most suitable configuration for all the working pairs with higher SCE and COP values under various operating conditions.
- The cascaded configuration B enables an electrical energy saving potential of over 60% to the VCR system depending on the operating condition, over that of its stand-alone configuration.

The present study identifies a significant potential for the low GWP refrigerants of R32 and R1234ze(E) for adsorption system-integrated commercial cooling applications. The findings of the present study are anticipated to pave way for further design and optimization of the hydrofluoro refrigerant-based adsorption cooling systems to contribute toward the climate goals. The present work can be further extended by identifying suitable adsorbent composites to lower the regeneration temperature requirements and enhance the heat and mass transfer properties. Furthermore, the blends of the high-performing refrigerant R1234ze(E) with various other refrigerants can be explored for the adsorption-based cooling applications.

Nomenclature

b_o	pre-exponential Tóth's constant (kPa)
C_p	specific heat capacity (kJ/kg K)
COP	coefficient of performance
h_{fg}	enthalpy of vaporization (kJ/kg)
ρ	density (kg/m^3)
P	pressure (Pa)
Q_{st}	heat of adsorption (kJ/kg)
R	gas constant (kJ/kg K)
T	temperature (K)
W	concentration of adsorbent (kg/kg)
m	mass (kg)
η	uptake efficiency
SCE	specific cooling energy (kJ/kg)

Subscripts

ads	adsorber
bulk	bulk adsorbent

cascade	cascading section
cond	condenser
des	desorber
eva	evaporator
hex	heat exchanger
hot	hot water
HT	heat transfer
inlet	inlet vapor to the adsorber
lift	temperature difference between the inlet and outlet of compressor
outlet	outlet vapor from the desorber
reg	regeneration
solid	skeletal density of adsorbent
thermal	thermal energy input
VCR	vapor compression refrigeration

References

1. IEA (2018). The future of cooling: opportunities for energy-efficient air conditioning. https://www.iea.org/reports/the-future-of-cooling (accessed 16 May 2024).
2. Sosa, J.E., Malheiro, C., Ribeiro, R.P. et al. (2020). Adsorption of fluorinated greenhouse gases on activated carbons: evaluation of their potential for gas separation. *Journal of Chemical Technology & Biotechnology* 95 (7): 1892–1905. https://doi.org/10.1002/jctb.6371.
3. UNEP (2016). *Kigali Amendment to the Montreal Protocol on Substances that Deplete the Ozone Layer*. https://treaties.un.org/doc/Treaties/2016/10/20161015%2003-23%20PM/Ch_XXVII-2.f-English%20and%20French.pdf (accessed 16 May 2024).
4. El Fadar, A., Mimet, A., and Pérez-García, M. (2009). Modelling and performance study of a continuous adsorption refrigeration system driven by parabolic trough solar collector. *Solar Energy* 83 (6): 850–861. https://doi.org/10.1016/j.solener.2008.12.003.
5. Islam, M.A., Mitra, S., Thu, K., and Saha, B.B. (2021). Study on thermodynamic and environmental effects of vapor compression refrigeration system employing first to next-generation popular refrigerants. *International Journal of Refrigeration* 131: 568–580. https://doi.org/10.1016/j.ijrefrig.2021.08.014.
6. Saha, B.B., El-Sharkawy, I.I., Chakraborty, A., and Koyama, S. (2007). Study on an activated carbon fiber–ethanol adsorption chiller: part I – system description and modelling. *International Journal of Refrigeration* 30 (1): 86–95. https://doi.org/10.1016/j.ijrefrig.2006.08.004.

7 Wang, L.W., Wu, J.Y., Wang, R.Z. et al. (2003). Study of the performance of activated carbon–methanol adsorption systems concerning heat and mass transfer. *Applied Thermal Engineering* 23 (13): 1605–1617. https://doi.org/10.1016/S1359-4311(03)00104-2.

8 Lin, S.H. and Lin, R.C. (1999). Prediction and experimental verification of HFC-134a adsorption by activated carbons. *Journal of Environmental Science and Health, Part A* 34 (1): 183–200. https://doi.org/10.1080/10934529909376830.

9 N.D. Banker, M. Prasad, P. Dutta, K. Srinivasan, Experimental results of an activated carbon-HFC 134a adsorption cooling system for thermal management of electronics, *Applied Thermal Engineering* 31 (2011) 1607–1612. https://doi.org/10.1016/j.applthermaleng.2011.01.042

10 Loh, W.S., Ismail, A.B., Xi, B. et al. (2012). Adsorption isotherms and isosteric enthalpy of adsorption for assorted refrigerants on activated carbons. *Journal of Chemical & Engineering Data* 57 (10): 2766–2773. https://doi.org/10.1021/je3008099.

11 Guahk, Y.T., Jeon, M., Moon, S. et al. (2021). Feasibility of regenerative adsorption of a hydrofluorocarbon (HFC-134a) using activated carbon fiber studied by the gaseous flow method. *Journal of Hazardous Materials* 411: 125009. https://doi.org/10.1016/j.jhazmat.2020.125009.

12 Ghazy, M., Askalany, A.A., and Saha, B.B. (2020). Maxsorb III/HFC404a as an adsorption pair for renewable energy driven systems. *International Journal of Refrigeration* 120: 12–21. https://doi.org/10.1016/j.ijrefrig.2020.06.017.

13 Jribi, S., Saha, B.B., Koyama, S. et al. (2013). Study on activated carbon/HFO-1234ze(E) based adsorption cooling cycle. *Applied Thermal Engineering* 50 (2): 1570–1575. https://doi.org/10.1016/J.APPLTHERMALENG.2011.11.066.

14 Ye, L., Islam, M.A., Rupam, T.H. et al. (2023). Study on the adsorption characteristics of Maxsorb III/HFO-1234ze(E) pair for adsorption refrigeration applications. *International Journal of Refrigeration* 146: 248–260. https://doi.org/10.1016/j.ijrefrig.2022.11.005.

15 Ghazy, M., Harby, K., Askalany, A.A., and Saha, B.B. (2016). Adsorption isotherms and kinetics of activated carbon/difluoroethane adsorption pair: theory and experiments. *International Journal of Refrigeration* 70: 196–205. https://doi.org/10.1016/j.ijrefrig.2016.01.012.

16 Habib, K., Saha, B.B., Rahman, K.A. et al. (2010). Experimental study on adsorption kinetics of activated carbon/R134a and activated carbon/R507A pairs. *International Journal of Refrigeration* 33 (4): 706–713. https://doi.org/10.1016/j.ijrefrig.2010.01.006 (accessed 20 July 2023).

17 Mariwala, R.K., Acharya, M., and Foley, H.C. (1998). Adsorption of halocarbons on a carbon molecular sieve. *Microporous and Mesoporous Materials* 22 (1–3): 281–288. https://doi.org/10.1016/S1387-1811(98)00109-7.

18 Elbassoussi, M.H. and Zubair, S.M. (2022). Performance evaluation of a cascading adsorption cooling system using a robust numerical model. *Arabian Journal for Science and Engineering* 47 (12): 16533–16550. https://doi.org/10.1007/s13369-022-07420-1.

19 Dakkama, H.J., Elsayed, A., AL-Dadah, R.K. et al. (2017). Integrated evaporator–condenser cascaded adsorption system for low temperature cooling using different working pairs. *Applied Energy* 185: 2117–2126. https://doi.org/10.1016/j.apenergy.2016.01.132.

20 Palomba, V., Vasta, S., Frazzica, A. et al. (2017). Analysis and testing of a novel cascaded adsorption-compression chiller for industrial applications. *12th IEA Heat Pump Conference.*

21 Saha, B.B., El-Sharkawy, I.I., Chakraborty, A. et al. (2006). Evaluation of minimum desorption temperatures of thermal compressors in adsorption refrigeration cycles. *International Journal of Refrigeration* 29 (7): 1175–1181. https://doi.org/10.1016/j.ijrefrig.2006.01.005.

22 Bell, I.H., Wronski, J., Quoilin, S., and Lemort, V. (2014). Pure and pseudo-pure fluid thermophysical property evaluation and the open-source thermophysical property library Coolprop. *Industrial and Engineering Chemistry Research* 53 (6): 2498–2508. https://doi.org/10.1021/ie4033999.

23 Saha, B.B., Chakraborty, A., Koyama, S. et al. (2008). Isotherms and thermodynamics for the adsorption of n-butane on pitch based activated carbon. *International Journal of Heat and Mass Transfer* 51 (7–8): 1582–1589. https://doi.org/10.1016/J.IJHEATMASSTRANSFER.2007.07.031.

24 Aristov, Y.I. (2014). Adsorption dynamics in adsorptive heat transformers: review of new trends. *Heat Transfer Engineering* 35 (11–12): 1014–1027. https://doi.org/10.1080/01457632.2013.863053.

25 Aristov, Y.I., Sapienza, A., Ovoshchnikov, D.S. et al. (2012). Reallocation of adsorption and desorption times for optimisation of cooling cycles. *International Journal of Refrigeration* 35 (3): 525–531. https://doi.org/10.1016/j.ijrefrig.2010.07.019.

26 Askalany, A.A., Saha, B.B., and Habib, K. (2014). Adsorption cooling system employing activated carbon/R32 adsorption pair. *MATEC Web of Conferences* 13: 06001. https://doi.org/10.1051/MATECCONF/20141306001.

27 Islam, Md.A. (2019). Thermophysical and adsorption characterization of functional and pore modified adsorbents. PhD thesis. Kyushu University. https://doi.org/hdl.handle.net/2324/2534480

28 Renaldi, R., Kiprakis, A., and Friedrich, D. (2017). An optimisation framework for thermal energy storage integration in a residential heat pump heating system. *Applied Energy* 186: 520–529. https://doi.org/10.1016/J.APENERGY.2016.02.067.

3

Optimal Utilization of Waste Biomass for the Development of Minimal Emission Sustainable Cooling Systems

Md. Amirul Islam[1,2] *and Bidyut Baran Saha*[1,3]

[1] *International Institute for Carbon-Neutral Energy Research (WPI-I²CNER), Multiscale Science and Engineering for Energy and the Environment Thrust, Kyushu University, Fukuoka, Japan*
[2] *Department of Electrical and Electronic Engineering, Bangabandhu Sheikh Mujibur Rahman Science & Technology University, Gopalganj, Bangladesh*
[3] *Department of Mechanical Engineering, Kyushu University, Fukuoka, Japan*

3.1 Introduction

Cooling systems have become an essential component of daily life to preserve food, beverages, and medicines, manufacture commodities and chemicals, keep machinery from overheating, and provide thermal comfort, PV panel cooling, ice manufacture, truck, ship, and airplane transportation of perishable products, data center cooling, and so on [1–6]. A conventional refrigeration system generates cooling at the cost of electricity consumption and harmful gas emissions, which contribute to global warming, high maintenance costs, health risks, noise pollution, and so on [7, 8]. In contrast, ACSs are energy efficient, environmentally friendly, and low maintenance; have a long lifespan and flexible installation even at remote or off-grid sites; and have a quiet and reliable operation [9–11]. However, the major drawbacks of ACS are that it is spacious, has a low cooling capacity, has a high initial cost, has limited adsorbent/refrigerant pairs availability, etc. [12]. Numerous researchers have synthesized AC from biomass sources such as coconut shells [13], bamboo [14], waste mangrove [15], rice husk [16], palm leaves [17], banana peel [18], corncob [19], walnut shells [20], coffee residue [21], and so on. Researchers have proposed various working fluids to pair with commercially ACs for adsorption cooling and other applications. Table 3.1 shows an overview of the feasible adsorbents employed for various types of available HFC adsorption.

Since, in this study, R32 adsorption was performed onto biomass-based ACs, the functional pairs in Table 3.1 include R32 adsorption performances with various adsorbents. Some of the high porosity carbons, such as spherical activated carbon

Advancements in Non-Conventional Cooling and Thermal Storage Strategies, First Edition.
Edited by Bidyut Baran Saha and Dibakar Rakshit.

Table 3.1 Adsorption characteristics of various types of adsorbents paired with HFC refrigerants.

Pairs	Description	Results	References
SAC/R32	R32 adsorption onto spherical phenol resin-based activated carbon was measured experimentally from 30 to 110 °C	$S_{BET} = 2992\,m^2/g$; $W_{max} = 2.23\,kg_{ref}/kg_{ads}$ @30 °C, 1.67 MPa; DA model parameters: $W_0 = 3.1344\,cm^3_{ref}/g_{ads}$, $E = 3.4981\,kJ/mol$, $n = 1.0217$, $k = 3.65$. Tóth model parameters: $C_0 = 3.9981\,kg_{ref}/kg_{ads}$, $b_0 = 2.20 \times 10^{-7}\,kPa$, $Q_{st} = 19.067\,kJ/mol$, $t = 1.8873$	[22]
Maxsorb III/R32	R32 adsorption onto activated carbon powder and fiber was measured experimentally from 25 to 75 °C	$S_{BET} = 3200\,m^2/g$; Tóth model parameters: $C_0 = 2.25\,kg_{ref}/kg_{ads}$, $b_0 = 1.71 \times 10^{-7}\,kPa$, $Q_{st} = 21.878\,kJ/mol$, $t = 1.07$	[23]
ACF/R32		$S_{BET} = 2200\,m^2/g$; Tóth model parameters: $C_0 = 2.08\,kg_{ref}/kg_{ads}$, $b_0 = 2.33 \times 10^{-7}\,kPa$, $Q_{st} = 21.872\,kJ/mol$, $t = 0.64$	[24]
Zeolite 5A/R32	R32 adsorption with Zeolite 5A was measured experimentally from 25 to 65 °C for gas separation	Tóth model parameters: $C_0 = 0.1526\,kg_{ref}/kg_{ads}$, $a_T = 5.13 \times 10^{-2}\,MPa^n$, $n = 0.13$	[25]
EcoSorb/R32	Commercially available EcoSorb was employed for R32 adsorption at 10, 30, and 50 °C	$S_{BET} = 1100\,m^2/g$; the maximum uptake is about $0.192\,kg_{ref}/kg_{ads}$ at 10 °C	[26]
Activated carbon composite/R32	Adsorption of R32 was performed onto Maxsorb (80%) + ionic liquid (10%) + PVA (10%) composite at 30, 50, and 70 °C	$S_{BET} = 2057\,m^2/g$; the maximum uptake is about $1.09\,kg_{ref}/kg_{ads}$ at 30 °C	[27]
Maxsorb III/R134a	HFC-134a adsorption onto commercially available activated carbon named Maxsorb III was performed from 30 to 60 °C	A maximum uptake of $1.945\,kg_{ref}/kg_{ads}$ was observed	[28]

Table 3.1 (Continued)

Pairs	Description	Results	References
AquaSorb 2000/R407C	HFC-407C adsorption was performed on commercially available activated carbon named AquaSorb 2000 from 25 to 75 °C	S_{BET} = 1050 m^2/g; the maximum uptake is about 0.42 kg_{ref}/kg_{ads} at 25 °C	[29]
AquaSorb 2000/R404A	HFC-404A adsorption was performed on commercially available activated carbon named AquaSorb 2000 from 25 to 75 °C	S_{BET} = 1050 m^2/g; the maximum uptake is about 0.5 kg_{ref}/kg_{ads} at 25 °C	[30]
ACF-A20/R507A	HFC-507A adsorption was performed at 5 and 40 °C	S_{BET} = 1930 m^2/g; the maximum uptake is about 1.19 kg_{ref}/kg_{ads}	[31]
Maxsorb III/R507A		S_{BET} = 3150 m^2/g; the maximum uptake is about 2.05 kg_{ref}/kg_{ads}	

(SAC) and Maxsorb III, show high adsorption capacity of R32 and other HFCs in Table 3.1.

The literature survey indicates that highly microporous AC has the capacity to adsorb a higher quantity of HFCs. Hence, one waste mangrove and one waste palm trunk-derived AC were investigated in this study. R32 was selected as the working fluid due to its acceptable GWP and lower environmental impact. Experimental data were collected at 30, 50, 70, and 90 °C adsorption temperatures using a high-accuracy TGA apparatus. Experimental data are fitted with DA and Tóth equations to propose a generalized equation for both samples. Specific cooling effect (SCE) and coefficient of performance (COP) analysis has also been performed. Finally, total equivalent warming impact (TEWI) analysis has been performed to compare the environmental friendliness of the ACSs.

3.2 Experimental

3.2.1 Waste Raw Precursors: Mangrove and Palm

Mangroves are found in tropical and subtropical coastal areas of intertidal zones, where land and sea meet. The Sundarbans in Bangladesh and India have the largest continuous mangrove forest in the world. Various species of mangroves

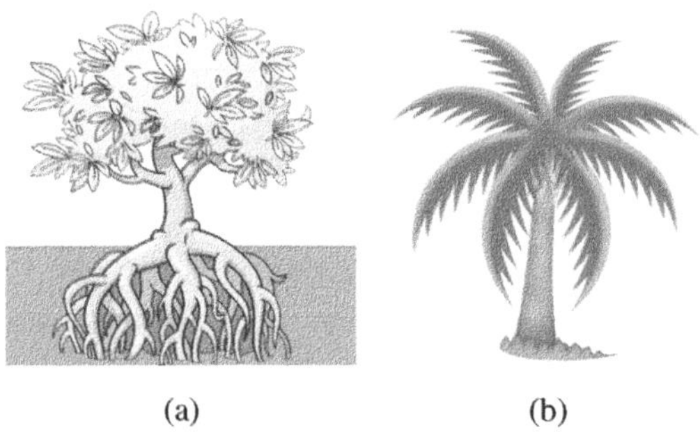

Figure 3.1 Snapshots of typical (a) mangrove and (b) palm tree found in the wild.

are also found in Indonesia, Malaysia, Thailand, Myanmar, Philippines, Sri Lanka, East Africa, America, Mexico, the Caribbean islands, coasts of Australia, Emirates, Qatar, Bahrain, Oman, Yemen, Solomon Islands, etc. Mangroves have special features such as salt filtration, aerial roots, viviparous seeds, and tolerance to salt and low oxygen levels. Therefore, they can survive in extreme climate conditions (Figure 3.1).

Palm trees are also grown primarily in tropical and subtropical climates associated with warm, sunny environments. India, Sri Lanka, Maldives, Indonesia, Malaysia, Thailand, Nigeria, Ghana, Cameroon, Madagascar, North America, Mexico, Costa Rica, Panama, Jamaica, Cuba, Saudi Arabia, United Arab Emirates, Iraq, Egypt, Spain, Italy, Morocco, Tunisia, Fiji, and Tahiti are the sanctuary of palm trees.

Dead branches and roots of these trees produce a huge amount of waste every year. Mostly, these wastes are rotted and mixed with the soil or are burnt. One of the best uses of these waste biomasses could be to utilize those as the precursor to produce AC. These biomasses have a high carbon content and are capable of producing high-grade AC. The precursors of this study were selected due to their wide availability, inexpensiveness, low impurity, and high carbon percentage and were collected from Indonesia.

3.2.2 Activated Carbon Synthesis

Initially, the raw sample was dried and crushed to powder form. This wood powder was then heated at 600 °C for one hour in a neutral environment with nitrogen flow at 100 ml/min rate. This step is called carbonization, while volatile water vapor and other volatile materials are washed out. The carbonized sample was then mixed with KOH. The amount of KOH was six times of the carbonized sample. The mixture was again heated in that neutral environment at 900 °C. This time, KOH reacted with the carbon and created pores. Remaining KOH and other chemical byproducts were then removed by washing several times with HCl. The acidic activated sample was then washed with deionized water until the pH level reached 7. The sample was then filtered, dried, and vacuum dried for several

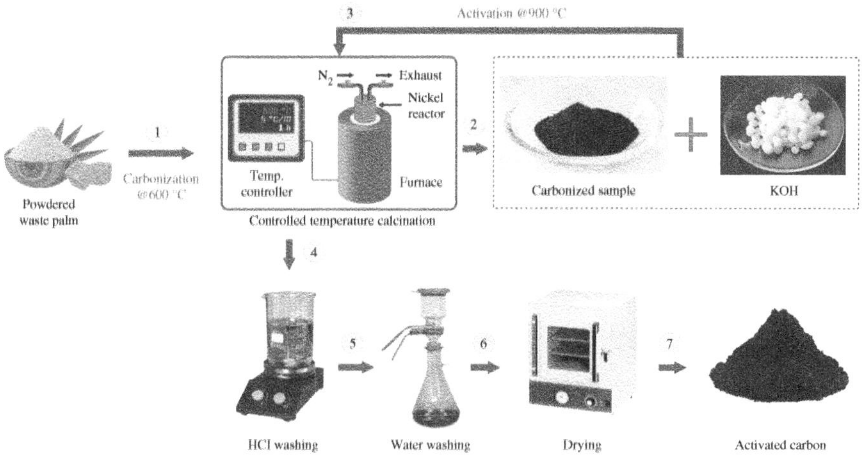

Figure 3.2 Chemical activation process of waste biomass into activated carbon. Source: Ref. [32]/Reproduced from Taylor and Francis.

hours. The synthesis process is illustrated in Figure 3.2 and the porous properties of the synthesized ACs at different conditions are presented in Table 3.2. Despite the different precursors, the surface areas of both samples were found to be almost identical. The micropore volume of the waste palm-derived sample showed a higher value than the mangrove precursor. However, the activation yield of mangrove-derived AC was about 10% higher than the palm-derived AC.

3.2.3 Adsorbate: R32

Widely used refrigerants in the current generation air conditioning and refrigeration systems are R134a, R404A, and R410A [32, 33]. These refrigerants have very high global warming potential and must be phased out by 2030. Therefore, R32 is selected as the adsorbate in this study. R32 is the latest generation refrigerant with a tolerable GWP (675 kg eq. CO_2) value according to the Montreal and Kyoto Protocol. In addition, R32 is more efficient than other refrigerants, which can assist in reducing energy use and electricity tariffs. Compared to other refrigerants, R32 is reasonably priced and more economically advantageous for producers and consumers. Heat transfer efficiency and other thermodynamic properties of R32 provide faster cooling and better performance compared to other systems. The properties of R32 refrigerants along with other widely used refrigerants are shown in Table 3.3.

3.2.4 Thermogravimetric Analyzer (TGA)

A high-precision thermogravimetric analyzer (TGA) apparatus was employed to acquire the adsorption isotherms of R32 onto waste biomass-derived ACs. The

Table 3.2 Properties of the studied waste mangrove and palm-derived samples [34].

Sample name	Total surface area S_{BET} (m^2/g)	External surface area S_{EXT} (m^2/g)	Micropore volume υ_{μ_pore} (cm^3/g)	Total pore volume υ_p (cm^3/g)	Average pore width w_p (nm)	Thermal conductivity λ (W/m K)	Activation yield (%)
M-AC-C600-A900-K6	2924	34.19	2.13	2.18	1.47	0.052	37.54
P-AC-C600-A900-K6	2927	62.52	2.41	2.51	1.68	0.044	27.27

Note: M, mangrove; P, palm: AC, activated carbon; C600, carbonization temperature 600 °C; A900, activation temperature 900 °C; K6 means that the carbonized sample to KOH ratio is 6 times.

Table 3.3 Properties comparison of R32 with current generation refrigerants.

Alternative names	R32, HFC32, Difluoromethane	R134a, 1,1,1, 2-Tetrafluoroethane, HFC134a	R410A, HFC410A
Chemical formula	CH_2F_2	CH_2FCF_3	CH_2F_2 (50%) C_2HF_5 (50%)
Molar mass (g/mol)	52.024	102.03	72.585
GWP (kg CO_2 equiv)	677	1300	1924
ODP	0	0	0
Critical temperature (°C)	78.105	101.1	71.342
Critical pressure (MPa)	5.782	4.06	4.901
Normal boiling point (°C)	−51.651	−26.074	−48.5
Triple point (°C)	−136.81	−103.3	−73.15
ASHRAE toxicity	No	No	No
ASHRAE safety group	A2L	A1	A1
ASHRAE flammability	Yes (mild)	No	No

Source: Refs. [32, 45, 46].

schematic in Figure 3.3 shows the major components of the TGA manufactured by Nippon Bell Co. Ltd. The magnetic suspension balance (MSB) is the most special part of the TGA which has an accuracy of ±30 μg. The MSB, sample holding bucket, heater unit, shields, and insulations are elaborately drawn in Figure 3.3.

The TGA apparatus is equipped with a rotary pump, a diaphragm pump, and a turbomolecular pump to evacuate various sections while necessary. These three pumps is capable of lowering the pressure to 1×10^{-5} Pa to obtain data with greater accuracy. In TGA, adsorption temperatures can be varied from 10 to 150 °C with a pressure swing up to 10 MPa pressure. The pressure of the adsorption cell is controlled using a Julabo oil bath.

Initially, the R32 gas is flown from the cylinder and liquefied in a refrigerant reservoir at a lower temperature (−5 °C). During the measurement, the gas cylinder is isolated by shutting the valve $V_{10,}$ and refrigerant is supplied from the liquid reservoir vessel. The temperature of the vessel is controlled to obtain a long range of low to high-pressure adsorption isotherms.

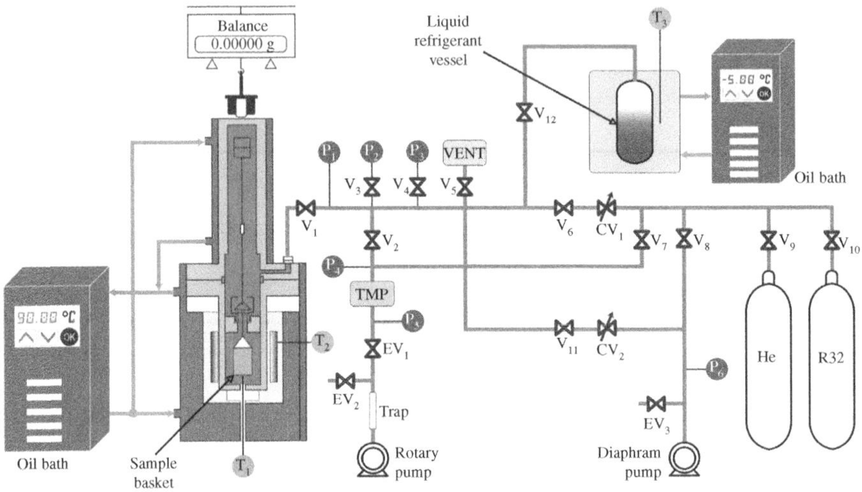

Figure 3.3 Schematic of the high-precision TGA apparatus for adsorption isotherm measurement. Here, EV_1 - EV_3 → automatic two-way valves; P_1 - P_6 → pressure sensors; T_1 - T_3 → temperature sensors; TMP → turbomolecular pump; V_1 - V_{12} → automatic two-way valves; CV_1 - CV_2 → flow control valves.

3.2.5 Measurement Procedure

A small amount of sample (20–100 mg) was put into the sample holding basket and then the basket was attached to the hook of the MSB, as shown in Figure 3.3. The basket was then surrounded by metal jackets, shields, and insulators to make the structure leakproof. Leakage tests were performed in both vacuum and high-pressure conditions to avoid experimental errors. Simultaneous heating (~150 °C) and vacuuming were performed for several hours to remove the gases and volatile materials. The mass decrement occurred at that time due to the removal of the volatile materials, and the built-in computer program automatically corrected the modified sample mass.

The preferred adsorption temperature was then set on the isothermal bath and waited until the temperature was stable. A very small amount of R32 was dosed from the refrigerant vessel to the tube by opening valve V_{12}. If the gas is overcharged, the vacuum pump removes the extra gas to match the target pressure. When the pressure was stable, valve V_1 was opened to let the dosed R32 gas flow into the adsorption chamber. At that time, a certain amount of R32 was adsorbed by the AC sample, and the MSB showed a mass change. In this way, one point of the adsorption was obtained. Hence, another dosing was performed to obtain the second point of the isotherm, and the process was repeated to get all the data points up to the saturation pressure.

3.2.6 Isotherm Models

The experimental data of R32 adsorption onto the samples were conducted at 30, 50, 70, and 90 °C with a pressure variation from 0 to 3 MPa. A practical system might require isotherms at other temperature and pressure regions. Hence, the experimental data were correlated with well-known modified DA and Tóth isotherm models. Once the fitting parameters of the models are evaluated, the adsorbent/adsorbate pair could be designed for any operating conditions.

The modified DA model [35, 36] is expressed by Eq. (3.1). Equations (3.2) and (3.3) define specific parameters of the model,

$$W = W_0 \exp\left[-\left\{\frac{RT\ln\left(\frac{P_s}{P}\right)}{E}\right\}^n\right] \tag{3.1}$$

Here,

$$W = CV_m \tag{3.2}$$

$$V_m = V_t \exp[\alpha(T - T_t)] \tag{3.3}$$

REFPROP database includes the specific volume of R32 (i.e. V_t) at triple-point (−136.81 °C) temperature which is 0.69966 cm^3/g. Moreover, the thermal expansion coefficient α was considered constant (0.0025 K) by many researchers [37–39]. Although, some researchers proposed that the value of α is inversely proportional to temperature (T) [40, 41]. However, the variation was not very significant in our operating temperature range and was considered constant.

In a practical situation, the thermal expansion of the adsorbate happens during adsorption. Hence, a volume correction must be performed in the case of the DA model. The correction in the specific volume of the adsorbed phase (V_m) is estimated using Eq. (3.3).

The DA model equation contains the term P_s, which is the saturation pressure of the adsorbate gas at adsorption temperature. However, the phase of the adsorbate is undefined above critical temperature. Therefore, the pseudo-saturated vapor pressure relation is employed to estimate the value of P_s, which can be defined by using Eq. (3.4)

$$\frac{P_s}{P_C} = \left(\frac{T_{ads}}{T_C}\right)^k \tag{3.4}$$

The exponential term k is a heterogeneity coefficient that can be determined from the correlation fitting of the modified DA model. Hence, four adjustable parameters (W_0, E, n, and k) were estimated from the experimental data fitting.

The Tóth model, which is also often used by researchers, can be expressed by the following equations:

$$C = C_0 \left[\frac{bP}{(1 + (bP)^t)^{\frac{1}{t}}} \right] \tag{3.5}$$

Here,

$$b = b_0 \exp\left(\frac{Q_{st}}{RT}\right) \tag{3.6}$$

Here, C and C_0 are the equilibrium and maximum adsorption uptake, respectively. Besides, P, T, t, b_0, R, and Q_{st} are the equilibrium pressure, adsorption temperature, heterogeneity factor, adsorption affinity at infinite temperature, universal gas constant, and isosteric heat of adsorption, respectively. When $t = 1$, the equation of the Tóth model represents Langmuir's model.

3.2.7 Isosteric Heat of Adsorption

During adsorption, adsorbate molecules collide with the adsorbent particles located on the pore walls. This impact produces heat which is referred to as the isosteric heat of adsorption (Q_{st}). By definition, Q_{st} is the quantity of heat that is released or absorbed when one mole of a substance is adsorbed onto a surface under constant temperature conditions. Q_{st} is a significant thermodynamic property that provides information on the interaction strength between adsorbent/adsorbate pairs. Q_{st} can be defined by the following well-known Clausius–Clapeyron Equation (3.7):

$$Q_{st} = -RT^2 \left[\frac{\partial \ln P}{\partial T}\right]_C \tag{3.7}$$

The abovementioned DA Equation (3.1) can be rearranged and can be expressed by Eq. (3.8),

$$P = \frac{P_S}{\exp\left[\frac{E}{RT}\left\{-\ln\left(\frac{W}{W_0}\right)\right\}^{\frac{1}{n}}\right]} \tag{3.8}$$

By taking ln on each side and then differentiating with respect to T, it results in Eq. (3.9),

$$\frac{\partial \ln P}{\partial T} = \frac{\partial \ln(P_S)}{\partial T} + \frac{E}{RT^2}\left\{-\ln\left(\frac{W}{W_0}\right)\right\}^{\frac{1}{n}} \tag{3.9}$$

After substituting Eq. (3.7) into Eq. (3.9), the expression of Q_{st} can be written by Eq. (3.10),

$$-Q_{st} = RT^2\frac{\partial \ln(P_S)}{\partial T} + E\left\{-\ln\left(\frac{W}{W_0}\right)\right\}^{\frac{1}{n}} \tag{3.10}$$

The surface coverage of the adsorbate can be defined by $\Theta = \frac{W}{W_0}$, and the above Eq. (3.10) can be rewritten as Eq. (3.11) [42],

$$Q_{st} = h_{fg} + E(-\ln\Theta)^{\frac{1}{n}} \tag{3.11}$$

Here, h_{fg} is the latent heat of vaporization in (kJ/kg). Above the critical temperature, the h_{fg} term can be approximated as kRT, and Eq. (3.11) can be written as Eq. (3.12),

$$Q_{st} = kRT + E(-\ln\Theta)^{\frac{1}{n}} \tag{3.12}$$

Rahman et al. [43] proposed a modification of the above model, which is defined by Eq. (3.13),

$$Q_{st} = h_{fg} + E\left\{(-\ln\Theta)^{\frac{1}{n}} + \frac{T\alpha}{n}(-\ln\Theta)^{\frac{1}{n}-1}\right\} \tag{3.13}$$

Above the critical temperature, Eq. (3.13) can be approximated as Eq. (3.14),

$$Q_{st} = kRT + E\left\{(-\ln\Theta)^{\frac{1}{n}} + \frac{T\alpha}{n}(-\ln\Theta)^{\frac{1}{n}-1}\right\} \tag{3.14}$$

Chakraborty et al. [44] further modified the model by including a nonideality term of the adsorbate and defined Q_{st} by Eq. (3.15):

$$Q_{st} = h_{fg} + E\left\{(-\ln\Theta)^{\frac{1}{n}} + \frac{T\alpha}{n}(-\ln\Theta)^{\frac{1}{n}-1}\right\} + Tv_g\frac{dP}{dT} \tag{3.15}$$

For temperatures above critical, the Chakraborty model is expressed by Eq. (3.16),

$$Q_{st} = kRT + E\left\{(-\ln\Theta)^{\frac{1}{n}} + \frac{T\alpha}{n}(-\ln\Theta)^{\frac{1}{n}-1}\right\} + Tv_g\frac{dP}{dT} \tag{3.16}$$

3.2.8 Specific Cooling Effect (SCE)

In an adsorption heat pump, the *SCE* refers to the produced cooling amount per kilogram of adsorbent material. *SCE* is a significant performance parameter of an ACS as it quantitatively assesses the cooling capacity relative to the amount of adsorbent material required. Higher values of *SCE* indicate a more efficient adsorption heat pump as more cooling can be produced with less amount of adsorbent. The *SCE* of an ACS can be determined using Eq. (3.17),

$$SCE = (C_{\max} - C_{\min})\left[\Delta H_{eva} - \int_{T_{eva}}^{T_{con}} c_{p_{ref}}\,dT\right] \tag{3.17}$$

3.2.9 Coefficient of Performance (COP)

The COP is a measure of the energy efficiency of an ACS. The COP of an ACS is cooling capacity over heat input, which is shown in Eq. (3.18),

$$COP = \frac{SCE}{Q_{total}} \tag{3.18}$$

The total heat input can be determined from Eq. (3.19),

$$Q_{total} = Q_{sensible} + Q_{latent} \tag{3.19}$$

The sensible heat mentioned in Eq. (3.19) depends on the specific heat capacity of the adsorbent, refrigerant, and adsorption uptake of the pair. The expression of sensible heat can be given by Eq. (3.20). Besides, the latent heat is dependent on Q_{st} and the uptake, as shown in Eq. (3.21),

$$Q_{sensible} = \underbrace{\int_{T_{f,ads}}^{T_{f,des}} C_{p_{ads}} dT}_{Pre\text{-heating+Desorption of adsorbent}} + \underbrace{W_{\max} \int_{T_{f,ads}}^{T_{i,des}} C_{p_{ref}} dT}_{Pre\text{-heating of refrigerant}} + \underbrace{\int_{T_{i,des}}^{T_{f,des}} WC_{p_{ref}} dT}_{\text{Desorption of refrigerant}} \tag{3.20}$$

$$Q_{latent} = Q_{st} \int_{W_{\min}}^{W_{\max}} dW \tag{3.21}$$

3.3 Results and Discussion

Experimentally obtained data and their fitting curves are plotted in Figure 3.4. The evaluated fitting parameters of both models are also presented in Table 3.3. The P-AC sample exhibits remarkably higher uptake than the M-AC sample due to its higher micropore volume, as mentioned earlier in Table 3.4. The maximum uptakes of R32 of both the studied samples were significantly higher than activated carbon powder (ACP), activated carbon fiber (ACF), and spherical phenol resin-based ACs by the previous researchers [22, 24]. Hence, the studied P-AC sample holds the current benchmark of R32 adsorption.

Modified DA models exhibit better fitting than the Tóth model according to the root-mean-square deviation (RMSD) values. Therefore, it would be more appropriate to use the modified DA equations and the corresponding fitting parameters for predicting the unknown isotherms of the studied samples.

The isosteric heat of adsorption (Q_{st}) for the models of Clausius–Clapeyron, Rahman et al., and Chakraborty et al. was estimated and plotted in Figure 3.5. For better readability, individual Q_{st} curves were also plotted and shown in the upper part of the same figure. Initially, Q_{st} was very high (for any adsorption temperature) and tended to decrease with increasing adsorption uptake. This happened because adsorbate molecules were first adsorbed in the narrower sites,

Table 3.4 Adjustable parameters of modified DA and Tóth models were determined using the RMSD method.

	Modified DA model parameters					Tóth model parameters				
Sample name	W_0 (cm³/g)	E (J/mol)	n	k	RMSD (%)	C_0 (kg/kg)	b_0 (kPa)	Q_{st} (J/mol)	t	RMSD (%)
M-AC-C600-A900-K6	2.373	4055.28	1.163	2.090	3.37	2.677	5.233×10^{-8}	23,748	3.243	8.76
P-AC-C600-A900-K6	2.598	3566.35	1.100	4.488	3.75	2.665	5.280×10^{-8}	23,663	3.801	4.99

and a strong interaction occurred between the surface particles and adsorbate. Hence, a higher Q_{st} was observed in the beginning. When the narrow sites are occupied, adsorbate molecules accumulate in larger pores. Therefore, the adsorption affinity starts to decrease, and Q_{st} gradually approaches the latent heat of vaporization.

The interaction between adsorbent particles and adsorbent molecules is stronger at lower adsorption temperature. Hence, the Q_{st} curve shifts the lower part of the y-axis when the adsorption temperature increases. However, at T_{ads} = 90 °C, which is above the critical temperature of R32, the interaction is exceptional. During this, the critical state R32 molecules interact with AC particles, and the curve shifts upward to the y-axis. The parameter k mentioned in Table 3.4 plays an important role, and the curve almost coincides with the Q_{st} curve of T_{ads} = 30 °C. This phenomenon is very interesting because although the temperature and pressure are high, the molecular interaction produces less heat at T_{ads} = 90 °C.

From Figure 3.5, it is clear that the Clausius–Clapeyron model underpredicts Q_{st} and is limited to ideal gases, whereas Rahman et al. and Chakraborty et al. considered the nonideality of the adsorbate and provided more accurate predictions of the isosteric heat of adsorption.

SCE at different evaporation temperatures as a function of desorption temperature was determined and drawn in Figure 3.6 for both M-AC and P-AC samples. A higher desorption temperature or higher evaporation temperature positively affects the SCE. The minimum desorption temperatures of this sample are also significantly low. For air-conditioning applications (T_{eva} = 20 °C), the SCE can be as high as 490 kJ/kg if the P-AC sample is employed. Even for refrigeration applications (T_{eva} = −5 °C), the SCE can be higher than 245 kJ/kg. Figure 3.6 also indicates a sharp rise in SCE up to the critical temperature.

The COP curves at different evaporation temperatures were drawn in Figure 3.7 for the two samples. The waste palm-derived sample shows higher COP than the

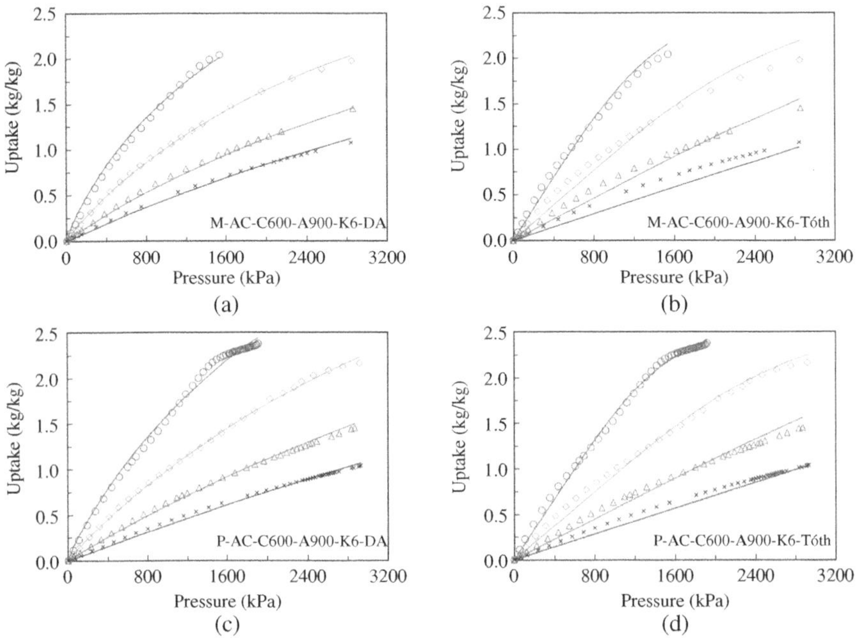

Figure 3.4 Experimental adsorption isotherm data points and correlated curves. Experimental isotherms are indicated using markers at adsorption temperatures: ○ 30 °C; ◇ 50 °C; Δ 70 °C; × 90 °C. Experimental data correlated with DA and Tóth models are shown using solid lines. Isotherm data of waste mangrove-derived samples are shown in (a) and (b) with fitted DA and Tóth models, respectively. Likewise, the isotherm data of waste palm trunk-derived samples are shown in (c) and (d) with fitted DA and Tóth models, respectively.

mangrove-derived sample. The COP sharply increases with desorption temperature until the critical point, and the slope is relatively lower after that. The COP of P-AC/R32 can be considered high for designing an efficient ACS.

3.4 Ecological Footprints

Whether it is a traditional vapor compression system or adsorption based, every cooling system produces GHGs directly or indirectly. Hence, estimating the TEWI of both systems is necessary to get an idea of their environmental impact. In order to assess TEWI, both direct and indirect global warming impacts need to be estimated.

Direct emission might occur due to the leakage of the refrigerant. Besides, indirect emission happens due to electricity consumption, manufacturing of the unit, disposal of refrigerants at the end of its lifetime, and so on.

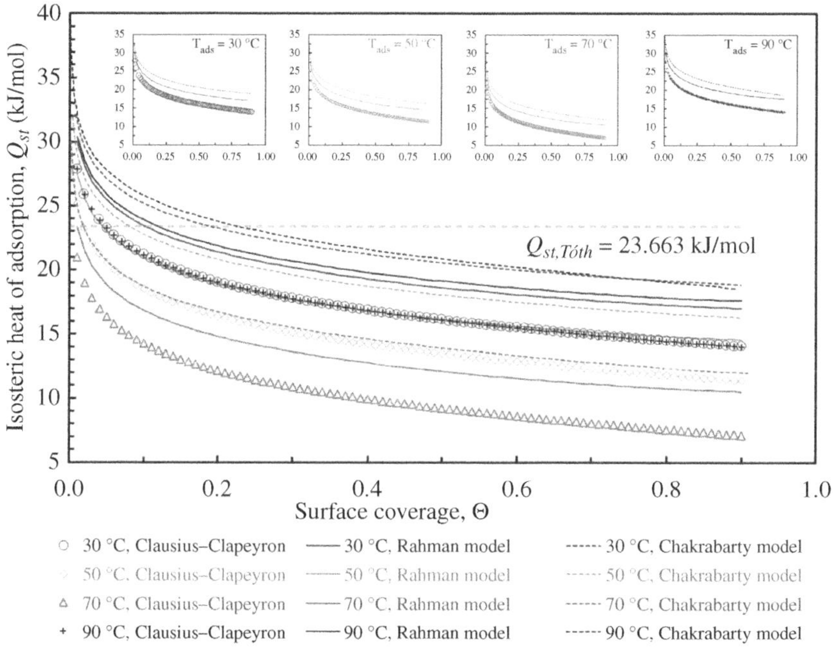

Figure 3.5 Isosteric heat of adsorption comparison of P-AC-C600-A900-K6 at different adsorption temperatures and surface coverage. Markers represent Q_{st} from the original Clausius–Clapeyron model, solid lines are for the Rahman et al. model, and dashed lines are for the Chakraborty et al. model.

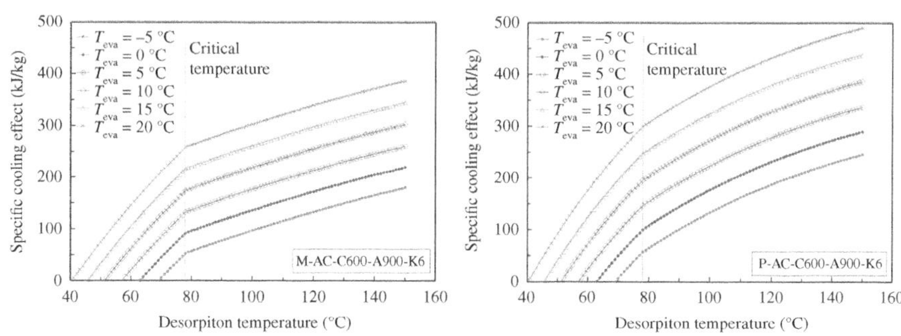

Figure 3.6 Specific cooling effect (SCE) comparison of M-AC and P-AC at different evaporation temperatures.

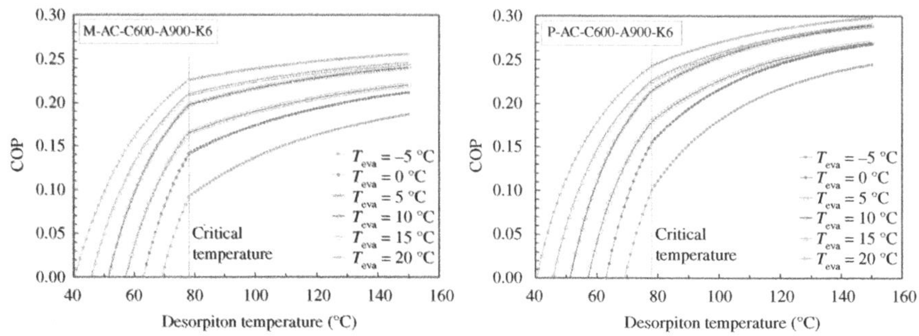

Figure 3.7 Coefficient of performance (COP) comparison of M-AC and P-AC at different evaporation temperatures.

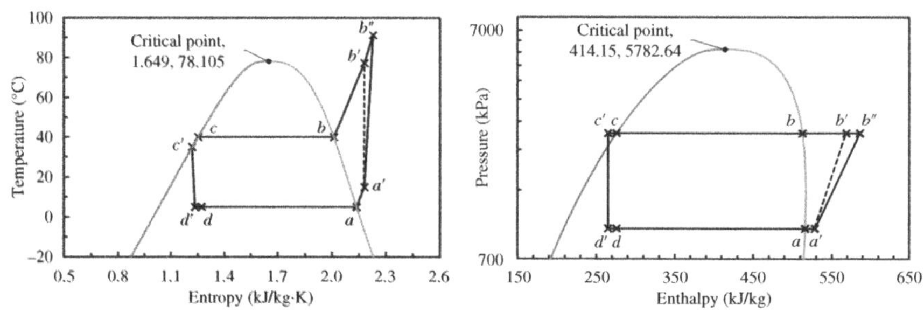

Figure 3.8 Temperature–entropy and pressure–enthalpy diagrams of vapor compression refrigeration cycle of R32 for conditions mentioned in Table 3.5.

TEWI provides an overall idea of the environmental impact of a refrigeration system and is commonly used to assess the sustainability of various cooling technologies. Lower TEWI value refers to lower environmental impact and greater sustainability.

This study assesses and compares the TEWI of a 20 kW ACS and a conventional refrigeration system. A simplified model is adopted for the assessment [9]. The assumptions and operating conditions are presented in Table 3.5.

In order to remove heat from the refrigerated space, the refrigerant is evaporated in the evaporator at low pressure. The change in enthalpy, entropy, and pressure depends on the properties of the refrigerant and the operating conditions. The properties can be obtained from the REFPROP database [45]. The temperature, pressure, enthalpy, and entropy of the refrigerant at different points were determined, and the refrigeration cycle is sketched in Figure 3.8. The COP of the system, energy consumption, and leaked refrigerant amount was estimated from the refrigeration cycle. Therefore, direct, indirect, and, finally, TEWI was assessed.

The result of the assessment is shown in Figure 3.9. The first two bars of the figure compares the indirect emission due to the power consumption between

Table 3.5 Assumptions, operating conditions, and system parameters for the environmental assessment.

Common parameters for both vapor compression and adsorption cooling system	
Cooling load	20 kW
Annual leakage rate of refrigerant	5%
GWP of refrigerant R32	677 kg CO_2 equiv
GWP for electricity generation	0.434 kg-CO_2/kWh [47]
System runtime	12 h/day
Required refrigerant amount	40 kg (approximately twice the cooling load) [9, 48]
GWP of per kg material production	Stainless steel: 2.13 kg-CO_2 equiv [49] Aluminum: 9.17 kg-CO_2 equiv [50] Copper: 4.97 kg-CO_2 equiv [50]
Evaporation temperature	5 °C
Condenser temperature	40 °C
Parameters applicable for VCRS	
Superheat	10 °C
Subcool	5 °C
Compressor's isentropic efficiency	70%
Product weight (indoor + outdoor unit)	100 kg [51, 52]
Approximate weight ratio of the raw materials to construct the VCRS	Stainless steel: 50% Copper: 20% Aluminum: 20% Plastic and others: 10%
System lifespan	15 years
Parameters applicable for ACS	
Energy consumption to drive pumps, sensors, valves, etc.	1.2 kW
Required adsorbent amount	70 kg
System weight	1200 kg [48]
Approximate weight ratio of the raw materials to construct the ACS	Stainless steel: 60% Aluminum: 10% Copper: 10% Others (plastic, refrigerant, adsorbent, heat exchanging water, etc.): 20%
System lifespan	25 years [48]

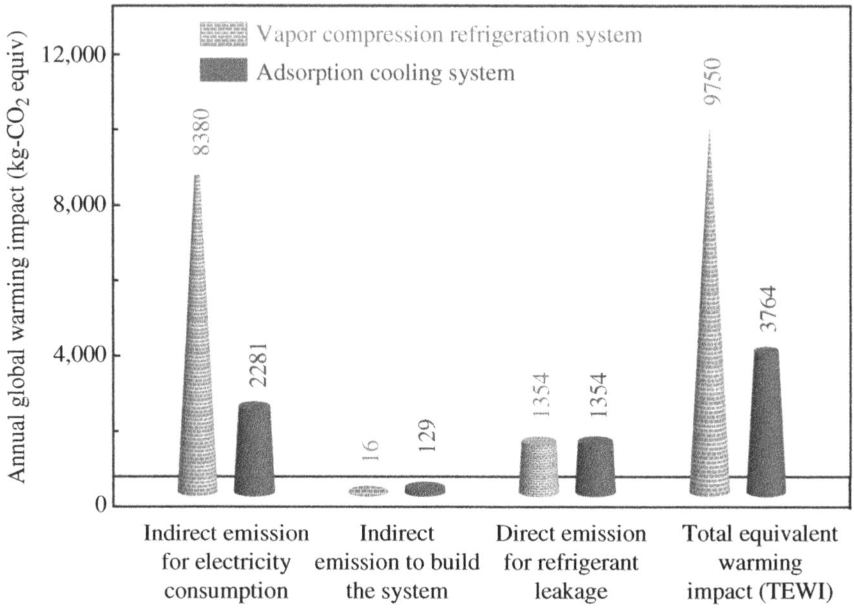

Figure 3.9 Indirect emission, direct emission, and total equivalent warming impact comparison between conventional and adsorption cooling systems at the same operating conditions.

vapor compression refrigeration system (VCRS) and ACS. A conventional cooling system utilizes more electric power than an ACS to drive the compressor. Besides, the ACS has no compressor and is generally powered by waste heat. However, electricity is required for ACS to drive the pumps, fans, sensors, etc. Hence, the ACS has a higher electrical COP, and the indirect emission is much lower. Secondly, the ACS is bulkier and requires more metal to build the system. Hence, the indirect emission of ACS due to metal production is higher than VCRS, as shown in Figure 3.9. Finally, direct emission is found to be the same due to the same operating conditions and leakage rate. In summary, if both VCRS and ACS are run by the same refrigerant, a small 20kW ACS can reduce about 5986 kg-CO_2 equiv emission. This 61% emission reduction can significantly contribute to saving the environment since millions of cooling units are operating in the whole world.

3.5 Conclusions

Adsorption of R32 onto waste mangrove and waste palm-derived ACs (M-AC and P-AC) were experimentally investigated in this research. Both samples were synthesized by the chemical activation method with the same operating conditions. Both the samples have very high but similar surface areas (2924 and

2927 m^2/g). However, P-AC has a slightly higher micropore volume than P-AC maybe because of the structural difference between the precursor materials. On the other hand, M-AC shows about 10% higher activation yield than the P-AC sample. Adsorption isotherms of R32 were experimentally obtained at 30, 50, 70, and 90 °C using a high-precision TGA apparatus. A record adsorption capacity of 2.677 kg/kg was achieved for the M-AC-C600-A900-K6/HFC32 pair. Between DA and Tóth isotherm models, the DA model shows better fitting with an RMSD error of less than 5%. Moreover, the models of Clausius–Clapeyron, Rahman et al., and Chakraborty et al. were used to determine the isosteric heat of adsorption (Q_{st}) of the pairs. The nonideality nature of the refrigerant gas was considered in the modified models, which provided a more appropriate trend of Q_{st} with surface coverage. Besides, the SCE and COP estimation revealed that the P-AC-C600-A900-K6/HFC32 pair was marginally better than the M-AC/R32 pair. In addition, the assesment showed that about 61% emission reduction is attainable compared to VCRS by employing R32 as the working fluid in ACS. All these results in this study indicate that a robust and sustainable ACS could be designed with either of these two pairs. However, P-AC/R32 would provide relatively higher performance than the M-AC/R32 pair.

Nomenclature

Symbol

A	activation temperature (°C)
AC	activated carbon
C	equilibrium adsorption uptake (kg/kg)
C_0	maximum adsorption uptake (kg/kg)
ACS	adsorption cooling system
b_o	equilibrium constant (kPa)
c_p	specific heat capacity (kJ/kg K)
COP	coefficient of performance
DA	Dubinin–Astakhov
E	adsorption characteristic energy (J/mol)
GWP	global warming potential (kg equiv CO_2)
HFC	hydrofluorocarbon
k	adjustable parameter of the modified DA equation
$K4$	carbonized sample to KOH ratio is 1 : 4
$K6$	carbonized sample to KOH ratio is 1 : 6
M	mangrove
$M\text{-}AC$	mangrove-derived activated carbon

MSB	magnetic suspension balance
n	exponential coefficient term of the DA equation
P	pressure (MPa)
P_C	critical pressure (MPa)
P_s	saturation pressure (MPa)
P-AC	palm-derived activated carbon
PTW	pressure-temperature-uptake
Q_{st}	isosteric heat of adsorption (J/mol)
R	universal gas constant (J/mol K)
RMSD	root-mean-square deviation (%)
t	heterogeneity factor
T	temperature (°C)
T_C	critical temperature (°C)
T_t	temperature at triple point (°C)
SCE	specific cooling effect (kJ/kg)
SAC	spherical activated carbon
TGA	thermogravimetric analyzer
TEWI	total equivalent warming impact (TEWI) (kg equiv CO_2)
VCRS	vapor compression refrigeration system
V_m	molar volume (cm^3/g)
V_t	specific volume of HFC32 at triple-point temperature (cm^3/g)
W	equilibrium volumetric uptake (cm^3/g)
W_0	maximum volumetric uptake, maximum volumetric adsorption capacity (cm^3/g)
ΔH_{eva}	enthalpy of vaporization (kJ/mol)

Greek Symbols

α	thermal expansion coefficient of the adsorbed refrigerant (K)
Θ	surface coverage
Δ	difference

Subscripts

ads	adsorption, adsorbent
con	condenser, condensation
eva	evaporation, evaporator
f	finished
i	initiated

max	maximum
min	minimum
ref	refrigerant

References

1 Aste, N., Del Pero, C., and Leonforte, F. (2017). Active refrigeration technologies for food preservation in humanitarian context – a review. *Sustainable Energy Technologies and Assessments* 22: 150–160. https://doi.org/10.1016/j.seta.2017.02.014.

2 Islam, M.A., Rupam, T.H., Hosan, S., Saha, B.B. (2022). Thermodynamic and environmental assessment of low-GWP alternative refrigerants for domestic cooling. *7th Natl. 1st Int. Conf. Refrig. Air Cond.*, Guwahati, India: pp. 1–11.

3 Isler-Kaya, A. and Karaosmanoglu, F. (2023). Life cycle assessment of a climate-friendly data center cooling device. *Energy and Buildings* 288: 113006. https://doi.org/10.1016/j.enbuild.2023.113006.

4 Zhao, Y., Zhang, X., and Xu, X. (2020). Application and research progress of cold storage technology in cold chain transportation and distribution. *Journal of Thermal Analysis and Calorimetry* 139: 1419–1434. https://doi.org/10.1007/s10973-019-08400-8.

5 Hassan, A.A., Hassan, H., Islam, M.A., and Saha, B.B. (2023). Thermoeconomic performance assessment of a novel solar-powered high-temperature heat pump/adsorption cogeneration system. *Solar Energy* 255: 71–88. https://doi.org/10.1016/j.solener.2023.03.017.

6 Wang, S.G. and Wang, R.Z. (2005). Recent developments of refrigeration technology in fishing vessels. *Renewable Energy* 30: 589–600. https://doi.org/10.1016/j.renene.2004.03.020.

7 Tassou, S.A. and Grace, I.N. (2005). Fault diagnosis and refrigerant leak detection in vapour compression refrigeration systems. *International Journal of Refrigeration* 28: 680–688. https://doi.org/10.1016/j.ijrefrig.2004.12.007.

8 Islam, M.A., Srinivasan, K., Thu, K., and Saha, B.B. (2017). Assessment of total equivalent warming impact (TEWI) of supermarket refrigeration systems. *International Journal of Hydrogen Energy* 42: 26973–26983. https://doi.org/10.1016/j.ijhydene.2017.07.035.

9 Islam, M.A. and Saha, B.B. (2020). TEWI assessment of conventional and solar powered cooling systems. In: *Solar Energy* (ed. H. Tyagi), 147–177. Springer Nature Singapore Pte Ltd. https://doi.org/10.1007/978-981-15-0675-8_9.

10 Pal, A., Rocky, K.A., and Saha, B.B. (2021). Thermodynamic analysis of promising biomass-derived activated carbons/CO_2 based adsorption cooling systems. *Journal of CO_2 Utilization* 46: 101457. https://doi.org/10.1016/j.jcou.2021.101457.

11 Singh, V.K. and Kumar, E.A. (2017). Experimental investigation and thermodynamic analysis of CO_2 adsorption on activated carbons for cooling system. *Journal of CO_2 Utilization* 17: 290–304. https://doi.org/10.1016/j.jcou.2016.12.004.

12 Islam, M.A. (2019). *Thermophysical and Adsorption Characterization of Functional and Pore Modified Adsorbents*. Ph.D. Thesis, Kyushu University, Japan. http://hdl.handle.net/2324/2534480 (accessed 23 June 2024).

13 Aljeboree, A.M., Alshirifi, A.N., and Alkaim, A.F. (2017). Kinetics and equilibrium study for the adsorption of textile dyes on coconut shell activated carbon. *Arabian Journal of Chemistry* 10: S3381–S3393. https://doi.org/10.1016/j.arabjc.2014.01.020.

14 Wang, Y.X., Ngo, H.H., and Guo, W.S. (2015). Preparation of a specific bamboo based activated carbon and its application for ciprofloxacin removal. *Science of the Total Environment* 533: 32–39. https://doi.org/10.1016/j.scitotenv.2015.06.087.

15 Pal, A., Uddin, K., Thu, K. et al. (2020). Thermophysical and adsorption characteristics of waste biomass-derived activated carbons. In: *Encyclopedia of Renewable and Sustainable Materials*, 617–628. Elsevier https://doi.org/10.1016/B978-0-12-803581-8.10832-X.

16 Vlaev, L., Petkov, P., Dimitrov, A., and Genieva, S. (2011). Cleanup of water polluted with crude oil or diesel fuel using rice husks ash. *Journal of the Taiwan Institute of Chemical Engineers* 42: 957–964. https://doi.org/10.1016/J.JTICE.2011.04.004.

17 Sidik, S.M., Jalil, A.A., Triwahyono, S. et al. (2012). Modified oil palm leaves adsorbent with enhanced hydrophobicity for crude oil removal. *Chemical Engineering Journal* 203: 9–18. https://doi.org/10.1016/J.CEJ.2012.06.132.

18 Alaa El-Din, G., Amer, A.A., Malsh, G., and Hussein, M. (2018). Study on the use of banana peels for oil spill removal. *Alexandria Engineering Journal* 57: 2061–2068. https://doi.org/10.1016/j.aej.2017.05.020.

19 J.O. Nwadiogbu, V.I.E. Ajiwe, P.A.C. Okoye, Removal of crude oil from aqueous medium by sorption on hydrophobic corncobs: equilibrium and kinetic studies, *Journal of Taibah University for Science* 10 (2016) 56–63. https://doi.org/10.1016/j.jtusci.2015.03.014.

20 Srinivasan, A. and Viraraghavan, T. (2010). Oil removal from water using biomaterials. *Bioresource Technology* 101: 6594–6600. https://doi.org/10.1016/j.biortech.2010.03.079.

21 Boudrahem, F., Soualah, A., and Aissani-Benissad, F. (2011). Pb(II) and Cd(II) removal from aqueous solutions using activated carbon developed from coffee residue activated with phosphoric acid and zinc chloride. *Journal of Chemical & Engineering Data* 56: 1946–1955. https://doi.org/10.1021/je1009569.

22 Sultan, M., Miyazaki, T., Saha, B.B. et al. (2018). Adsorption of difluoromethane (HFC-32) onto phenol resin based adsorbent: theory and experiments. *International Journal of Heat and Mass Transfer* 127: 348–356. https://doi.org/10.1016/j.ijheatmasstransfer.2018.07.097.

23 Askalany, A.A. and Saha, B.B. (2015). Experimental and theoretical study of adsorption kinetics of difluoromethane onto activated carbons. *International Journal of Refrigeration* 49: 160–168. https://doi.org/10.1016/j.ijrefrig.2014.10.009.

24 Askalany, A.A., Saha, B.B., Uddin, K. et al. (2013). Adsorption isotherms and heat of adsorption of difluoromethane on activated carbons. *Journal of Chemical & Engineering Data* 58: 2828–2834. https://doi.org/10.1021/je4005678.

25 Yancey, A.D., Corbin, D.R., and Shiflett, M.B. (2022). Difluoromethane (HFC-32) and pentafluoroethane (HFC-125) sorption on Linde Type A (LTA) zeolites for the separation of azeotropic hydrofluorocarbon refrigerant mixtures. *Langmuir* 38: 1937–1953. https://doi.org/10.1021/acs.langmuir.1c02904.

26 Sosa, J.E., Malheiro, C., Ribeiro, R.P.P.L. et al. (2020). Adsorption of fluorinated greenhouse gases on activated carbons: evaluation of their potential for gas separation. *Journal of Chemical Technology and Biotechnology* 95: 1892–1905. https://doi.org/10.1002/jctb.6371.

27 Yagnamurthy, S., Rakshit, D., Jain, S. et al. (2021). Adsorption of difluoromethane onto activated carbon based composites: thermophysical properties and adsorption characterization. *International Journal of Heat and Mass Transfer* 171: 121112. https://doi.org/10.1016/j.ijheatmasstransfer.2021.121112.

28 Saha, B.B., Habib, K., El-Sharkawy, I.I., and Koyama, S. (2009). Adsorption characteristics and heat of adsorption measurements of R-134a on activated carbon. *International Journal of Refrigeration* 32: 1563–1569. https://doi.org/10.1016/j.ijrefrig.2009.03.010.

29 El-sharkawy, M.M., Askalany, A.A., Harby, K., and Ahmed, M.S. (2016). Adsorption isotherms and kinetics of a mixture of pentafluoroethane, 1,1,1,2-tetrafluoroethane and difluoromethane (HFC-407C) onto granular activated carbon. *Applied Thermal Engineering* 93: 988–994. https://doi.org/10.1016/j.applthermaleng.2015.10.077.

30 Ghazy, M., Askalany, A.A., Harby, K., and Ahmed, M.S. (2016). Adsorption isotherms and kinetics of HFC-404A onto bituminous based granular activated carbon for storage and cooling applications. *Applied Thermal Engineering* 105: 639–645. https://doi.org/10.1016/j.applthermaleng.2016.03.057.

31 Loh, W.S., Bin Ismail, A., Xi, B. et al. (2012). Adsorption isotherms and isosteric enthalpy of adsorption for assorted refrigerants on activated carbons. *Journal of Chemical & Engineering Data* 57: 2766–2773. https://doi.org/10.1021/je3008099.

32 Islam, M.A., Mitra, S., Thu, K., and Saha, B.B. (2021). Study on thermodynamic and environmental effects of vapor compression refrigeration system employing first to next-generation popular refrigerants. *International Journal of Refrigeration* 131: 568–580. https://doi.org/10.1016/j.ijrefrig.2021.08.014.

33 Islam, M.A., Rupam, T.H., Hosan, S. et al. (2023). Thermodynamic and environmental assessment of low-GWP alternative refrigerants for domestic cooling. *Journal of The Institution of Engineers (India): Series C* 104: 377–383. https://doi.org/10.1007/s40032-023-00927-y.

34 Islam, M.A., Pal, A., Baran Saha, B. et al. (2022). Thermophysical characteristics of novel biomass-derived activated carbon as a function of synthesis parameters. *Heat Transfer Engineering* 43: 1694–1707. https://doi.org/10.1080/01457632.2021.2001743.

35 Do, D.D. (1998). *Adsorption Analysis: Equilibria and Kinetics*. London: Imperial College Press and Distributed by World Scientific Publishing Co. https://doi.org/10.1142/p111.

36 Saha, B.B., Jribi, S., Koyama, S., and El-Sharkawy, I.I. (2011). Carbon dioxide adsorption isotherms on activated carbons. *Journal of Chemical & Engineering Data* 56: 1974–1981. https://doi.org/10.1021/je100973t.

37 Jribi, S., Miyazaki, T., Saha, B.B. et al. (2017). Equilibrium and kinetics of CO_2 adsorption onto activated carbon. *International Journal of Heat and Mass Transfer* 108: 1941–1946. https://doi.org/10.1016/j.ijheatmasstransfer.2016.12.114.

38 Pal, A., El-Sharkawy, I.I., Saha, B.B. et al. (2016). Experimental investigation of CO_2 adsorption onto a carbon based consolidated composite adsorbent for adsorption cooling application. *Applied Thermal Engineering* 109: 304–311. https://doi.org/10.1016/j.applthermaleng.2016.08.031.

39 Ozawa, S., Kusumi, S., and Ogino, Y. (1976). Physical adsorption of gases at high pressure. IV. An improvement of the Dubinin–Astakhov adsorption equation. *Journal of Colloid and Interface Science* 56: 83–91. https://doi.org/10.1016/0021-9797(76)90149-1.

40 Saha, B.B., Koyama, S., El-Sharkawy, I.I. et al. (2007). Evaluation of adsorption parameters and heats of adsorption through desorption measurements. *Journal of Chemical & Engineering Data* 52: 2419–2424. https://doi.org/10.1021/je700369j.

41 Rahman, K.A., Loh, W.S., Yanagi, H. et al. (2010). Experimental adsorption isotherm of methane onto activated carbon at sub- and supercritical temperatures. *Journal of Chemical & Engineering Data* 55: 4961–4967. https://doi.org/10.1021/je1005328.

42 Hazwani, F., Azahar, M., Saha, B.B., Harata, A. (2017). Study on the isosteric heat and its effect on the adsorption systems for various sorption pairs. 1–7.

43 Rahman, K.A., Chakraborty, A., Saha, B.B., and Ng, K.C. (2012). On thermodynamics of methane + carbonaceous materials adsorption. *International Journal of Heat and Mass Transfer* 55: 565–573. https://doi.org/10.1016/j.ijheatmasstransfer.2011.10.056.

44 Chakraborty, A., Saha, B.B., Koyama, S., and Ng, K.C. (2006). On the thermodynamic modeling of the isosteric heat of adsorption and comparison with experiments. *Applied Physics Letters* 89: 171901. https://doi.org/10.1063/1.2360925.

45 Lemmon, E.W., Bell, I.H., Huber, M.L., and McLinden, M.O. (2018). *NIST Standard Reference Database 23: Reference Fluid Thermodynamic and Transport Properties-REFPROP, Version 10.0*. National Institute of Standards and Technology https://doi.org/10.18434/T4JS3C.

46 NIST (2023). *NIST Chemistry WebBook, SRD 69*. https://webbook.nist.gov/cgi/inchi/InChI%3D1S/CH2F2/c2-1-3/h1H2 (accessed 23 June 2024).

47 Electricity Rewiew Japan (2023). The Federation of Electric Power Companies of Japan, 2023. https://www.fepc.or.jp/english/library/electricity_eview_japan/index.html (accessed 29 April 2023).

48 (2023). *Technical Specification of 10 kW Bry-Chill Adsorption Chiller*. Bry-Air (Asia) Pvt. Ltd https://www.bryair.com/products/desiccant-adsorption-chillers/ (Accessed 23 June 2024).

49 Chisalita, D.-A., Petrescu, L., Cobden, P. et al. (2019). Assessing the environmental impact of an integrated steel mill with post-combustion CO_2 capture and storage using the LCA methodology. *Journal of Cleaner Production* 211: 1015–1025. https://doi.org/10.1016/j.jclepro.2018.11.256.

50 Kosai, S. and Yamasue, E. (2019). Global warming potential and total material requirement in metal production: identification of changes in environmental impact through metal substitution. *Science of the Total Environment* 651: 1764–1775. https://doi.org/10.1016/j.scitotenv.2018.10.085.

51 Hitachi Room Aircon Product Brochure (2018). http://www2.tochi.hitachi.co.jp/ra.drowing/zumen/JT90J2E7.PDF (accessed 23 June 2024).

52 General – Product Catalogue 2018 – Air Conditioners Lineup (2018). https://www.fujitsu-general.com/shared/g-eu/pdf-geur-support-ctlg-3eg019-1801e-01.pdf (accessed 23 June 2024).

4

Functionalization on Metal–Organic Frameworks to Enhance Water Adsorption Uptakes and Kinetics for Cooling Applications

Bo Han and Anutosh Chakraborty

School of Mechanical and Aerospace Engineering, Nanyang Technological University, Singapore

4.1 Introduction

4.1.1 Background

The use of cooling and heating power is increasing worldwide because of the fast technological developments and population growth. Traditionally, these needs have been met by burning fossil fuels in power plants, while the heating and cooling needs have been met with the aid of boilers and chillers. Electricity generation is involved in these processes; cooling and heating cause serious environmental pollution. Therefore, it is more important to explore technology transfer for cooling/heating and TES applications. In recent years, the investigation and commercialization on adsorption heat transfer (AHT) systems for cooling production [1] and TES have been successfully developed.

The adsorption cooling (AC) uses the operation principle of (i) evaporation–adsorption of water onto porous adsorbents and (ii) desorption–condensation of water vapor from the adsorbents to generate cooling power. As shown in Figure 4.1a, the AC system consists of two or more adsorption reactors or beds, an evaporator, and a condenser. The detailed working mechanism of the AC system is well established in the literature [2]. Due to evaporation/boiling of the water vapor, energy transfer occurs from the evaporator to the adsorption bed and the energy is stored there as the isosteric heat of adsorption. In addition, the heat of adsorption can be applied in the heat pump application. On the other hand, by supplying the bed with renewable energy, the water vapor that was adsorbed during the previous cycle (due to the porous adsorbent) is desorbed. Transferring the desorbed water vapor to the condenser, where the heat from condensation is dissipated, cools the water. By using a U-tube to reflux the condensed water back to the evaporator, the pressure differential between the two is kept constant. The refrigerant change (Δq)

Advancements in Non-Conventional Cooling and Thermal Storage Strategies, First Edition.
Edited by Bidyut Baran Saha and Dibakar Rakshit.

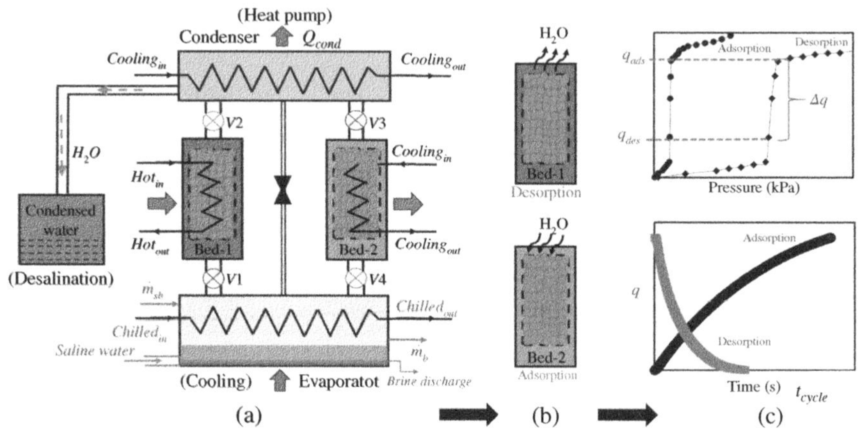

Figure 4.1 (a) Conceptual illustration of the AHT system that generates cooling/heating and TES effects, (b) conceptual illustration of the adsorption and desorption processes, and (c) conceptual illustration of the isotherms and kinetics to comprehend water transfer ($\Delta q = q_{ads} - q_{des}$) and cycle duration ($t_{cycle}$).

every adsorption/desorption cycle determines the overall efficiency of the air conditioner [2]. However, as demonstrated in Figure 4.1b and c, the Δq value is dependent on adsorption/desorption kinetics. The mass and energy balances for all of the chiller's components are used to build the thermodynamic framework for AC [2]. Desalinated water can be kept in storage along with the condensed water [3].

Traditionally, various types of zeolites [4] and silica gel [5] have been used as adsorbents to fabricate thermal compressors for AC systems. The main disadvantage of AC is the bulky size of the compressor, which is mainly due to the poor conversion of the adsorbate in each adsorption–desorption cycle. The low capacity of the porous adsorbent will cause a small difference in adsorbate (hence water) transfer (Δq) during cooling/heating pump operation. Therefore, to meet the cooling/heating requirements, the adsorption bed requires more adsorbents, which makes the adsorption device large as compared to the conventional vapor compression chiller. MOFs with V-shaped (or "S-shaped") adsorption isotherms have been found to be the most suitable materials for heat-operated (60–70 °C) AC systems [6]. The combined hydrophilic/hydrophobic behaviors of MOFs provide higher water transfer with fast kinetics so as to improve the overall performances of the adsorption chiller.

Metal clusters and organic ligands make up the crystalline mesoporous–microporous hybrid structures known as MOFs [7]. The most common metal materials are bivalent (Cu, Mg, etc.), trivalent (Al, Fe, Cr, etc.), or tetravalent (Zr, Ti, etc.), and the organic ligand with multidentate chains is typically utilized to connect the metal clusters [8]. MOFs can be designed and modified more easily than commonly utilized zeolites or silica gels [9]. Additionally, MOFs often feature large pores, a big surface area, stable pore structure, and a large pore size. In fact, MOFs are widely used in a variety of applications. Adsorbents having hydrophobic characteristics, for instance, are useful for dehumidification in low pressure areas [10]. On the other hand, it has been demonstrated that adsorbents with hydrophilic properties are appropriate for cooling and heat pumps [11, 12]. It is important to highlight that MOFs with good heat transfer performance include aluminum fumarate, MOF-801 (Zr), and MIL-101 (Cr) [13].

4.1.2 Research Motivation

Recently, UiO-66 (Zr) and MIL-125 (Ti) [14] have recently drawn interest in the fields of desalination, heat pumps, and adsorption-aided cooling. With a face-centered cubic structure, UiO-66 (Zr) possesses a large Brunuer–Emmet–Teller (BET) area (approximately $1200\,m^2/g$) and pore volume ($0.49\,cm^3/g$) [15]. In addition, a "S-shaped isotherm" appropriate for heating applications is produced by the water adsorption of UiO-66 (Zr). The high water content of UiO-66 (Zr) results in a very slow rate of water adsorption in the Henry zone [16]. Similar to the parent MIL-125 (Ti), the primary issue with MIL-125 (Ti) is its long hydrophobic length ($P/P_s \approx 0.35$) and very poor water transfer under the AC condition. MIL-125 (Ti) MOFs were also discovered in several investigations to be unstable with water adsorption [17]. The use of MOFs in applications involving heat transformation is prohibited by their considerable hydrophobicity and hydrothermal instability. At the start of adsorption, water molecules interact negatively with the porous walls of MOFs due to the parent MOFs' long hydrophobic length. This issue can be solved by changing the structure of MOFs.

The hydrophilicity/hydrophobicity of MOFs can be tuned via ligand functionalization. For instance, Gordeeva et al. [18] decreased the step pressure of the original MIL-125 (Ti) by grafting amine ($-NH_2$) functional groups onto it, and they were able to attain a COP (around 0.77) in AC application. Other functional groups with hydrophilic/hydrophobic characteristics exist in addition to $-NH_2$. For instance, haloalkanes, a class of organic molecules with links between carbon and fluorine, chlorine, bromine, and iodine, significantly influence the development of micropores in MOFs [19]. It is thought that the pyridine group has a significant role in modifying the MOF + water system's isotherm's form [20]. A substantial attraction for polar molecules like water is also produced by the hydrogen bonding that is frequently present in hydroxyl-containing MOFs [21]. Ammonium, another functional group, has drawn a lot of interest in attempts to change the MOF structure [2]. These functional groups can alter the pore structure of MOFs, which will change the isotherms and kinetics of water adsorption.

We are aware of no other publication that goes into great detail into the kinetics of water adsorption on pristine and modified (functionalized) MOFs.

4.1.3 Research Objectives and Scope

The overall goal of this chapter is to create novel, useful adsorbents and to research how they interact with water when used in air-conditioning systems. Thermodynamic analyses of water adsorption under equilibrium and dynamic settings on the natural and functionalized MOFs UiO-66 (Zr) and MIL-125 (Ti) are presented in this section. The studies took place at a range of 30 and 60 °C in temperature. Calculations were made about the MOFs' porosity, thermal stability, and shape. To measure isotherms and kinetic parameters, Langmuir type isotherms and kinetic equations were fitted to the experimental dewatering/drainage data. The Van't Hoff equation was used to get the isosteric heat of adsorption (Q_{st}) in the coordinate system of temperature–pressure absorption. The presentation of a thermodynamic model illustrating sensible and latent heat transport. It also includes the assessment of AC and TES performance using isotherms and kinetic data. The following are the aims and range:

(1) Fundamental knowledge of MOF synthesis and functionalization. Different MOFs have been created using the hydrothermal reaction technique. The inclusion of various functional groups changes the initial MOF.
(2) XRD, TGA, SEM, and N_2 adsorption are used to assess the porosity and thermal characteristics of intact and functionalized MOFs.
(3) Using absorption/discharge models based on temperature, an experimental study on the equilibrium and quality conditions of AC and TES systems is carried out.
(4) Different MOFs/composites + water-based AC functions are assessed in terms of SCP (specific cooling power) and COP using experimentally known adsorption kinetic and isotherm data.
(5) The adsorption system's TES performance is assessed.

4.2 Thermodynamic Frameworks of AHT System

4.2.1 Isotherm and kinetic fitting

The experimental data for the water vapor uptakes on MOFs exhibit "S-shaped" isotherms, and the following isotherm equation fits these data [22],

$$\theta = \frac{K\left(\frac{P}{P_s}\right)^m}{1+(K-1)\left(\frac{P}{P_s}\right)^m} \tag{4.1}$$

where q is the uptake under an equilibrium state, $\theta = q/q^o$ denotes the surface coverage, and q^o is the limiting uptake. The pre-exponential coefficient is denoted by α, while the heterogeneous coefficient is denoted by m. The adsorption equilibrium constant is $K = \alpha \exp[m(Q_{st}^* - h_{fg)}/RT]$, where h_{fg} stands for the evaporation–enthalpy, Q_{st}^* stands for the isosteric heat at zero coverage, and R is the constant for all gases.

$$K = k_{ads}/k_{des}$$

where k_{ads} is the adsorption rate coefficient and k_{des} is the desorption rate coefficient, is the formula given by the Langmuir analogy [23]. When $K = k_{ads}/k_{des}$ is substituted into Eq. (4.1) under experimental equilibrium conditions, the surface coverage becomes

$$k_{ads}(1-\theta)\left(\frac{P}{P_s}\right)^m = k_{des}\theta\left[1-\left(\frac{P}{P_s}\right)^m\right] \tag{4.2}$$

A_s is the adsorbent's BET specific surface area, and $k_{ads} = W\beta A_s$. The formula for β, which resembles the sticking coefficient, is $\beta = [\beta_0 \exp(E_a/RT)]$. Here, β_0 denotes the frequency factor or pre-exponential with a unit value for monolayers and E_a stands for the activation energy. Since M represents the adsorbate molecular weight, W determines the rate of adsorbate molecule collision with the adsorbent surface per unit time and is given by $W = MP/\sqrt{2\pi MRT}$. Consequently, it is possible to determine the water vapor uptake over time under nonequilibrium conditions by

$$\frac{dq}{dt} = k_{ads}(1-q)\left(\frac{P}{P_s}\right)^m - k_{des}q\left[1-\left(\frac{P}{P_s}\right)^m\right] \tag{4.3}$$

$$\frac{dq}{dt} = k_{ads}\left(\frac{P}{P_s}\right)^m - q\left\{k_{ads}\left(\frac{P}{P_s}\right)^m + k_{des}\left[1-\left(\frac{P}{P_s}\right)^m\right]\right\} \tag{4.4}$$

This leads to the definition of the time constant as $\frac{1}{\tau} = k_{ads}\left(\frac{P}{P_s}\right)^m + k_{des}\left[1-\left(\frac{P}{P_s}\right)^m\right]$. Equation (4.4) consequently becomes

$$\frac{dq}{dt} = k_{ads}\left(\frac{P}{P_s}\right)^m - q\frac{1}{\tau} \tag{4.5}$$

4.2.2 Energy Balance

Adsorption–desorption processes for air-conditioning, heat pumps, and ocean desalination are described in many studies using thermodynamic models. Hence, a straightforward lumped heat capacity model was selected. The primary drawback of this model is the assumption that the temperature of the adsorption bed is merely a function of time and independent of position (both in terms of position and work required for storage). An external cooling system (coolant from copper tubes in the evaporator) or cooling water from the TES (for use in heat recovery reasons) evaporates the water vapor in the evaporator. As a result, the water vapor at low pressure is continually evaporated and flows into the adsorption bed.

The heat transfer from the chilled water and water vapor from the evaporator to the adsorption bed generates the energy flow in the evaporator. The mass–thermal interaction between the evaporator and the adsorber is primarily responsible for determining the energy balance in the evaporator, which is defined by

$$(Mc_p)_e \frac{dT_e}{dt} = -h_{fg} M_s \frac{dq_a}{dt} + (UA)_e \frac{1}{N} \sum_{i=1}^{N} (T_e - T_i) - h_f M_s \frac{dq_d}{dt} \tag{4.6}$$

where the adsorption and desorption rates – calculated from the isotherm and kinetic models – are $\frac{dq_a}{dt}$ and $\frac{dq_d}{dt}$, respectively. The total heat capacity of the copper-tube heat exchanger and the water inside the evaporator is represented by $(Mc_p)_e$ on the left-hand side of Eq. (4.6). The latent heat of evaporation is therefore defined by the first item on the right-hand side of Eq. (4.6); the enthalpy of the condensate liquid is indicated by the second term, and the load generated by the supply of chilled water is represented by the third term, depending on the need for cooling. The following equation provides the chilled water temperature at each section of the heat exchange tube (Ti):

$$\frac{V_e(\rho c_p)_{cw}}{N} \frac{dT_i}{dt} = (\dot{m} c_p)_{cw} (T_{i-1} - T_i) + \frac{(UA)_e}{N} (T_e - T_i) \tag{4.7}$$

The heat of adsorption is produced by the water vapor transmitted to the bed and is eliminated by delivering cooling water (at 30 °C) to the heat exchanger tubes. On the other hand, the transfer of hot water to the bed causes the mass change of water vapor to occur. Following that, the desorbed water vapor enters the condenser. The following equation represents these adsorbent bed occurrences quantitatively,

$$(Mc_p)_b \frac{dT_{b,k}}{dt} = \psi . M_s \frac{dq_{b,k}}{dt} \Delta H - \frac{(UA)_{b,k}}{N} \sum_{j=1}^{N} (T_{b,k} - T_j) \tag{4.8}$$

where the adsorption or desorption phase is indicated by k. The flags $\psi = 1$ and $\psi = -1$ are valid for adsorption and desorption, respectively. The heat capacity of the bed made up of porous adsorbents contained in a tube-fin heat exchanger is described on the left-hand side of Eq. (4.8). While the second term in Eq. (4.8) describes the heat transfer from the bed to hot water (during desorption) or from cold water to the bed (during adsorption), the first term on the right-hand side of Eq. (4.8) denotes the latent heat exchange resulting from the adsorption enthalpy. It should be noted that the temperature of the adsorption bed and the pressure of the evaporator both affect how much water vapor is absorbed. The condenser pressure and the desorption temperature are determined in accordance with the outflow rate (e.g. the water load during desorption). The temperature T_j of the heat transfer medium (i.e. the hot and cold water) is determined by

$$\frac{V_b(\rho c_p)_f}{N}\frac{dT_j}{dt} = (\dot{m}c_p)_f(T_{j-1} - T_j) + \frac{(UA)_{b,k}}{N}(T_b - T_j) \tag{4.9}$$

The total heat capacity of the heat transfer fluid in tubes is shown by the quantity "$V_b(\rho c_p)_f$" in Eq. (4.9). The adsorbed water vapor is desorbed from the adsorbent material and delivered to the condenser by feeding hot fluid to the heat exchange tube. Condensation heat is simultaneously released into the surrounding area. To maintain the difference in pressure between the evaporator and condenser, the condensed water goes back to the evaporator through an expansion valve. It is expected that a specific volume of condensate can be stored on the condenser chamber's surface. The sensible heat rejection from the condenser to the radiator (second term on the right side of Eq. (4.10)) and latent heat transfer from the absorbed water vapor (first term on the right side of Eq. (4.10)) account for the condenser's energy balance, respectively. It is calculated that the condenser temperature T_c is

$$(Mc_p)_c\frac{dT_c}{dt} = -h_{fg}M_s\frac{dq_d}{dt} + \frac{(UA)_c}{N}\sum_{i=1}^{N}(T_i - T_c) + h_f M_s\frac{dq_d}{dt} \tag{4.10}$$

As a result, $(Mc_p)_c$ on the left-hand side of Eq. (4.10) denotes the total heat capacity made up of the condenser's heat-exchanging tubes and the condensed water with water vapor. The condenser-heat exchanger tube's energy balance is

$$\frac{V_c(\rho c_p)_c}{N}\frac{dT_i}{dt} = (\dot{m}c_p)_{cw}(T_{i-1} - T_i) + \frac{(UA)_c}{N}(T_c - T_i) \tag{4.11}$$

The mass interaction throughout the adsorption/desorption operation phase as well as the switching duration influences the mass balance of the refrigerant (water). As evaporation proceeds, the amount of water in the evaporator decreases, changing the mass balance to $\frac{dM_e}{dt} = M_s\frac{dq_a}{dt} + M_s\frac{dq_d}{dt}$. Energy that can be used to improve the system's energy recovery capabilities is stored in the TES. The energy balance of the thermal storage system for mass recovery in the TES system is represented by the notation $(Mc_p)_{ts}\frac{dT_{ts}}{dt} = \alpha \cdot M_s\frac{dq_{ts}}{dt}\Delta H$. The finite difference method is used to numerically solve the differential equation during the simulation. For a large number of different parts, the total number of counts was calculated in equal increments. The temperature is controlled using a control volume that contains a number of nearby fins and MOF sorbent. The design presupposes unidirectional heat transfer through the fins. The basic difference scheme estimates the diffusion term in the equation, while the flow difference scheme substitutes the convective term. A collection of subroutines calculates the isotherms and kinetics of adsorption. The tolerance is set to 1×10^{-6}, and double precision is employed. Calculations are also made for the

Table 4.1 Important variables for calculating AHT performance parameters.

Parameters	Equations and values [32]
$(\dot{m}_w)_{bed}$	0.75 kg/s for desorption and 0.8 kg/s for adsorption beds
$\dot{m}_{chill}$	0.6 kg/s
$\dot{m}_w$	0.7 kg/s for condenser tubes
$(Mc_p)_{bed}$	$(9\,kg \times 386\,J/(kg\,K) + 41\,kg \times c_{p,w}\,J/(kg\,K))$
$(Mc_p)_{cond}$	$(23\,kg \times 386\,J/(kg\,K) + 6\,kg \times c_{p,w}\,J/(kg\,K))$
$(Mc_p)_{evap}$	$(8\,kg \times 386\,J/(kg\,K) + 35\,kg \times c_{p,w}\,J/(kg\,K))$
M_s (MOFs-mass)	20 kg
$(UA)_{evap}$	$(2600\,W/(m^2\,K) \times 1.31\,m^2)/1000\,kW/K$
$(UA)_{cond}$	$(4100\,W/(m^2\,K) \times 1.26\,m^2)/1000\,kW/K$
$(UA)_{bed}$	$(1080\,W/(m^2\,K) \times 2.70\,m^2)/1000\,kW/K$ (for adsorption bed)
	$(1160\,W/(m^2\,K) \times 3.70\,m^2)/1000\,kW/K$ (for desorption bed)
V_{bed}	$1.05 \times 10^{-3}\,m^3$
V_{evap}	$0.75 \times 10^{-3}\,m^3$
V_{cond}	$0.72 \times 10^{-3}\,m^3$

temperature of the adsorption bed and the uptake of water vapor. The simulation's key variables are continuously updated by numerical combination. Only the initial circumstances are used to start the adsorption system, and the subsequent simulation results are derived from the transient conditions to the steady state. Table 4.1 displays the parameters [24, 25] for the energy and mass equations.

Performance metrics like SCP and COP are computed under cyclic steady-state settings. Cycle time (t_{cycle}), which includes switching and adsorption–desorption operation time, is the foundation of the computation. The formulas for the energy needed for evaporation (Q_e) and desorption (Q_h) are as follows:

$$Q_e = \int_0^{t_{cycle}} \frac{(\dot{m}c_p)_{cw}\,(T_{cw,i} - T_{cw,o})dt}{t_{cycle}} \tag{4.12}$$

$$Q_h = \int_0^{t_{cycle}} \frac{(\dot{m}c_p)_{hw}\,(T_{hw,i} - T_{hw,out})dt}{t_{cycle}} \tag{4.13}$$

SCP and COP, two criteria of AC performances, are defined by

$$SCP = \int_0^{t_{cycle}} \frac{(\dot{m}c_p)_{cw}}{M_s} \frac{(T_{cw,i} - T_{cw,o})dt}{t_{cycle}} \tag{4.14}$$

$$COP = \frac{Q_e}{Q_h} \tag{4.15}$$

4.3 Experimental

4.3.1 MOF Synthesis and Functionalization

4.3.1.1 UiO-66 (Zr)

UiO-66-based MOFs were created using two distinct techniques (hence referred to as methods A and B). With certain modifications made by various chemical ligands, the UiO-66 synthesis technique [20, 26] was employed as a guide.

Method A: Hydrothermal reaction process. First, 2.7 mmol of terephthalic acid and 2.7 mmol of zirconium oxychloride octahydrate ($ZrOCl_2{\cdot}8H_2O$, 98%) were put to a stainless-steel autoclave. Second, 106 ml of *N,N*-dimethylformamide (DMF, 98%) was added to the mixture together with 5 ml of formic acid (HCOOH, 96%). The reactor was sealed and heated for 24 hours at 120 °C after stirring at room temperature for 30 minutes. Filter paper was then used to separate the mixture. Following separation, the product was washed twice with ethanol and three times with DMF (10,000 rpm each time). The synthetic material was then dried for eight hours at 80 °C.

Method B: Under circumstances of ambient pressure, UiO-66-based MOFs are synthesized. First, 30 ml of an acidic solution (27 ml of formic acid plus 3 ml of distilled water) is combined with 2.2 mmol of zirconyl chloride octahydrate and 2.2 mmol of terephthalate-based acid. At a temperature of 120 °C, the mixture is agitated in a reflux for three hours. The suspension was quenched in 25 °C cold water and then rotated to separate it. Purification was done by giving the preparation two 20-minute rinses in ethanol and water (8000 rpm each time). The product was then dried by heating it for six hours at 80 °C.

The above mentioned terephthalic acid include terephthalic acid of parent UiO-66 (Zr) ($C_8H_6O_4$, 98%); 2,5-pyridinedicarboxylic acid ($C_7H_5NO_4$, 98%) is N-UiO-66 (Zr); OH-UiO-66 (Zr) is 2-hydroxyterephthalic acid ($C_8H_6O_5$, 97%), NH_2-UiO-66 (Zr), and 2-aminoterephthalic acid ($C_7H_7NO_4$, 99%), as shown in Figure 4.2. In this case, intact UiO-66(Zr), NH_2-UiO-66(Zr), and method A's preparation of OH-UiO-66(Zr) and method B's preparation of N-UiO-66(Zr) were used to create these compounds. Every material needs to be activated. For at least six hours, the UiO-66(Zr)-based MOF is heated to 155 °C (method A) or 110 °C (method B) in a vacuum oven. The purification procedure is intended to get rid of any remaining water vapor, chemicals, or unreacted acids.

4.3.1.2 MIL-125 (Ti)

MIL-125 (Ti)-MOFs were made using methods A, which contained DMF, and B, which contained formic acid.

Method A: A hydrothermal reaction approach is used to create intact MIL-125 (Ti). In a stainless steel autoclave, 10 mmol of titanium (IV) isopropoxide

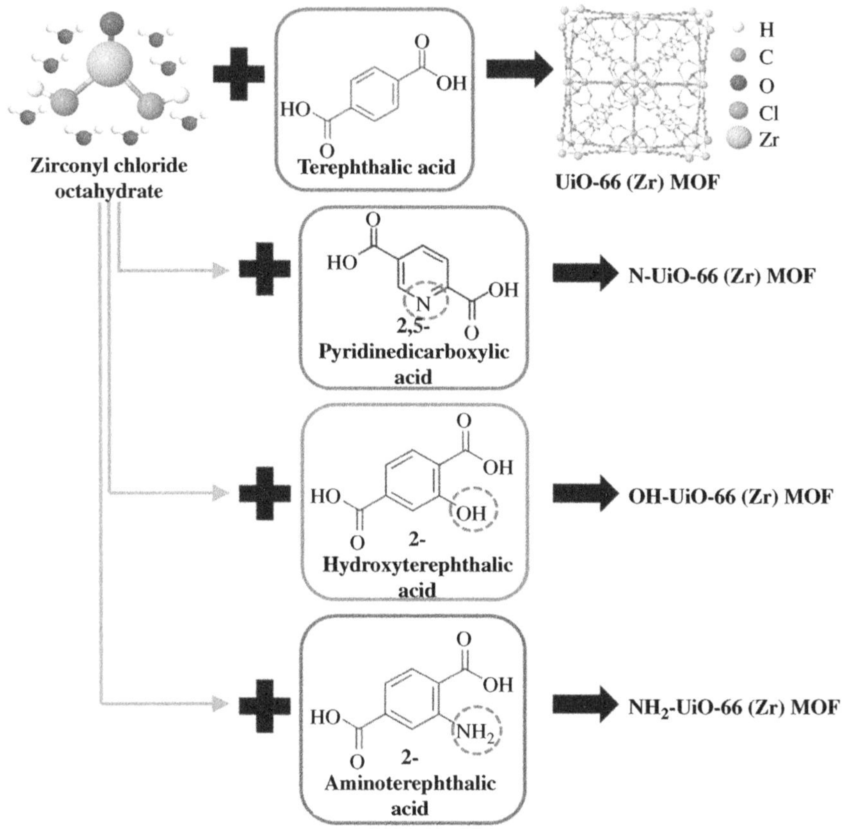

Figure 4.2 Parent and functionalized UiO-66 (Zr) MOFs' schematic depiction.

($TiOCH(CH_3)_2]_4$), 97%) and 15 mmol of terephthalic acid ($C_8H_6O_4$, 98%) were first added. Then, the mixture was given 45 ml of *N*,*N*-dimethylformamide (DMF, 98%) and 5 ml of dry methanol (CHOH, 99%). The reactor was sealed and heated for 16 hours at 150 °C after stirring at ambient temperature for 20 minutes. A centrifuge was then used, to separate the mixture. Following separation, the obtained material was subjected to two separate washes with DMF and methanol (10,000 rpm, 15 minutes each). The finished white material was overnight dried at 80 °C. MIL-125 (Ti) had modifications during preparation. The stainless-steel autoclave was first filled with 4.0 mmol of titanium (IV) isopropoxide and 8.0 mmol of terephthalate-based acid. The mixture was then supplemented with 25 ml of dry methanol and 25 ml of DMF. The reactor was sealed and heated for 16 hours at 150 °C after stirring at ambient temperature for 20 minutes. Following the same steps as before, workable MIL-125 (Ti) was extracted and purified.

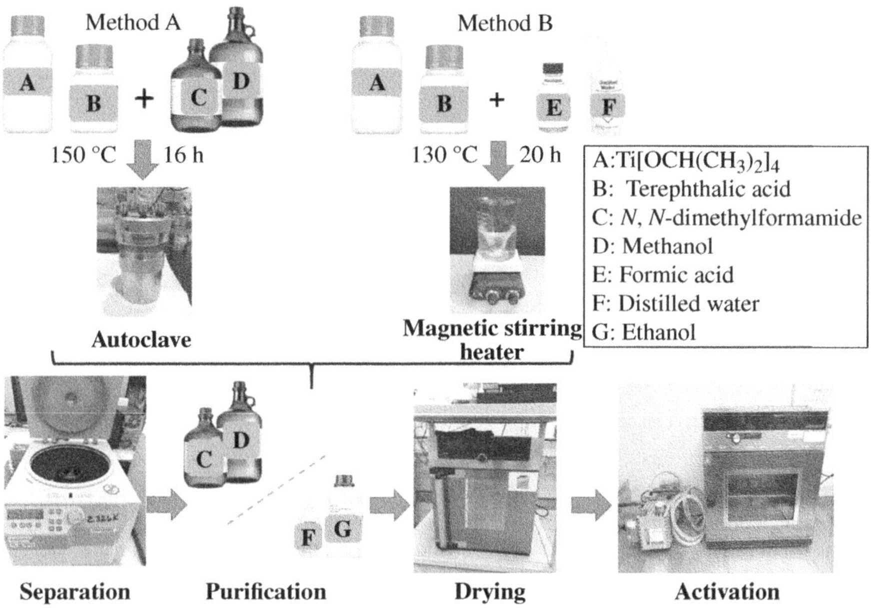

Figure 4.3 MIL-125 (Ti) (original and functionalized MOF synthesis method).

Method B: Using formic acid and ambient pressure, technique B produces functionalized MIL-125 (Ti). First, 50 ml of acid solution (45 ml of formic acid plus 5 ml of distilled water) was combined with 3 mmol of titanium (IV) isopropoxide and 6 mmol of terephthalate-based acid. The mixture was then agitated for 20 hours at a reflux temperature of 130 °C, quenched in cold water at 25 °C, and extracted using a centrifugal spinner. Finally, the synthesized material was purified. Ethanol and water were cleaned (7000 rpm, twice for 30 minutes each). The moist samples were then heated for an overnight period at 80 °C. Figure 4.3 displays a schematic representation of the synthesis procedure.

The terephthalate-based acid includes terephthalic acid ($C_8H_6O_4$, 98%) for parent MIL-125 (Ti), bromoterephthalic acid ($C_8H_5BrO_4$, 95%) for Br-MIL-125 (Ti), 2,5-pyridinedicarboxylic acid ($C_7H_5NO_4$, 98%) for N-MIL-125 (Ti), 2-aminoterephthalic acid ($C_7H_7NO_4$, 99%) for NH_2-MIL-125 (Ti), 2-hydroxyterephthalic acid ($C_8H_6O_5$, 97%) for OH-MIL-125 (Ti), and 2-nitroterephtgalic acid ($C_8H_5NO_6$, 99%) for NO_2-MIL-125 (Ti) (Figure 4.4). In this case, technique A is used to construct the original MIL-125 (Ti), NH_2-MIL-125 (Ti), OH-MIL-125 (Ti), and NO_2-MIL-125 (Ti), while method B is used to fabricate Br-MIL-125 (Ti) and N-MIL-125 (Ti). It should be emphasized that all synthetic materials call for an activation step. The vacuum over is used to heat MIL-125 (Ti) MOFs for 12 hours at 160 °C (method A) or 120 °C (method B). Any remaining

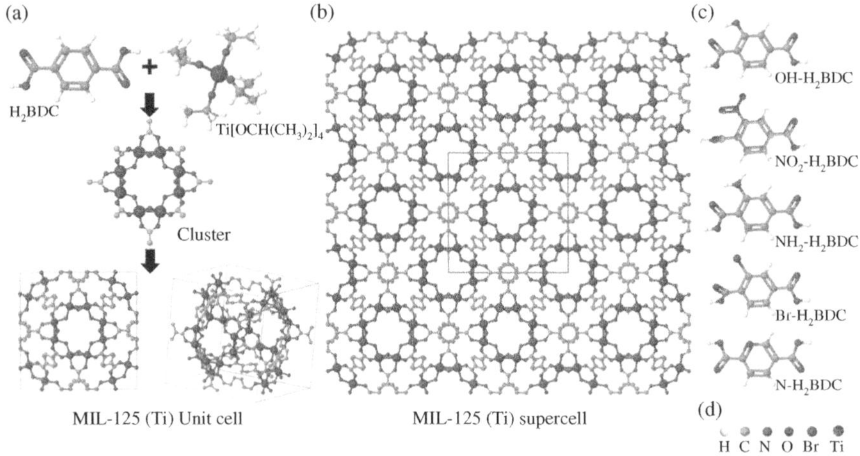

Figure 4.4 The unit cell of the crystal MIL-125 (Ti), a metal cluster ($Ti[OCH(CH_3)_2]_4$), an organic modulator (H_2BDC), and modified modulators with different functional groups are shown in (a), (b), and (c), respectively. For the sake of simplicity, the H atoms are not displayed here.

reagent, unreacted acid, or water vapor is intended to be further eliminated during this activation procedure. Figure 4.4 displays the molecular structures of the functionalized ligands as well as the unit cell of the original MOF.

4.3.2 MOF Characterization

First, it was done by using XRD analysis to examine the crystallography of MOFs. Utilizing a parallel X-ray mirror, a Cu radiation preamplifier, and these conditions, XRD was carried out at 40 mA and 40 kV (with direct light and observer). Depending on the peak, 1/6- or 1/4-degree divergence was seen throughout experiments. Second, SEM was used to examine the morphology, particle size, and microstructure of MOFs. Prep coating was used during sample preparation to increase the powder sample's conductivity. TGA was also used to examine the thermal stability of MOFs. Original and modified MOFs were dried at 80 °C under vacuum before being exposed to an environment with 80% relative humidity for 24 hours. The hydrated MOF was then placed in the TGA apparatus, which heated it from room temperature (30 °C) to approximately 600 °C at a rate of 10 °C per minute while supplying a steady stream of dry N_2. The MOF standard weight that varies with temperature and time should be tracked. The N_2 adsorption–desorption examination at −196 °C was also used to derive the MOF's pore volume, pore diameter, BET surface area, and pore size distribution (PSD) from the N_2 adsorption study. All of the MOFs were cleaned and degassed at 160 °C for three hours before the experiment to get rid of any remaining gas. The entire pore volume of the materials can be determined under saturated states ($P/P_s = 1$), which is equal to the total volumetric amount of liquid N_2 adsorbed. The measurement starts from Henry's area ($P/P_s \rightarrow 0$). $V_{liq} = \frac{P_a V_{ads} V_m}{RT}$, where V_{ads} is the volume of N_2 adsorbed, P_a is the ambient pressure, V_m is the molar volume of liquid of N_2 with a value of 3.47×10^{-5} m^3/mol, and R and T are the gas constant and temperature, respectively, can be used to compute the total pore volume. Additionally, the average pore size (r_p) of the sample can be calculated using the formula $r_p = \frac{2V_{liq}}{A_s}$, where A_s is the BET surface area of the adsorbent sample, which can be inferred from the N_2 isotherm using the BET equation.

4.3.3 Water Adsorption Experiments

A gravimetric device (AQUADYNE *DVS*™) is used to measure the amount of water adsorbed/desorbed by MOF samples at various pressures ($0 < P/P_s < 1$) and temperatures (25–60 °C), as illustrated in Figure 4.5. The microbalance, balanced sampling, humidifier, sample cell, pump, and temperature and humidity measurement are the six basic parts of this system. The sample cell held by the microbalance was covered with the purified MOF sample before the experiment began. Over the control room is a microbalance with a 5 g (±0.1 μg) maximum capacity. Continuous lines link the sample cell and microbalance, enabling TGA

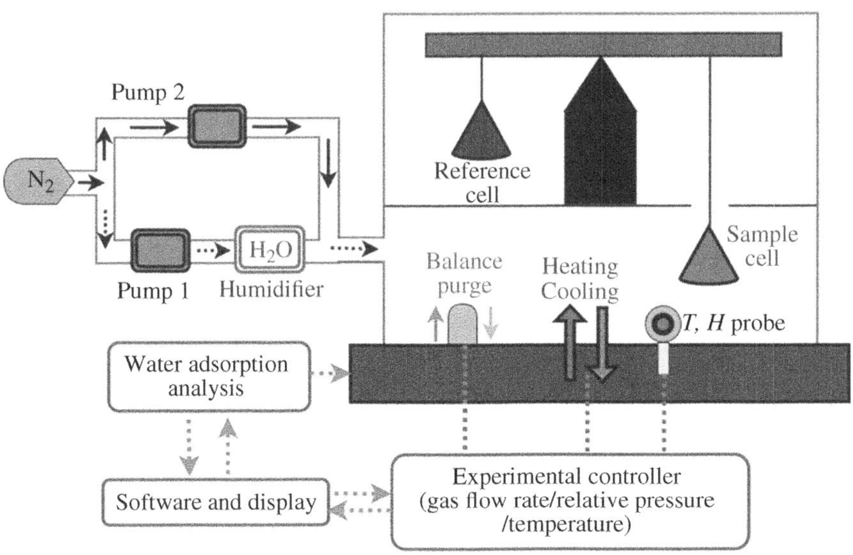

Figure 4.5 A gravimetric analyzer with a schematic diagram for measuring water uptakes on functionalized MOFs over a wide temperature range (25–80 °C) is shown.

to directly affect the amount absorbed during the adsorption/desorption process. A microbalance is used to measure the mass gain or loss of the MOF sample, and the data collected can be computed using the adsorption isotherm. Under isothermal settings, tests on the adsorption of water were conducted over time at various relative humidity (RH) ratios or relative pressures. The TGA has a built-in Class A RTD (resistance temperature detector) thermometer that may be used to regulate the heater and create an isothermal atmosphere. Additionally, the temperature ranges from 25 to 60 °C. The adsorption unit was evacuated using a vacuum pump at the start of the experiment to maintain the TGA's internal pressure.

Two crucial elements in the adsorption experiment are regulating the room temperature and the relative pressure. The internal heater, which is positioned directly into the reaction chamber, warms the adsorbent material before the test. First, the heat flow from the radiant heater and the cooling air flow from the chamber are measured in order to regulate the temperature. It should be noted that, in comparison to the experimental set, the temperature of the adsorption chamber needs to be raised by an additional 15 °C in order to prevent condensation. To maintain microbalance buoyancy effects and remove lingering MOF adsorbent, all adsorption equipment should have a steady supply of dry nitrogen that is 99.999% pure. Second, the flow of dry nitrogen gas into the space is adjusted to manage the relative humidity, which is then measured using a humidity probe. When the RH is below the acceptable level, extra water moisture is added to the adsorption medium to promote the nitrogen flow. On the other side, more dry nitrogen will enter the vacuum chamber if the RH is higher than anticipated. As a result, the RH value should be changed to suit the experiment's requirements. MOFs increase in mass during the adsorption process when the adsorbent absorbs water vapor. Weight variations on the microbalance can be used to identify and document significant changes in the adsorbate–adsorbent system. $M_s(P_0)$ is the initial known mass of the MOFs adsorbent + water vapor adsorbate, and $M_s(P_i)$ is the total sample mass of the adsorbent as a function of pressure. This formula yields the mass of the water vapor adsorbate as $m_{a,i} = M_s(P_i) - M_s(P_0)$. In addition, $q_i = m_{a,i}/M_s$ is used to calculate the gravimetric water uptake. The created program keeps track of all values and may be used to examine adsorption isotherms, kinetics, and calorimetry under steady-state and dynamic settings.

4.4 Results and Discussion

4.4.1 MOF Characterization

Figure 4.6 displays the PXRD (power X-ray diffraction) patterns of the primary and functional UiO-66 MOFs. The modified MOF exhibits comparable XRD

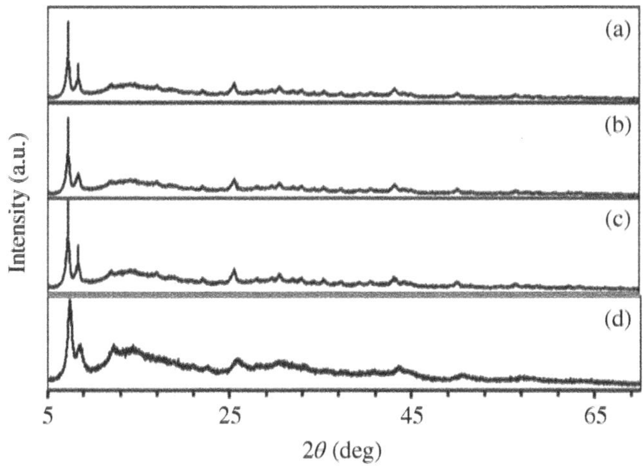

Figure 4.6 XRD results of original and functionalized UiO-66 MOFs. Here, (a) UiO-66, (b) N-UiO-66, (c) NH_2-UiO-66, and (d) OH-UiO-66.

patterns that are synchronized with the standard profile to the original UiO-66 (Zr) [15]. As a result, the MOFs are correctly constructed, and the incorporation of functional groups has no impact on their structural integrity. About 7.5° is where UiO-66's apparent peak can be seen. Modified MOFs showed smaller variations in maximum, which could have been brought on by extra functional group additions. Further evidence that the synthesis process affects the crystallography of the MOFs comes from the fact that the N-UiO-66 MOF has a few extra peaks when compared to other MOFs. Therefore, the stable structures are formed by all UiO-66-based MOFs.

Figure 4.7 displays the parent and functionalized MIL-125 (Ti) PXRD models. Synthesized MIL-125 (Ti)'s XRD trends match those of the reference sample [23], proving that the product was properly synthesized.

4.4.1.1 SEM

Figure 4.8 displays SEM plots. The sample has an octahedral shape and is around 1 m in size. All of the nanocrystals exhibit clear agglomeration behavior, which is comparable to MOF-801, another Zr-based MOF.

The SEM pictures for the operational MIL-125 (Ti) MOFs are displayed in Figure 4.9. The MIL-125 (Ti) particles have a thin disk shape and a size of around 1 μm, and all of the samples exhibit agglomeration characteristics. These morphological traits are congruent with the work that has been published [27]. The functionalized materials contain some irregular polyhedral microcrystallites, particularly for Br-MIL-125 (Ti) (Figure 4.9e) and N-MIL-125(Ti) (Figure 4.9f), as compared to the parent MOFs depicted in Figure 4.9a. Varied synthesis

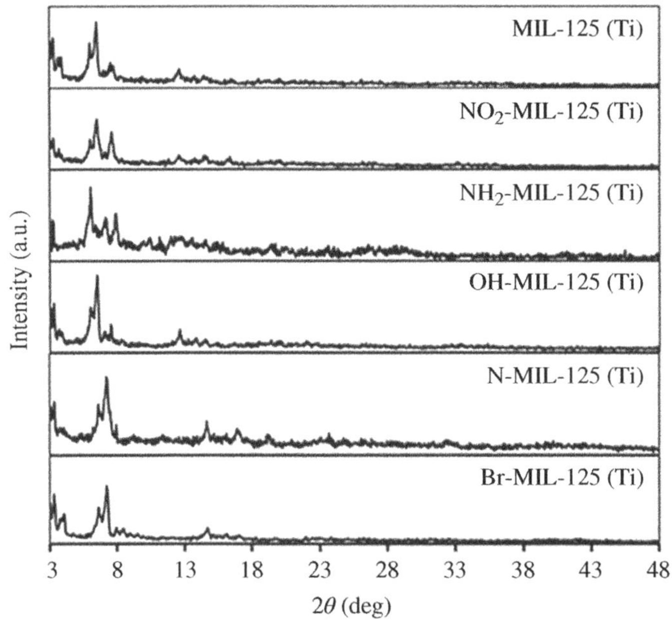

Figure 4.7 Parent and functionalized MIL-125 (Ti) MOF XRD profiles.

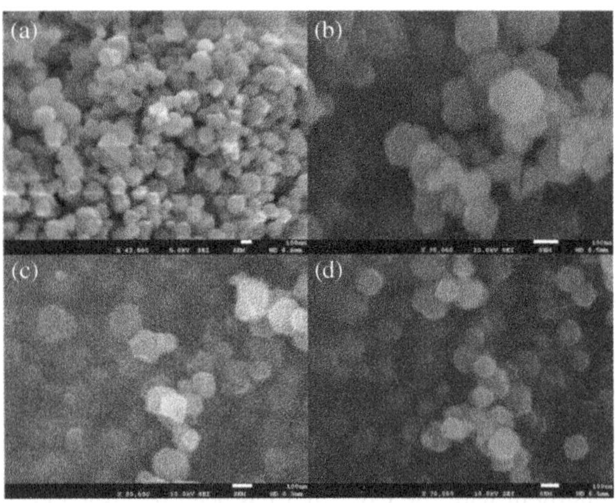

Figure 4.8 SEM images of original/-N/-NH_2/-OH-based UiO-66 MOFs: (a) UiO-66 (Zr), (b) N-UiO-66 (Zr), (c) OH-UiO-66(Zr), and (d) NH_2-UiO-66 (Zr) MOFs.

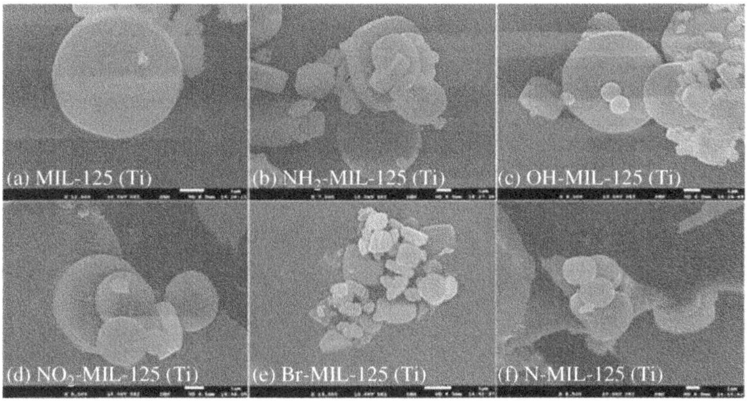

Figure 4.9 SEM images for the MIL-125 (Ti)-based MOFs: (a) MIL-125 (Ti), (b) NH_2-MIL-125 (Ti), (c) OHMIL-125 (Ti), (d) NO_2-MIL-125 (Ti), (e) Br-MIL-125 (Ti), (f) N-MIL-125 (Ti).

techniques and the addition of terephthalic acid may be to blame for this behavior (thus, varied crystal growth processes are seen). The SEM plots demonstrate that (i) the morphologies of the functionalized MOFs are preserved and (ii) the insertion of functional group additions does not affect the structural stability.

4.4.1.2 TGA

Figure 4.10 displays the thermal stability of unfunctionalized and functionalized UiO-66 MOFs. TGA curves with multiple steps offer crucial details regarding MOF porosity, temperature resistance, and even water dynamics. For all MOFs, material loss typically takes place between 30 and 500 °C. At temperatures between 45 and 80 C, water molecules are typically removed to cause the initial weight loss. The UiO-66 (Zr) MOF commences the initial phase of weight loss together with other instances at the lowest temperature of 45 °C, demonstrating that the physical link between the UiO-66 MOF and water molecules is the weakest and the addition of functional groups. It improved UiO-66's interaction with water molecules. The fact that N-UiO-66 MOF experiences the greatest weight loss (19%) during the first stage suggests that it produces the most pores of any functionalized material, which is in good agreement with the results of the pore size analysis. More water molecules are then released from the adsorbents' inner pore as a result of the deeper dehydration process. The second weight loss phase primarily takes place between 102 and 135 °C, which is near to the water boiling point. At this point, functionalized UiO-66 MOF weight loss of another 10% reveals a large number of micropores within the samples. The MOF structure disintegrates when the temperature rises, breakdown occurs, and the final weight loss results. In all three stages of weight loss, it is seen that the first stage experiences the greatest amount of weight loss, which suggests that the majority of water vapors are largely adsorbed in the outer cages of the UiO-66-based

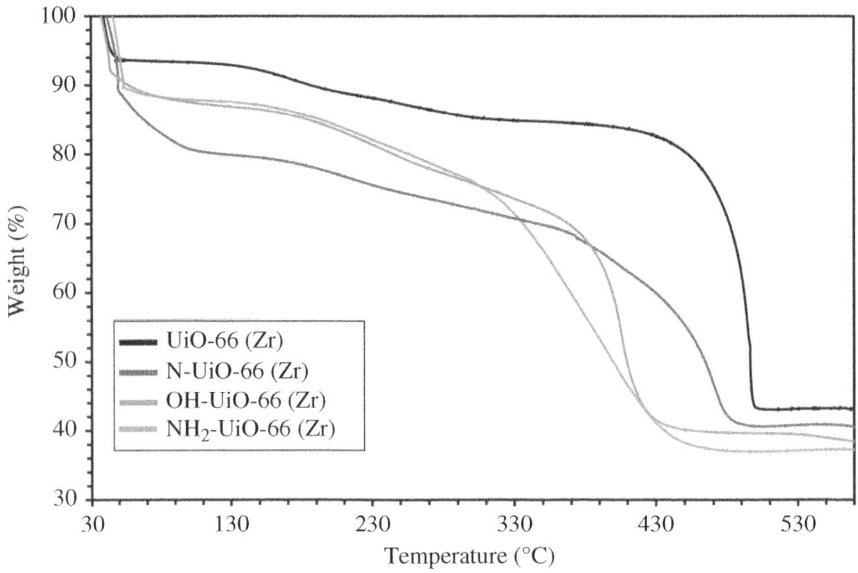

Figure 4.10 TGA analysis was performed on the original and functionalized UiO-66 (Zr)-based MOF adsorbents to determine their thermal stabilities.

MOF structure. Additionally, the critical temperatures for UiO-66(Zr), N-UiO-66, OH-UiO-66, and NH2-UiO-66 at the point of greatest weight loss are 510.7, 479.2, 429.6, and 450.1 °C, respectively. As a result, the implantation of many functional groups slightly reduces the thermal stability of UiO-66 MOF. Functionalized UiO-66 MOFs, however, are thermally stable substances.

Figure 4.11 displays the TGA charts for the parent and functionalized MIL-125 (Ti) MOFs. The pore architectures of the MOFs and the interaction between the MOF and the water are the key determinants of the TGA profiles. For temperatures ranging from 25 to 550 °C, the weight loss profiles of MIL-125 (Ti) MOFs are typically observed in three phases. Between 40 and 60 °C, these phenomena are seen, and they are associated with the early weight loss. More water vapor is then discharged from the MOFs' micropores as the dehydration process progresses. In relation to the boiling point of water, the second stage of weight loss is primarily seen between 80 and 120 °C. Finally, as the temperature rises to 350 °C, the MOF structure disintegrates, the adsorbent decomposes, and the last stage of weight loss takes place.

4.4.1.3 N_2 Adsorption

Table 4.2 lists the various adsorbents' porous properties. Figure 4.12 displays the N_2 adsorption isotherms for UiO-66-based adsorbents.

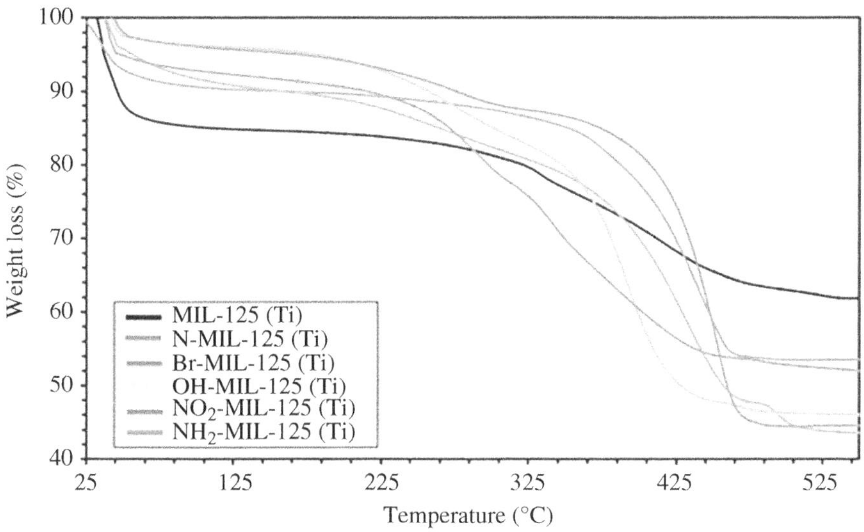

Figure 4.11 Understanding the thermal stabilities of MIL-125 (Ti)-based MOFs using TGA graphs.

Table 4.2 Surface characteristics of the original and functionalized UiO-66 (Zr)-based MOFs, including the BET surface area, average pore radius, and pore volume.

Samples	BET surface area (m^2/g)	Average pore radius (Å)	Pore volume (cm^3/g)
UiO-66 (Zr)	1296.8	10.74	0.54
N-UiO-66 (Zr)	1162.9	10.68	0.46
NH_2-UiO-66 (Zr)	979.2	10.87	0.44
OH-UiO-66 (Zr)	646.0	11.25	0.33

Figure 4.13 displays the PSD curves for the original and functionalized UiO-66 MOFs. The UiO-66 (Zr)-based MOFs are discovered to have a high degree of microporousness. However, NH_2 functional groups in UiO-66 (Zr) MOFs have fewer micropores than other functional groups.

N_2 adsorption is used to measure the functionalized MOFs' pore size, breadth, and specific surface area. The results are shown in Table 4.3.

Figure 4.14 displays the N_2 adsorption isotherms for all MIL-125 (Ti)-based MOFs. NH_2-MIL-125 (Ti) > NO_2-MIL-125 (Ti) > N-MIL-125 (Ti) > MIL-125 (Ti) > Br-MIL-125 (Ti) > OH-MIL-125 (Ti) could be used to summarize the limiting N_2 adsorption capacity.

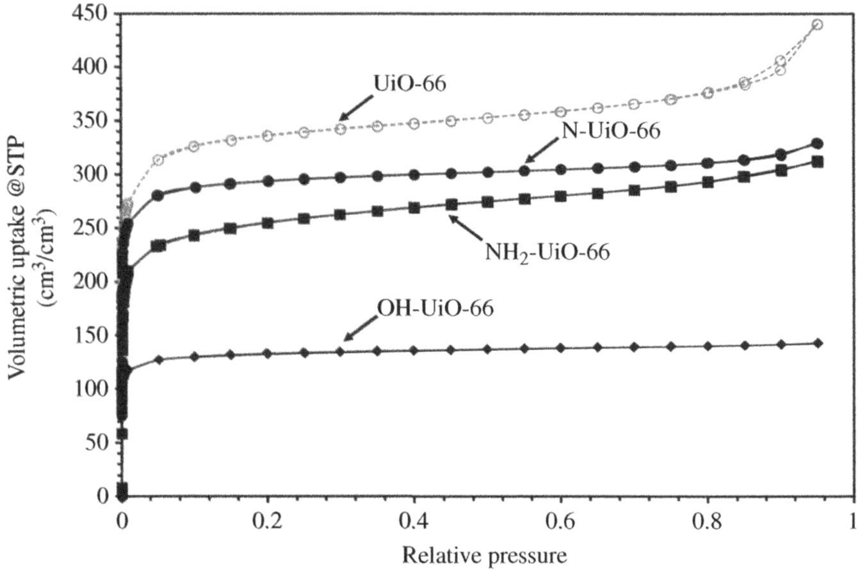

Figure 4.12 N_2 adsorption isotherms for UiO-66-based MOFs.

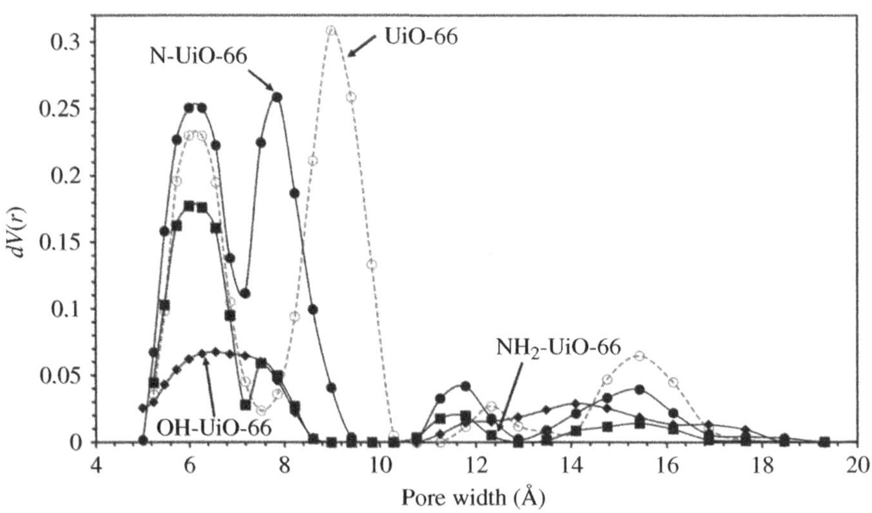

Figure 4.13 PSD of UiO-66-based MOFs employing DFT (density functional theory) analysis.

Table 4.3 N_2 adsorption analysis of the MIL-125 (Ti)-based MOFs.

Samples	BET surface area (m^2/g)	Average pore size (Å)	Pore volume (cm^3/g)
MIL-125	1260.2	11.76	0.86
NH_2-MIL-125	1457.6	11.58	0.91
NO_2-MIL-125	1380.7	11.32	0.90
N-MIL-125	1289.2	11.57	0.89
Br-MIL-125	953.2	11.49	0.82
OH-MIL-125	791.2	11.23	0.79

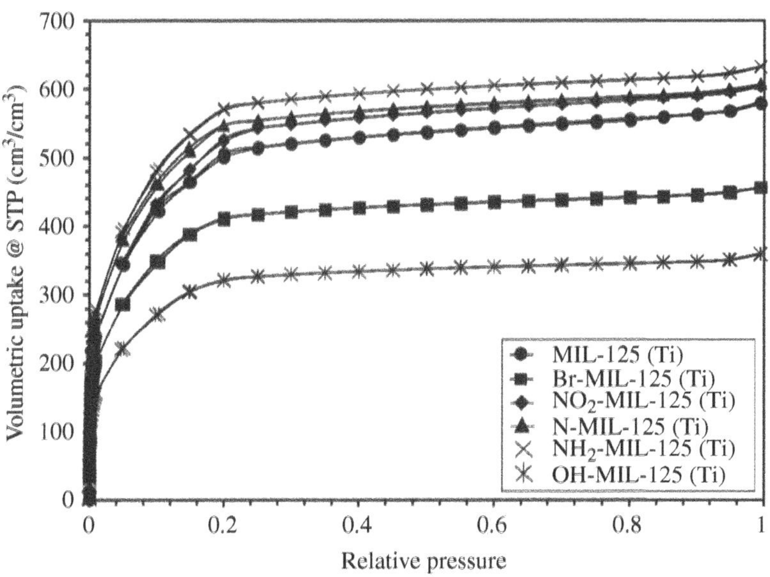

Figure 4.14 N_2 adsorption isotherms for MIL-125 (Ti)-based MOFs.

Figure 4.15 displays the PSD curves for the parent and functionalized MIL-125 (Ti) MOFs. The functionalized MIL-125 (Ti) MOFs are very microporous, according to the PSD curves, which display the frequency of pore distribution for pore widths ranging from 2.0 to 16 Å. The supermicropore area exhibits a larger pore distribution frequency for functionalized MIL-125 (Ti) MOFs, indicating that the structure of these MOFs contains more micropores.

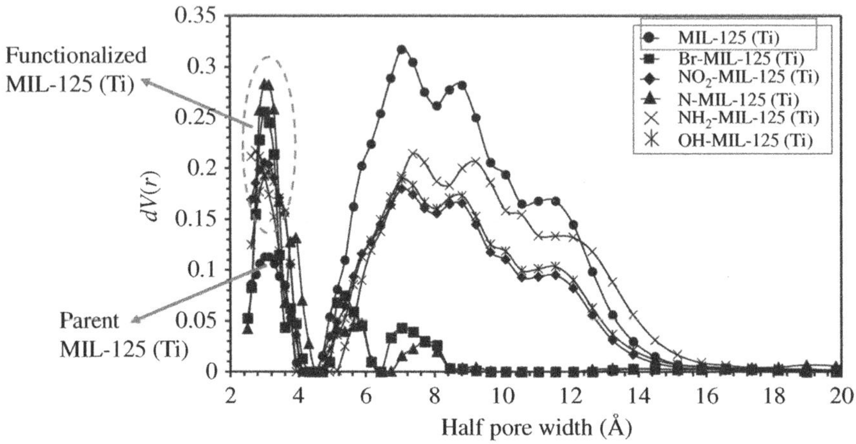

Figure 4.15 PSD of functionalized MIL-125 (Ti) employing DFT analysis and original MIL-125 (Ti) curve shown for comparison purpose.

4.4.2 MOF–Water Adsorption Performance

4.4.2.1 Isotherms

Figure 4.16a and b show the amount of water that is adsorbed on the functionalized (–N, –OH, $-NH_2$) and original UiO-66 MOFs at 25 °C for pressures ranging from Henry's region to the capillary condensation zone. For reliability, water adsorption studies are run at least three times under identical circumstances. It is discovered that there is a moderate amount of hysteresis behavior, which is brought on by the incomplete desorption process with leftover water molecules stuck in the inner pores of the MOFs. As a result, the functionalized materials are represented by the colored dotted lines, while the parent UiO-66 is represented by the black solid line. A typical "S-shaped" isotherm with a high limiting water absorption (q^o) of 0.49 kg/kg (or g/g of adsorbent), as depicted in Figure 4.16, is produced by the parent material UiO-66. However, UiO-66 has a very lengthy step pressure area ($0 < P/P_s < 0.4$) and the low-pressure region only absorbs a very small amount of water vapor. It is notable that all modified UiO-66 materials exhibit the revised isotherms and begin to trap water molecules in the low-pressure region after the addition of various functional group additions.

The water adsorption isotherm shape of NH_2-UiO-66 (Zr) is comparable to that of the parent compound UiO-66 (Zr). It does, however, offer a hydrophobic length that is shorter ($P/P_s = 0.20$). The OH-UiO-66 + water adsorption system's isotherm resembles that of MOF-801, another zirconium-based adsorbent [28], which has a limiting absorption of 0.34 g/g (of adsorbent), which is similar to a change from a "Type I" to a "Type V" isotherm. Additionally, extensive water molecule adsorption

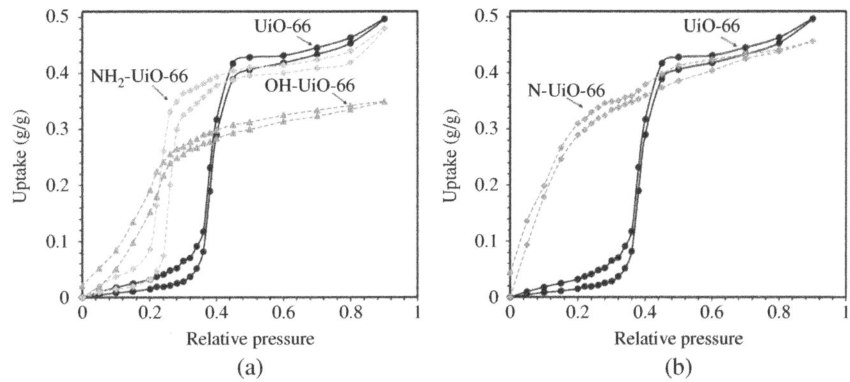

Figure 4.16 The water adsorption–desorption isotherm at 25 °C for the functionalized MOFs NH_2-UiO-66 and OH-UiO-66 is shown in (a) and (b), respectively. For comparison purposes, the original UiO-66 isotherm is displayed as a black solid line.

occurs on N-UiO-66 due to the strong hydrophilic adsorbent–adsorbate interaction at $P/P_S = 0.10$ and a "type I" adsorption isotherm is discovered, which is in good accord with other findings [20]. The maximum water intake of all functionalized materials is lower than that of the parent material, UiO-66 (Zr), and this is an important point to remember. Functional groups, such as amino acids, may result in a decreased pore volume of MOFs and prevent the water molecules from adhering to the pore site at the high-pressure area, which is a plausible explanation for this behavior. The N_2 adsorption analysis is consistent with this hypothesis. The hydrophilicity of all functionalized UiO-66 MOFs increases, leading to greater water uptake in the low-pressure region. The usage of UiO-66-based MOFs in low temperature heat transfer applications, such as adsorption chillers, is made possible by the decrease of the step pressure region, which points to a bigger water uptake-removal differential (Δq).

Figure 4.17a and b display water adsorption–desorption isotherms on various MIL-125 (Ti) MOFs at 30 °C for pressures ranging from the saturation condition ($P/P_S \rightarrow 1$) to the low-pressure region (Henry's region, $P/P_S \rightarrow 0$), where the solid line represents the adsorption process and the dashed line represents the desorption path. Figure 4.17 shows that functionalization has significant impact on the water adsorption behaviors (such as the hydrophobic length, limiting uptake, and isotherm type) of the original MOF.

4.4.2.2 Water Transfer (Δq)

For the temperatures of 25, 35, 45, 60, 75, and 80 °C, the quantity of water uptake on the parent and functionalized (–N, –OH, $-NH_2$) UiO-66 MOFs is depicted in Figure 4.18a–d, where the adsorption–desorption process of a thermal compressor

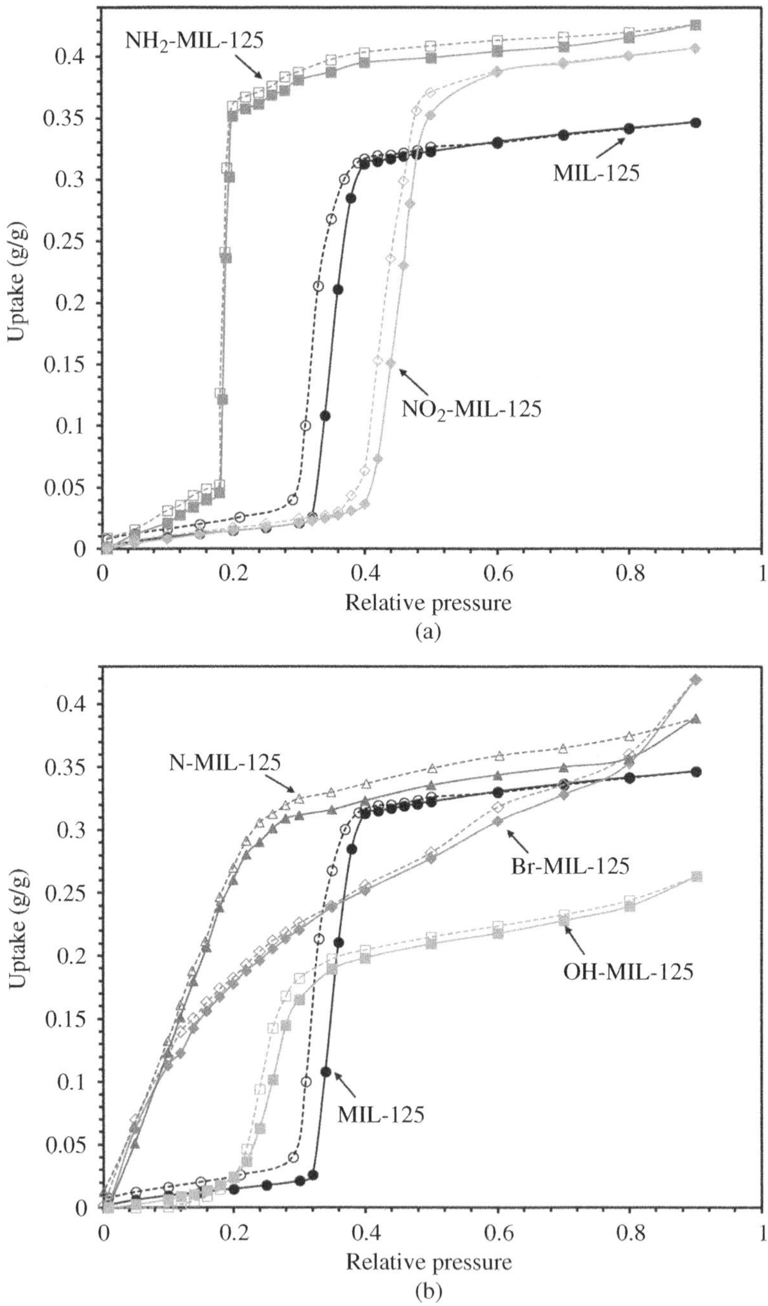

Figure 4.17 MIL-125 (Ti) functionalized water adsorption/desorption isotherms at 30 °C. The original MIL-125 (Ti) isotherms are displayed here for comparison.

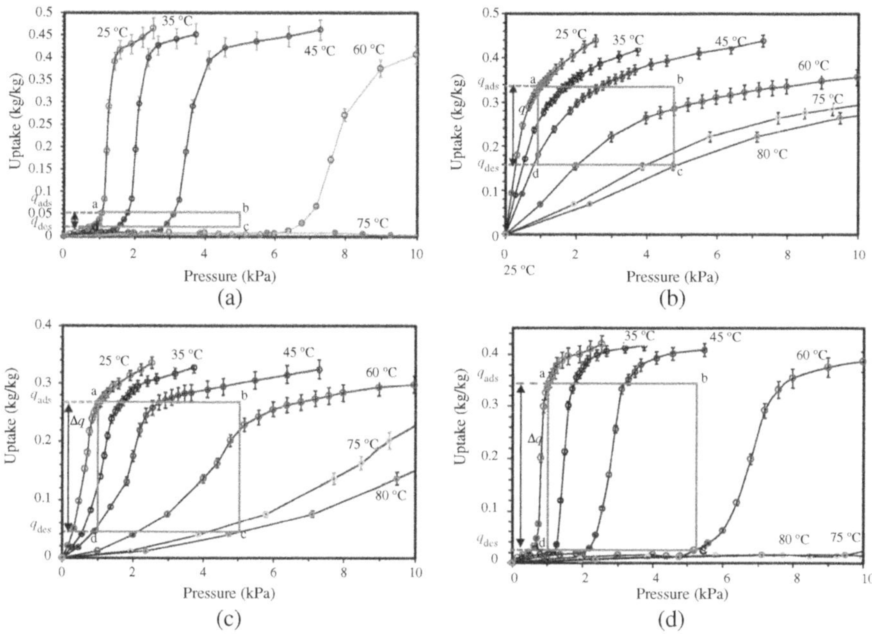

Figure 4.18 Adsorption isotherms of water vapor for temperature ranges of 25–80 °C on (a) UiO-66 (Zr), (b) N-UiO-66 (Zr), (c) OH-UiO-66(Zr), and (d) NH_2-UiO-66 (Zr) MOFs. Here, a denotes desorption and c denotes adsorption.

(adsorption bed) is presented. The amount of water transfer Δq ($= q_{ads} - q_{des}$) each cooling cycle of an AHT may be computed from the pressure variation of 1–5 kPa and temperature variation of 25–60 °C/80 °C in an ideal thermal compression system (Figure 4.18a–d). Under adsorption chiller operating conditions (adsorption-triggered evaporation at 25 °C and 1 kPa; desorption-activated condensation 60–80 °C with 5 kPa), Δq ($= q_{ads} - q_{des}$) per thermal compression cycle of the original UiO-66 (Zr) MOF + water system is found to be the lowest (Figure 4.18a) due to the strong hydrophobic behavior in UiO-66 (Zr) at lower evaporation pressure (≤1 kPa). As a result, even at greater desorption/regeneration temperatures (>80 °C), very less cooling power is produced.

Functional groups are added to the MOF structure to get the better Δq outcomes. Figure 4.18b demonstrates that the N-UiO-66 (Zr) MOF exhibits a somewhat higher Δq at relatively higher desorption temperatures (about 80 °C) because it reflects Type I isotherm curves. As a result, at the desorption temperature of 80 °C, N-UiO-66 (Zr) MOF offers substantial Δq. *Δq* is quite modest at 60 °C, though. While utilizing hydroxyterephthalic acid/aminoterephthalic acid in place of terephthalic acid, Figure 4.18c and d demonstrate the type V isotherms when the functional groups –OH and $-NH_2$ are added to the original UiO-66 (Zr) structure. As a result, both OH-UiO-66 (Zr) and NH_2-UiO-66 (Zr) MOFs exhibit a short hydrophobic length at 298 K (25 °C) with relative pressures of 0.1–0.3. At a desorption temperature of 60 °C, Δq for NH_2-UiO-66 is discovered to be 0.31 kg/kg. The $-NH_2$ functional group gives H mole, which is the major factor. Additionally, due to electrostatic interactions, the amino group contributes a partially negatively charged "O" of the water molecule as well as a positive charge for interactions with negatively charged materials. OH-UiO-66 (Zr) MOFs have a somewhat higher Δq (0.18 kg/kg at 80 °C desorption temperature) than the parent UiO-66 (Zr) MOFs. Because it is an electron-donating group, the hydroxyl group (–OH) increases the interactions between water and MOF. As a result, adding –OH makes the UiO-66 (Zr) adsorbent more hydrophilic, which results in significantly higher Δq values when operating an adsorption chiller or heat pump.

Table 4.4 displays the isotherm fitting parameters for both the initial and functionalized UiO-66 (Zr) + water adsorption systems.

Figure 4.19a–f depict the water adsorption on parent (Figure 4.19a) and functionalized MIL-125 (Ti) MOFs at pressures up to the saturation point and temperatures ranging from 30 to 60 °C. The water transfer (Δq) per adsorption–desorption cycle can be calculated using the isotherm data, which are plotted under equilibrium conditions. The adsorption phase at 30 °C and 1 kPa, as well as the desorption phase at 5 kPa and 60 °C, is therefore used to determine Δq. For a thorough understanding, the experimentally determined water adsorption isotherm and kinetic data are presented in this section.

The MIL-125 (Ti) MOFs display type I water adsorption isotherms when N and Br functional groups are present (Figure 4.19d and e). As a result, the original MOFs' hydrophilic behavior is greatly enhanced. As a result, the low-pressure area experiences a dramatic rise in water adsorption. An extremely high temperature (for desorption) is necessary to remove water vapor from these MOFs' surfaces. The cooling, heating, and desalination effects are not effectively produced when the heating source temperature is 60 °C or lower. Second, MIL-125 (Ti)'s

Table 4.4 For the parent and functionalized UiO-66 adsorbents + water systems, curve fitting parameters of the adsorption isotherms were used.

System	q^o (kg/kg)	Q^*_{st} (kJ/kg)	m	α
UiO-66 + water	0.45	2566.5	8.0	4.41×10^{-1}
N-UiO-66 + water	0.38	2723.4	2.2	9.15×10^{-1}
OH-UiO-66 + water	0.19	2974.1	2.1	2.93×10^{-2}
NH_2-UiO-66 + water	0.41	2601.8	6.3	9.49×10^{-1}

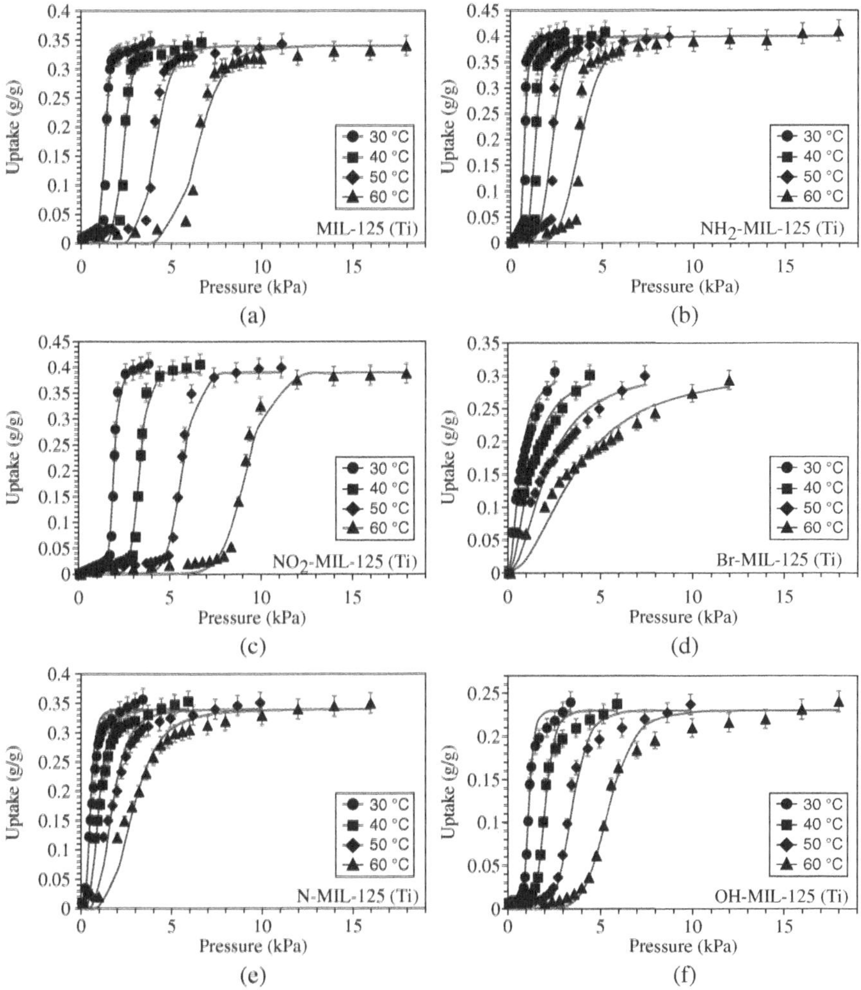

Figure 4.19 Adsorption isotherms of water vapor at different temperatures for MOFs based on UiO-66 (Zr): (a) UiO-66 (Zr), (b) N-UiO-66 (Zr), (c) OH-UiO-66(Zr) and (d) NH_2-UiO-66 (Zr) MOFs. Here, the red lines show the results of the fitting, while the black markers show the experimental data.

OH and NH_2 functional groups (Figure 4.19b and f) also make the compound more hydrophilic and allow for greater water vapor capture at low relative pressures (P/P_s up to 0.3). More water vapor is transported (Δq) every cooling, heating, and desalination cycle as a result of the S-shaped isotherms (Figure 4.19b and f). Third, due to the NO_2-MIL-125 (Ti) group's long hydrophobic length, the water

Table 4.5 (Using a fitting model and experimental data) adsorption isotherm parameters for the parent and functionalized MIL-125 (Ti) MOFs + water systems.

Adsorbent	Q_{st}^* (kJ/kg)	m	q^o (kg/kg)	α
MIL-125 (Ti)	2638.5	9.0	0.34	1.22×10^{-2}
NH_2-MIL-125 (Ti)	2698.4	6.5	0.40	2.98×10^{-3}
NO_2-MIL-125 (Ti)	2584.2	13.2	0.39	2.01×10^{-3}
N-MIL-125 (Ti)	2730.6	4.0	0.34	8.36×10^{-1}
Br-MIL-125 (Ti)	2715.1	3.3	0.30	5.49×10^{-4}
OH-MIL-125 (Ti)	2669.7	7.8	0.23	3.06×10^{-1}

transfer (Δq) under the chosen working conditions (adsorption at 1 kPa and 30 °C; desorption at 5 kPa and 60 °C) is practically negligible.

Table 4.5 provides the isotherm parameters for functionalized MIL-125 (Ti) MOF + water systems, including maximum water uptake (q^o), surface-structural heterogeneity factor (m), pre-exponential coefficient (α), and isosteric heat of adsorption at zero surface coverage (Q_{st}^*).

4.4.2.3 Kinetics

The plots of water absorption and desorption vs time for the parent and functionalized (–N, –OH, $-NH_2$) UiO-66 MOFs are both given in Figure 4.20a–d, and they indicate how the functional group also impacts the amount of time needed for thermal compression of the adsorption-assisted heat transfer system. As a result, the kinetics are described in terms of the adsorption and desorption conditions of an AC system (1 kPa and 25 °C for evaporation, and 5 kPa and 60 °C for condensation). Due to the faster movement of water molecules at higher temperatures and pressures as compared to the adsorption rates, which are relatively slower, it is discovered that the desorption rates for the provided MOF–water systems are greater and take less time to achieve the equilibrium condition. The adsorption kinetics of N-UiO-66 and OH-UiO-66 MOFs are approximately 1.4–1.5 times faster than those of original UiO-66 (Zr). Additionally, it is discovered that NH_2-UiO-66 has a 1.2 times greater adsorption absorption rate. The UiO-66 structure develops additional micropores with the addition of hydrophilic functional group additions, resulting in a more intense MOF + water interaction and faster adsorption kinetics. The water adsorption kinetics' overall performance might be summed up as OH-UiO-66 > N-UiO-66 > NH_2-UiO-66 > UiO-66.

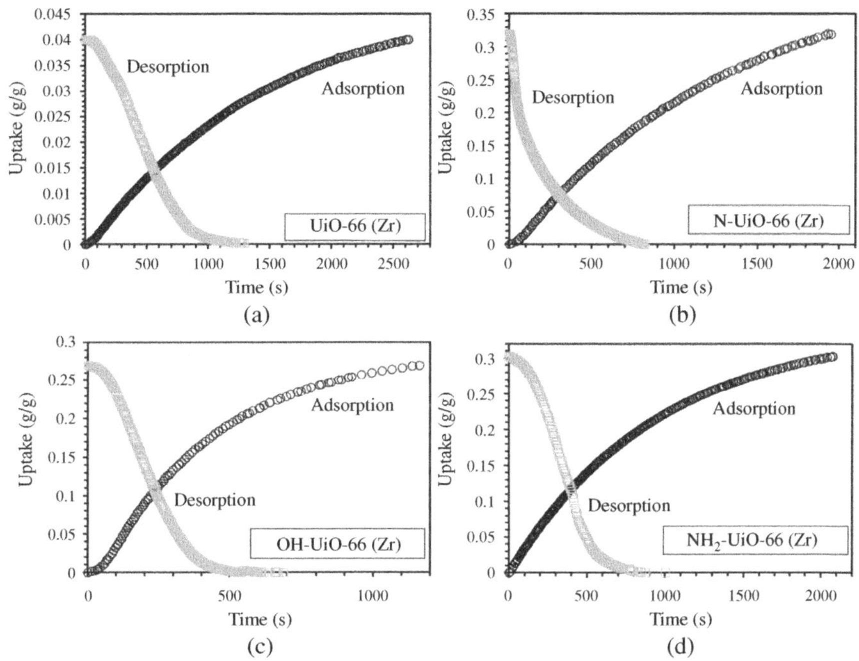

Figure 4.20 Adsorption kinetics for the water-soluble MOFs UiO-66 (Zr), N-UiO-66 (Zr), OH-UiO-66 (Zr), and NH_2-UiO-66 (Zr) under the adsorption and desorption conditions of an AC process (adsorption: 25 °C with 1 kPa and desorption: 60 °C with 5 kPa).

From the isotherm and kinetic data, it can be shown that more micropores are produced inside the MOF structure with the implantation of the functional group (–N, –OH, –NH_2). As a result, the redesigned MOF and water systems have greater interatomic and electrostatic interaction potentials. By adding electrical charge to the functionalized MOFs, the usage of functional group dramatically improves the electron interaction. Second, the functional group also lowers the MOFs' pore apertures. According to the PSD results, functionalized MOFs have fewer pores than their parent UiO-66 (Zr) MOFs do in terms of pore volume, surface area, and width.

Table 4.6 provides the fitted kinetic parameters. Calculating the water uptake on functionalized UiO-66 (Zr) as a function of pressure, temperature, and time using isotherm and kinetic parameters can be utilized to predict the mass transfer properties of the AHT system.

The cycle time (τ_{cycle}) of a batch-operated AHT is calculated using the water adsorption/desorption rates on porous adsorbents. Figure 4.21 displays the

Table 4.6 Fitting parameters of kinetic equations for the UiO-66-based adsorbents + water systems.

System	k_{ads} (1/s)	E_a (30 °C)	β_0	k_{des} (1/s)	E_a (60 °C)
UiO-66 + water	0.645	43,200	1	0.0017	43,000
N-UiO-66 + water	0.135	46,800	1	0.0055	47,600
OH-UiO-66 + water	0.334	43,100	1	0.0062	46,400
NH_2-UiO-66 + water	0.774	42,050	1	0.0019	41,000

Table 4.7 Adsorption kinetic parameters for the parent and functionalized MIL-125 (Ti) MOFs + water systems (using experimental data and Eq. (4.12)).

Adsorbent	$k_{ads\,@\,30\,°C}$ (1/s)	E_a(30 °C)	β_0	$k_{des\,@\,60\,°C}$ (1/s)	E_a (60 °C)
MIL-125 (Ti)	0.736	42,800	1	0.0013	43,550
NH_2-MIL-125 (Ti)	0.786	43,000	1	0.0019	43,100
NO_2-MIL-125 (Ti)	1.028	42,200	1	0.0014	43,500
N-MIL-125 (Ti)	1.273	41,500	1	0.0037	41,000
Br-MIL-125 (Ti)	1.690	40,050	1	0.0027	41,200
OH-MIL-125 (Ti)	1.709	39,560	1	0.0034	40,000

kinetic graphs for the adsorption/desorption of water on the unmodified and modified MIL-125 (Ti) MOFs. Improved adsorption kinetics are found for the functionalized MOFs, where the reasons have been explained above for the UiO-66 (Zr) type MOFs.

Table 4.7 displays the fitting kinetic parameters. The water absorption on functionalized MIL-125 (Ti) as a function of pressure, temperature, and time may be calculated using the isotherm and kinetic parameters, and the results can be utilized to predict the mass transfer properties of the AHT system.

4.4.2.4 Cyclic Stability

Under cyclic operating circumstances that include both adsorption and desorption, the transfer of water vapor on the produced NH_2-UiO-66 (Zr) is carried out using a gravimetric analyzer. The temperature swings from 25 to 70 °C during the experiment, while the relative pressure (P/P_S) varies from 0.1 to 0.9. Figure 4.22 depicts the water vapor's cyclic adsorption and desorption on NH_2-UiO-66 (Zr) MOFs. As a result, the results are displayed for more than 300 cycles, with a cycle length of 10 minutes. It is noted that there is no overt degradation and that the trends of water uptake and desorption values are nearly the same for each cycle.

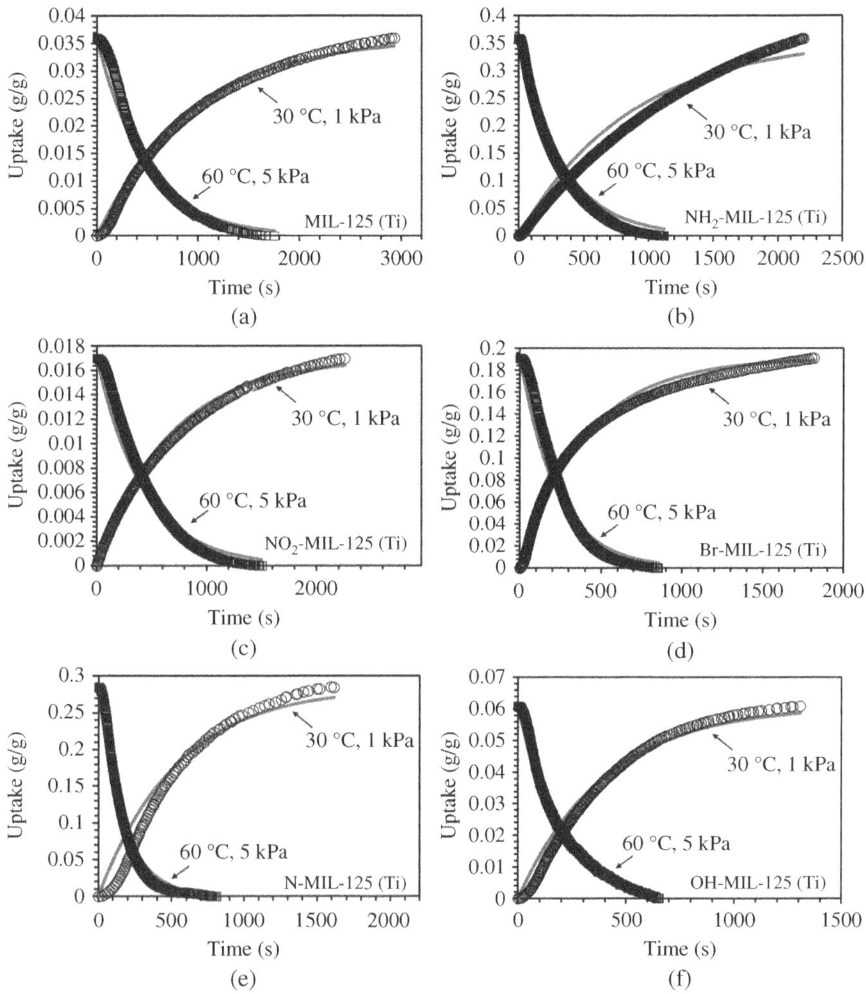

Figure 4.21 Volume of water that MIL-125 (Ti)-based MOFs adsorb under dynamic settings: (a) MIL-125 (Ti), (b) NH_2-MIL-125 (Ti), (c) NO_2-MIL-125 (Ti), (d) Br-MIL-125 (Ti), (e) N-MIL-125 (Ti), (f) OH-MIL-125 (Ti). Here, the red lines represent the results of the fitting, while the black markers represent the experimental data.

4.4.3 MOF-Assisted AC and TES Performances

4.4.3.1 AC (Adsorption Chiller) Performances

The simulation is run for cooling water and chilled water inlet temperatures of 30 and 14.8 °C using the adsorption chiller modeling. Figure 4.23 displays the COP and SCP of the parent and functionalized UiO-66 + water-assisted

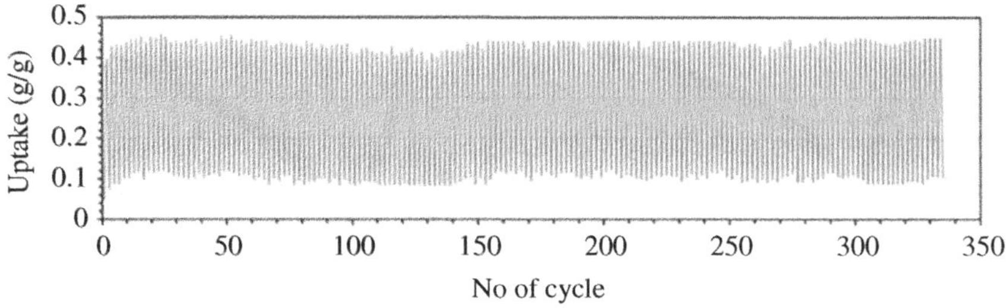

Figure 4.22 NH_2-UiO-66 (Zr) MOFs and water exhibit cyclic adsorption–desorption behaviors.

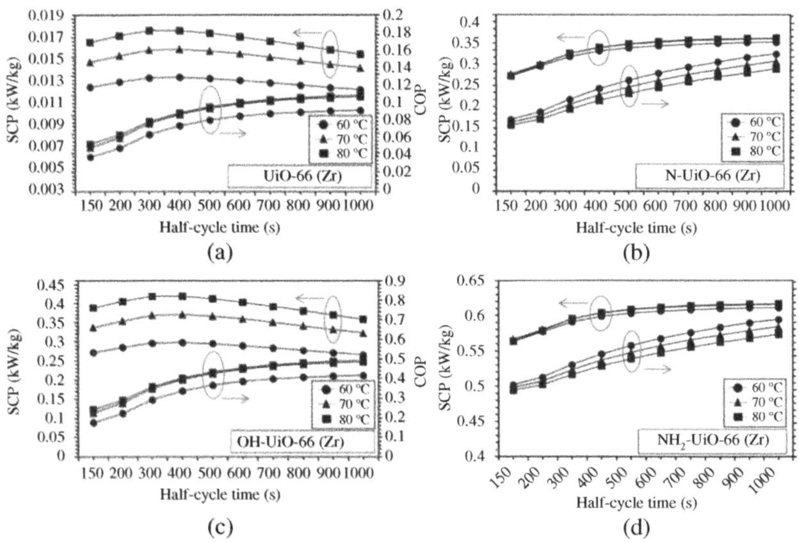

Figure 4.23 COP and SCP (in terms of kW/kg) for (a) UiO-66, (b) N-UiO-66, (c) OH-UiO-66, and (d) NH_2-UiO-66 and water-assisted adsorption chillers under varied regeneration temperatures (60–80 °C).

adsorption chiller at various half-cycle times (t_{cycle}) for the heat source (hot water) temperature range of 60–80 °C and the heat sink (cooling water) temperature of 30 °C. The COP increases monotonically with t_{cycle}, as depicted in Figure 4.25. The fundamental cause is that shorter cycle times result in incomplete desorbed water vapor due to inefficient heating energy use, which lowers the uptake-removal differential (Δq) every AC cycle and lowers COP. On the other hand, a longer cycle time allows the heat energy to be fully utilized, resulting in a larger COP and the complete release of water vapor from the adsorbents. As a result, the functional UiO-66 (Zr) has a greater COP than the UiO-66 (Zr) parent.

The SCP exhibits a non-monotonic behavior when the cycle period changes, in contrast to the COP (Figure 4.23). The SCP grows sharply for the parent UiO-66 (Zr) MOFs up to 400 seconds, but the slope flattens off for the half-cycle duration over 500 seconds. At shorter half-cycle times (150–400 seconds) for N-UiO-66 and OH-UiO-66 + water-assisted adsorption chillers, insufficient cooling and heating of the beds result in a diminished degree of adsorption and desorption effects, which lowers Δq. The lower SCP is as a result obtained. All UiO-66 (Zr)-based adsorbents approach their adsorption limit at longer half-cycle times since more heat energy is needed for desorption purposes. Additionally, due to the N-UiO-66/OH-UiO-66 adsorbent and water's strong interaction, adsorption may also take place at extended desorption durations. Due to these restrictions, the SCP will be poor and undesirable. Therefore, a longer half-cycle duration (t_{cycle} > 500 seconds) leads to a declining trend in SCP.

The regeneration temperature affects the adsorption chiller's efficiency as well. Greater q and faster kinetics cause both COP and SCP to rise with greater regeneration temperature. The highest COP for UiO-66, N-UiO-66, OH-UiO-66, and NH_2-UiO-66 is determined to be 0.15, 0.47, 0.42, and 0.82, respectively, for the hot water inlet temperature of 60 °C. As a result, it is discovered that the COP of all functionalized adsorbents is greater than that of the original UiO-66 (Zr) MOFs. This is attributed to the fact that OH-UiO-66 (Zr) and NH_2-UiO-66, when compared to other functionalized materials, exhibit remarkably quick water adsorption kinetics and could reach around 80% of the equilibrium state in 700 seconds. As a result, no appreciable increase in COP is seen for operation situations with longer cycle times (700–1000 seconds), as the uptake–offtake difference for each AC cycle is not significantly reduced. In addition, it is seen that all functionalized materials have greater COP than their parent material, UiO-66 (Zr). The primary causes are faster kinetics and a greater uptake–offtake disparity for each adsorption–desorption interval. The functionalized UiO-66 (Zr) MOFs fully utilize the provided heating energy to drive the AHT system, resulting in greater SCP and COP. This is due to the enhanced hydrophilic behavior in the low pressure area as well as the quicker adsorption–desorption rates.

Figure 4.23 also shows the impact of the regeneration temperature (60–80 °C) on the AC performance. While the SCP virtually stays the same, the COP for the parent UiO-66 (Zr) and NH_2-UiO-66 (Zr) MOFs drops when the regeneration temperature rises. The hydrophobic length of these two adsorbents is less than 0.4 (P/P_s < 0.4), which results in "S-shaped" isotherms for water adsorption. As a result, higher desorption temperatures (>70 °C) do not result in a significant increase in water vapor transfer per AC cycle, but higher hot water inlet temperatures do result in a greater consumption of heating energy.

On the other hand, higher regeneration temperatures are needed to achieve complete desorption processes with N-UiO-66 and OH-UiO-66 MOFs because

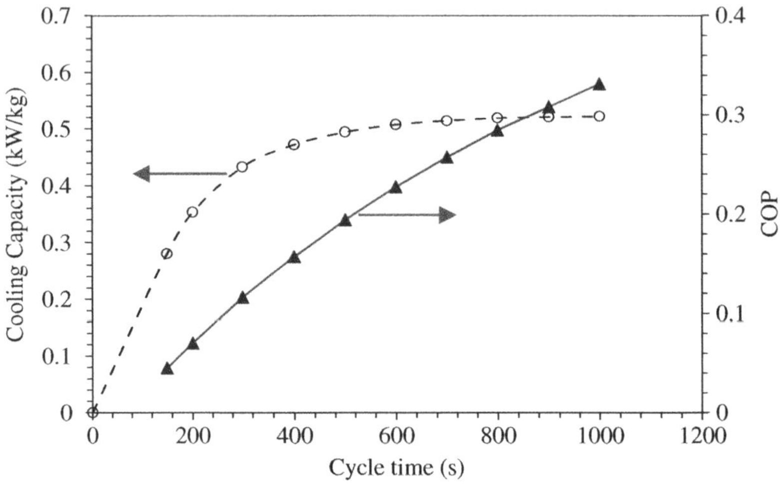

Figure 4.24 NH_2-MIL-125 (Ti) MOF at various half-cycle times for adsorption-assisted cooling applications. Here, the simulation is run at the 70, 30, and 14.8 °C heating, cooling, and chill water inlet temperatures.

they exhibit hydrophilic behavior in the low-pressure zone. As a result, higher hot water inlet temperature is where the higher COP and SCP are found. The order of NH_2-UiO-66 > OH-UiO-66 > N-UiO-66 > UiO-66 for the overall performance of various UiO-66 (Zr) MOFs + water-based adsorption chillers could be used.

Figure 4.24 displays the performance metrics for NH_2-MIL-125 (Ti) + water-assisted AC, including Q_{cool} and COP. The findings at the desorption temperature of 70 °C are displayed. As the cycle period lengthens, both the COP and cooling capacity grow. Both the adsorption and desorption stages are not fully exploited at a cycle period of 200 seconds or less, and very less water vapor is produced throughout each cycle. As a result, it is discovered that the cooling effect and COP are negligible. Additionally, the water transfer per cycle is fully utilized for the heat transformation process at very long cycle times (>1000 seconds), and the energy for desorption increases. As a result, the quantity of unnecessary energy input for desorption is increased, which could reduce the AHT system's cooling performance. For a thorough understanding, the supporting documents also show the NH_2-MIL-125 (Ti) MOF for adsorption-assisted cooling and heat pump applications at various half-cycle times (150–1000 seconds) and regeneration temperatures (60/70/80 °C) (Table 4.8).

4.4.3.2 TES Performances

TESD is defined as the amount of thermal energy stored in adsorption bed per unit volume or mass. In addition, TESD also reflects the charge/discharge rate of

Table 4.8 NH_2-MIL-125 (Ti) MOF for varied half-cycle times in adsorption-aided cooling and heat pump applications.

	Tregen = 60 °C		Tregen = 70 °C		Tregen = 80 °C	
Half-cycle time (s)	COP	Q_{cool} (kW/kg)	COP	Q_{cool} (kW/kg)	COP_h	Q_{heat}(kW/kg)
150	0.033	0.012	0.045	0.280	0.033	0.255
200	0.038	0.020	0.070	0.353	0.053	0.332
300	0.048	0.029	0.116	0.433	0.091	0.416
400	0.065	0.034	0.157	0.472	0.126	0.459
500	0.082	0.036	0.194	0.494	0.158	0.483
600	0.098	0.037	0.227	0.507	0.187	0.498
700	0.109	0.038	0.257	0.514	0.213	0.506
800	0.118	0.039	0.284	0.519	0.237	0.512
900	0.127	0.039	0.308	0.521	0.260	0.515
1,000	0.132	0.039	0.331	0.522	0.280	0.516

Chiller water inlet, cooling water inlet, and regeneration temperature are 14.8, 30, and 60–80 °C in this instance.

heat to/from a TES system [29]. In particular, the TESD of the MOF adsorbent used in the adsorption system is calculated by $TESD = \rho_{MOF} \cdot \int_{q_{des}}^{q_{ads}} Q_{st}(q) \cdot dq$, where ρ_{MOF} is the density of the MOF, $Q_{st}(q)$ is the isosteric heat of adsorption as a function of the uptake [30], and q_{ads} and q_{des} are the water uptake and offtake, respectively. It is also reported that the density of the functionalized MOFs would be approximately double that of the original density. For example, by implanting amino functional groups into the UiO-66 MOF, the density of NH_2-UiO-66 was increased from 0.43 to 0.93 g/cm^3 [31]. The synthesis procedures of the original and modified MOFs in our work are comparable to those found in the literature. Therefore, it is reasonable to consider the density of the modified MOFs as 0.8 g/cm^3 (which is double that of the original MOF). Based on this assumption, Figure 4.25 shows the TESD of the original and functionalized MOFs at the optimum Δq. Due to the enhanced adsorbent–adsorbate interaction (Q_{st}) as well as the improved water capacity, the functionalized MOFs show higher TESD compared to the original MOF. For example, at RH = 0.9, the TESD of the original UiO-66, OH-UiO-66, NH_2-UiO-66, and N-UiO-66 is 0.54 MJ/l (equivalent to 2.57 kJ/g), 0.86 MJ/l (+58.8%, equivalent to 2.97 kJ/g), 1.06 MJ/L (+95.5%, equivalent to 2.60 kJ/g), and 0.77 MJ/l (+41.8%, equivalent to 2.72 kJ/g).

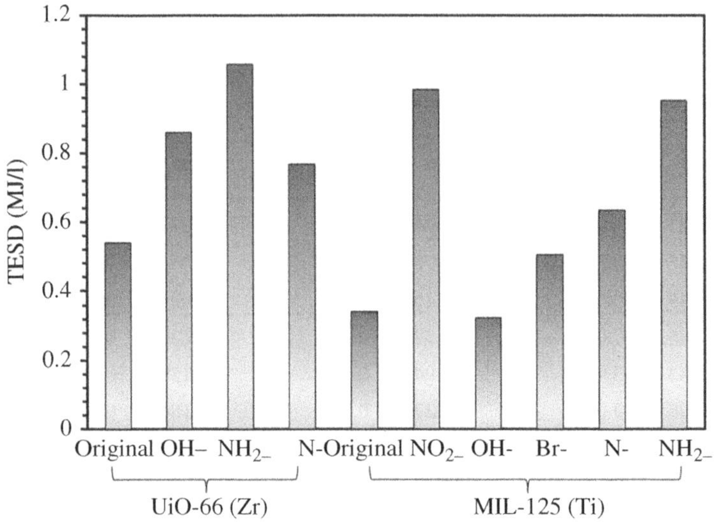

Figure 4.25 TESD of the original and functionalized MOFs.

4.5 Conclusion

The MOFs UiO-66 (Zr) and MIL-125 (Ti) are modified with various functional groups in this chapter. XRD, TGA, SEM, and N_2 adsorption are used to characterize both the original and functionalized MOFs. The original and modified MOFs are the subject of theoretical and experimental studies on water adsorption. The functionalized MOFs are observed to have promising stability characteristics and increased microporosity. The modified materials are shown to have faster adsorption–desorption kinetics and more water vapor transfer per AC cycle than the parent MOF adsorbents. The Langmuir-based equation is fitted to the water adsorption kinetics and isotherms, and the performance of AC is examined using the critical parameters. The performance of the adsorption chillers using the original and modified MOF adsorbent–water pairings is effectively predicted, and the results are provided in terms of SCP and COP. With regard to various regeneration temperatures, cold water inlet temperatures, and half-cycle times, the impacts of operating conditions on the AC performance are examined. The improved MOF-based adsorption chiller shows superior results. According to the findings, the functionalized MOFs outperform the original MOF in terms of enhanced AC performance, more water transfer per operating cycle, faster water adsorption/desorption kinetics, and higher TESD. As a result, the simulation's findings were supported by data on water adsorption isotherms and kinetics and experimentally verified MOFs. The finding offers a potential blueprint for designing and optimizing chillers that use both parent and modified MOFs to adsorb water. To validate these novel findings, additional study should focus on creating useful AC studies. A greater water uptake–offtake difference with faster kinetics per AC cycle could also be achieved with the original MOFs using other modification approaches, like pore size enlargement.

References

1 Yang, G., Xia, Z., Wang, R. et al. (2006). Research on a compact adsorption room air conditioner. *Energy Conversion and Management* 47 (15-16): 2167–2177.

2 Cmarik, G.E., Kim, M., Cohen, S.M., and Walton, K.S. (2012). Tuning the adsorption properties of UiO-66 via ligand functionalization. *Langmuir* 28 (44): 15606–15613.

3 Han, B. and Chakraborty, A. (2023). Evaluation on the performances of adsorption desalination employing functionalized metal-organic frameworks (MOFs). *Applied Thermal Engineering* 218: 119365.

4 Myat, A., Choon, N.K., Thu, K., and Kim, Y.-D. (2013). Experimental investigation on the optimal performance of Zeolite–water adsorption chiller. *Applied Energy* 102: 582–590.

5 Chen, C., Wang, R., Xia, Z. et al. (2010). Study on a compact silica gel–water adsorption chiller without vacuum valves: design and experimental study. *Applied Energy* 87 (8): 2673–2681.

6 Han, B. and Chakraborty, A. (2020). Advanced cooling heat pump and desalination employing functional UiO-66 (Zr) metal-organic frameworks. *Energy Conversion and Management* 213: 112825.

7 Kitagawa, S. (2014). Metal–organic frameworks (MOFs). *Chemical Society Reviews* 43 (16): 5415–5418.

8 Yuan, S., Feng, L., Wang, K. et al. (2018). Stable metal–organic frameworks: design, synthesis, and applications. *Advanced Materials* 30 (37): 1704303.

9 Han, B. and Chakraborty, A. (2022). Synthesis and characterization of various MOFs and MOFs-zeolite composites: water adsorption and rapid cooling production. In: *Rapid Refrigeration and Water Protection: Next Generation Adsorbents*, 29–85. Springer.

10 Zhang, Y., Wu, L., Wang, X. et al. (2020). Super hygroscopic nanofibrous membrane-based moisture pump for solar-driven indoor dehumidification. *Nature Communications* 11 (1): 1–11.

11 Solovyeva, M., Gordeeva, L., Krieger, T., and Aristov, Y.I. (2018). MOF-801 as a promising material for adsorption cooling: equilibrium and dynamics of water adsorption. *Energy Conversion and Management* 174: 356–363.

12 Han, B. and Chakraborty, A. (2021). Tailoring zirconium-based metal organic frameworks for enhancing hydrophilic/hydrophobic characteristics: simulation and experimental investigation. *Journal of Molecular Liquids* 341: 117381.

13 Akiyama, G., Matsuda, R., Sato, H. et al. (2012). Effect of functional groups in MIL-101 on water sorption behavior. *Microporous and Mesoporous Materials* 157: 89–93.

14 Luz, I., Llabrés i Xamena, F., and Corma, A. (2010). Bridging homogeneous and heterogeneous catalysis with MOFs:"Click" reactions with Cu-MOF catalysts. *Journal of Catalysis* 276 (1): 134–140.

15 Jeremias, F., Lozan, V., Henninger, S.K., and Janiak, C. (2013). Programming MOFs for water sorption: amino-functionalized MIL-125 and UiO-66 for heat transformation and heat storage applications. *Dalton Transactions* 42 (45): 15967–15973.

16 Hossain, M.I., Cunningham, J.D., Becker, T.M. et al. (2019). Impact of MOF defects on the binary adsorption of CO_2 and water in UiO-66. *Chemical Engineering Science* 203: 346–357.

17 Kim, S.-N., Kim, J., Kim, H.-Y. et al. (2013). Adsorption/catalytic properties of MIL-125 and NH2-MIL-125. *Catalysis Today* 204: 85–93.

18 Gordeeva, L.G., Solovyeva, M.V., and Aristov, Y.I. (2016). NH2-MIL-125 as a promising material for adsorptive heat transformation and storage. *Energy* 100: 18–24.

19 Biswas, S. and Van Der Voort, P. (2013). A general strategy for the synthesis of functionalised UiO-66 frameworks: characterisation, stability and CO_2 adsorption properties. *European Journal of Inorganic Chemistry* 2013 (12): 2154–2160.

20 Waitschat, S., Fröhlich, D., Reinsch, H. et al. (2018). Synthesis of M-UiO-66 (M = Zr, Ce or Hf) employing 2, 5-pyridinedicarboxylic acid as a linker: defect chemistry, framework hydrophilisation and sorption properties. *Dalton Transactions* 47 (4): 1062–1070.

21 Peng, Q., Guo, J., Zhang, Q. et al. (2014). Unique lead adsorption behavior of activated hydroxyl group in two-dimensional titanium carbide. *Journal of the American Chemical Society* 136 (11): 4113–4116.

22 Sun, B. and Chakraborty, A. (2014). Thermodynamic formalism of water uptakes on solid porous adsorbents for adsorption cooling applications. *Applied Physics Letters* 104 (20): 201901.

23 Sun, B. and Chakraborty, A. (2015). Thermodynamic frameworks of adsorption kinetics modeling: dynamic water uptakes on silica gel for adsorption cooling applications. *Energy* 84: 296–302.

24 Zhang, L. (2000). A three-dimensional non-equilibrium model for an intermittent adsorption cooling system. *Solar Energy* 69 (1): 27–35.

25 Teo, H.W.B., Chakraborty, A., and Han, B. (2017). Water adsorption on CHA and AFI types zeolites: modelling and investigation of adsorption chiller under static and dynamic conditions. *Applied Thermal Engineering* 127: 35–45.

26 Garibay, S.J. and Cohen, S.M. (2010). Isoreticular synthesis and modification of frameworks with the UiO-66 topology. *Chemical Communications* 46 (41): 7700–7702.

27 Guo, X., Huang, H., Ban, Y. et al. (2015). Mixed matrix membranes incorporated with amine-functionalized titanium-based metal-organic framework for CO_2/CH_4 separation. *Journal of Membrane Science* 478: 130–139.

28 Kim, H., Yang, S., Rao, S.R. et al. (2017). Water harvesting from air with metal-organic frameworks powered by natural sunlight. *Science* 356 (6336): 430–434.

29 Hall, M.R. (2010). *Materials for Energy Efficiency and Thermal Comfort in Buildings*. Elsevier.

30 Han, B., Chakraborty, A., and Saha, B.B. (2022). Isosteric heats and entropy of adsorption in Henry's law region for carbon and MOFs structures for energy conversion applications. *International Journal of Heat and Mass Transfer* 182: 122000.

31 Dhainaut, J., Avci-Camur, C., Troyano, J. et al. (2017). Systematic study of the impact of MOF densification into tablets on textural and mechanical properties. *CrystEngComm* 19 (29): 4211–4218.

5

Ammoniated Salt-Based Gas–Solid Sorption Devices

Rakesh Sharma[1] *and E. Anil Kumar*[2]

[1] *Department of Mechanical Engineering, National Institute of Technology Raipur, Raipur, India*
[2] *Department of Mechanical Engineering, Indian Institute of Technology Tirupati, Tirupati, India*

5.1 Introduction to Gas–Solid Sorption Pair of Ammonia and Metal Halides

As the world population is increasing rapidly, the need for energy and thermal comfort is rising day by day. The significant contribution to the world's energy demand is fulfilled by fossil fuels (coal and petroleum). However, it is well known that the fossil fuel reserves are exhaustible, and with growing demands of energy, it is predicted to be unsustainable. Also, the consumption of fossil fuels pollutes the environment. The exhaust gases emitted by the burning of fossil fuels are the primary cause of global warming and respiratory diseases. To provide thermal comfort, the present world's community largely relies upon a conventional refrigeration system (electricity driven) which operates on the energy produced by fossils fuels. The refrigerants employed in the conventional refrigeration system cause weakening of the ozone layer due to which harmful UV enters the atmosphere which causes severe skin diseases. Thus, it is immensely required to investigate the alternative sources of energy which are sustainable in nature and technology which does not have any adverse impact on the environment. One such technology is based on the interaction of gases with solids or liquids. The fascinating nature of technologies based on gas–solid/liquid interaction is their ability to be thermally driven and can be operated using solar energy or waste heat. The exchange of heat during the interaction between the gas and solid or liquid forms the basis for the development of devices for various applications like refrigeration system, heating system, and energy storage [1–5]. The limited possibilities of gas–liquid pairs and their restriction in applicability turn the attention of researchers toward gas–solid pairs.

Advancements in Non-Conventional Cooling and Thermal Storage Strategies, First Edition.
Edited by Bidyut Baran Saha and Dibakar Rakshit.

Since the gas undergoes reaction with the solid in chemisorption rather than attached by weak van der Waals force like in physisorption, the degree of heat exchange involved during exothermic adsorption and endothermic desorption is higher in chemisorption than that during physisorption. Also, the adsorption capacity is higher in chemisorption. Therefore, the chemisorption-based gas–solid pairs are favorable for the development of energy conversion or storage devices. Among different chemisorption pairs based on H_2, NH_3, H_2O, or CO_2 gas, metal alloys which absorb H_2 gas are very expensive, and the product formed from the gas–solid pair of hydrogen–metal alloys, i.e., metal hydride, is very reactive. While in the case of water vapor-based pairs, evacuation below atmospheric pressure is required for the evaporation of water and also the formation of a liquid layer on the surface of metallic salt or oxide hampers their water adsorption capacity. Instead, the ammonia adsorption capacity of metal halides is very high, and metal halides are comparably inexpensive. Also, the ammonia gas reacts with a variety of available metal halides within an extensive range of working temperature from −50 to 300 °C [6, 7]. The reversible reaction between ammonia and metal halide salts follows the following equations:

$$\text{Salt} + x\text{NH}_3 \leftrightarrow \text{Salt} \cdot x\text{NH}_3 + \Delta H_1 \tag{5.1}$$

$$\text{Salt} \cdot x\text{NH}_3 + y\text{NH}_3 \leftrightarrow \text{Salt}(x + y)\text{NH}_3 + \Delta H_2 \tag{5.2}$$

where x and y are moles of ammonia. The reaction equilibrium of ammonia and halide salt is monovariant which implies that the equilibrium pressure and temperature are interdependent [7]. By increasing the equilibrium temperature, the pressure of the ammoniated salt complex compound can be increased, and vice versa. The equilibrium pressure and temperature are related by the Van't Hoff equation which is as follows:

$$\ln P_e = \frac{\Delta H}{RT} - \frac{\Delta S}{R} \tag{5.3}$$

where P_e is the equilibrium pressure of NH_3 gas over halide salt in Pa, ΔH is the ad-/desorption enthalpy in kJ/mol of NH_3 gas, ΔS is the ad-/desorption entropy in kJ/mol K, R is the universal gas constant in kJ/mol K, and T is the temperature in K.

Also, the use of ammonia as a refrigerant is highly favored because of its zero ozone depletion potential and global warming potential. Among various natural refrigerants, the latent heat of evaporation of ammonia is next to water but comparatively higher than that of methanol and ethanol. As the boiling point of ammonia is well below (−33 °C) at atmospheric conditions, it is not required to work under vacuum condition in the case of ammonia evaporation which is essential while using water as a refrigerant. Ammonia is toxic but not below the concentration of 250 ppm to cause eye or skin irritation and 500–1000 ppm to cause breathing problems. However, being lighter than air, ammonia is easily vented to the atmosphere. Also, due to pungent smell, the leakage of ammonia is easily identified. Thus, the gas–solid pair of ammonia and metal halide offers a promising choice to develop sorption-based energy conversion or storage devices or heating/cooling application, solar energy utilization, and industrial decarbonization.

5.2 Ammoniated Salt-Based Adsorption System

The adsorption refrigeration system consists of an evaporator, a condenser, and an adsorber container (Figure 5.1). The cooling effect is produced in the evaporator by the latent heat of refrigerant during its phase change. The evaporated vapor is then adsorbed by the solid adsorbent contained in the adsorber which results in the release of adsorption heat to the ambient. The adsorber is then heated up to the sufficient desorption temperature by taking heat input from an available heat source which may be either solar energy or waste heat. Once the desorption temperature is reached, the refrigerant is desorbed from the adsorbent by consuming desorption reaction enthalpy and then condensed in the condenser. The condensed refrigerant is throttled back to the evaporator for the next cycle. Unlike the absorption refrigeration system in which the absorption and release of refrigerants by the liquid absorbent take place separately in an absorber and a generator, respectively, the solid adsorbent undergoes temperature swing (and consequently pressure swing) between ambient and desorption conditions to adsorb and desorb the refrigerant, respectively, in the adsorption refrigeration system.

The adsorption cooling system based on the working pair of ammonia and halide salt is firstly proposed by Rockenfeller in his patent [8]. It has been suggested by Rockenfeller that the temperature as low as −65 °C can be achieved by using a gas–solid pair of ammonia and halide salt. Although even before this, few researchers had recognized the affinity of halide salt toward ammonia

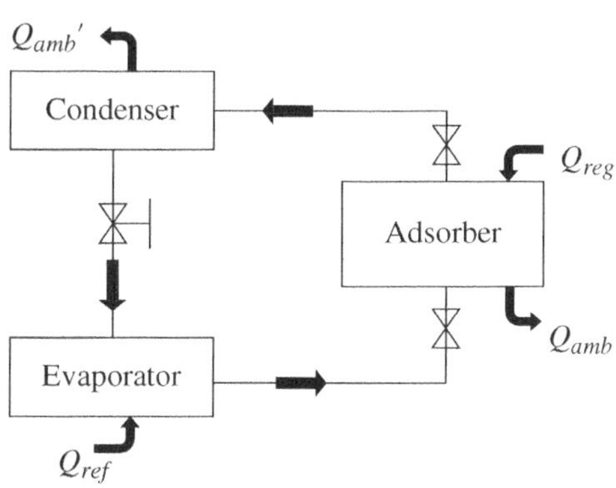

Figure 5.1 Schematic diagram of the adsorption refrigeration system.

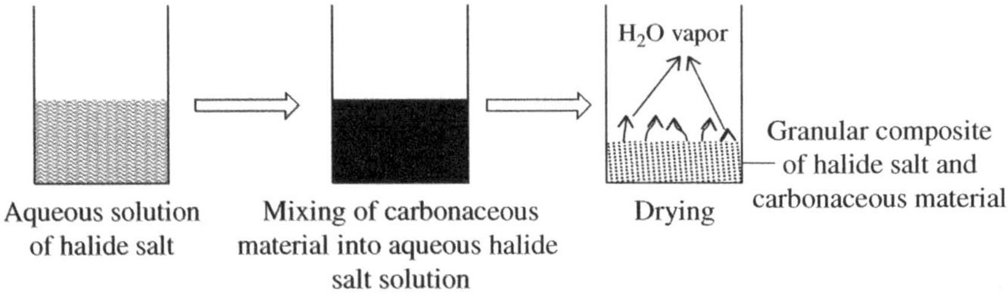

Figure 5.2 Method of impregnation to prepare a composite adsorbent.

and performed the thermochemical characterization of ammoniated halide salt [9, 10]. The ammoniated salt adsorption system has been given more attention in the period starting from 1990s. In the early 1990s, the work was mainly focused on the reaction kinetics of ammonia and halide salt [11, 12]. It has been understood that the importance of heat and mass transfer is vital for the designing of practical systems. The problems of swelling and agglomeration occur during ammonia adsorption of halide salts which reduces the diffusion of ammonia gas. Also, the heat transfer through the reactive packed bed of ammoniated halide salt is adversely affected by the low effective thermal conductivity. As the thermal conductivity of reactive halide salt is low, Coste et al. [13] proposed in the year 1986 that the rate of gas–solid reaction can be enhanced by adding graphite to the reactive salt by testing a composite mixture of calcium chloride and graphite with methylamine as a reactive gas. Mauran et al. [14] suggested the methods of mixing expanded graphite and studied the heat transfer through the reactive beds with the determination of effective thermal conductivity and permeability during the reaction between ammonia and a composite of $MnCl_2$ and expanded graphite. The composite adsorbents (solidified or granular) of halide salt and porous carbonaceous material are favorably employed to investigate the sorption system rather than using halide salts in the basic granular form [14]. The composite adsorbent improves the diffusion of ammonia gas and enhances the effective thermal conductivity of reactive packed beds.

The method of impregnation (Figure 5.2) to develop composite adsorbents using expanded graphite is suggested by Mauran et al. in his US patent [15]. Impregnation is the process in which a porous carbonaceous material is mixed with the halide salt solution and then the mixture is dried to remove the solvent which is usually water. The drying is carried out in a vacuum condition to allow the halide salts get penetrated into the pores of the carbonaceous material. Expanded natural graphite (ENG) is the most common, and inexpensive porous carbonaceous material used to prepare composite adsorbents other than activated carbon, activated carbon fibers (ACFs), carbon nanotubes (CNTs), multi-walled carbon nanotubes (MWCNTs), and carbon nanofibers (CNFs). Table 5.1 shows the measured values

Table 5.1 Effective thermal conductivity and permeability of composite halide salt adsorbents.

Composite adsorbent	Effective thermal conductivity (W/m K)	Permeability(m^2)
$CaCl_2$ + ENG	10.2–48.6 [16]	$10^{-12} - 5.3 \times 10^{-16}$ [17]
(consolidated)	7.05–9.2 [18]	—
	0.87–1.62 [19]	$3.86 \times 10^{-11} - 8.95 \times 10^{-13}$ [19]
	0.85–1.70 [20]	$3.16 \times 10^{-11} - 8.61 \times 10^{-13}$ [20]
	0.87–3.01 [21]	$7.15 \times 10^{-11} - 1.45 \times 10^{-13}$ [21]
$CaCl_2$ + ENG (granular)	0.45–0.58 [18]	—
$CaCl_2$ + ENG treated with sulfuric acid (consolidated)	23.5–88.1 [22]	$9.31 \times 10^{-10} - 3.05 \times 10^{-14}$ [22]
$MnCl_2$ + ENG	10.8–22.9 [16]	$10^{-12} - 4.9 \times 10^{-15}$ [17]
(consolidated)	0.79–1.69 [19]	$9.43 \times 10^{-11} - 3.02 \times 10^{-13}$ [19]
	0.82–2.07 [21]	$9.61 \times 10^{-11} - 1.42 \times 10^{-13}$ [21]
$BaCl_2$ + ENG	9.63–30.7 [16]	$10^{-12} - 1.1 \times 10^{-14}$ [17]
(consolidated)	0.62–1.84 [19]	$4.73 \times 10^{-11} - 1.01 \times 10^{-13}$ [19]
	0.62–1.72 [21]	$9.02 \times 10^{-11} - 1.96 \times 10^{-13}$ [21]
NaBr + ENG	0.77–1.61 [19]	$9.05 \times 10^{-10} - 2.80 \times 10^{-11}$ [19]
(consolidated)	0.88–1.75 [21]	$8.13 \times 10^{-11} - 3.37 \times 10^{-13}$ [21]
$NiCl_2$ + ENG	0.72–1.84 [19]	$3.87 \times 10^{-10} - 1.80 \times 10^{-13}$ [19]
(consolidated)	0.72–2.01 [21]	$8.55 \times 10^{-11} - 1.98 \times 10^{-13}$ [21]
$FeCl_3$ + ENG	0.89–2.09 [19]	$8.48 \times 10^{-11} - 1.87 \times 10^{-13}$ [19]
(consolidated)	0.92–2.04 [21]	$1.21 \times 10^{-11} - 1.14 \times 10^{-13}$ [21]
NH_4Cl + ENG	0.87–2.13 [19]	$4.60 \times 10^{-11} - 4.13 \times 10^{-12}$ [19]
(consolidated)	0.87–1.68 [21]	$5.92 \times 10^{-11} - 1.41 \times 10^{-13}$ [21]
$SrCl_2$ + ENG	0.87–1.87 [19]	$6.50 \times 10^{-10} - 2.76 \times 10^{-13}$ [19]
(consolidated)	0.86–2.37 [21]	$6.62 \times 10^{-11} - 1.47 \times 10^{-14}$ [21]

ENG, expanded natural graphite [17, 19–22].

of effective thermal conductivity and permeability of composite halide salt adsorbents prepared by the impregnation method using ENG by various researchers [16–22]. In the meantime, another method to make composite adsorbents is suggested as intercalation which provides stronger bond, but the preparation is slightly more challenging than impregnation [23, 24]. The impregnation is found to be a more simple and effective method for preparing composites [25, 26].

Meanwhile, researchers started analyzing the systems based on ammoniated salt reactions and studied extensively about cooling and heating systems with various methods to improve heat transfer and the performance of the system [27–29]. The studies involving the numerical modeling and second law analysis of chemical heat pump based on ammonia–halide salt pairs have also gained momentum in the late 1990s [30–32]. Various techniques like advanced cycles, thermal wave process, porous material for composites, and design of heat exchanger have been discussed to improve the performance of the sorption system [33]. A collective work by renowned scientists working with different gas–solid pairs, like zeolite–water, activated carbon–methanol, activated carbon–ammonia, halide salt–ammonia, metal hydride–hydrogen, water–ammonia, and LiBr–water, has been performed with the support of European Union in which sorption cooling and heating systems based on different working pairs are compared [34].

The measurement of sorption characteristics is equally important to understand the adsorption/desorption behavior for the better designing of a thermochemical system [35–40]. Hysteresis is observed between adsorption and desorption due to which the desorption of ammonia takes place at pressure lower than that of adsorption at a particular temperature [35, 41–46]. The decrease in the dynamic adsorption capacity of halide salts is observed with an increase in temperature [47, 48], while the high driving pressure difference results in the improvement of the dynamic capacity [49, 50]. The adsorption isotherms of a binary salt mixture having a common cation are observed to be in between the isotherms of two independent components [51, 52]. The ammonia adsorption characteristics of composite adsorbents are measured, and significant reduction in the swelling and agglomeration of salt adsorbent is observed in the course of adsorption by mixing the salt with a porous matrix [53–57]. Recently, the solution to the sorption hysteresis is given by Gao et al. [58, 59] by mixing together the reactive salt adsorbents.

Progressively, researchers have started testing and developing prototypes of the ammoniated salt-based adsorption refrigeration system for several applications. Wang et al. [60] conducted an experiment to investigate the adsorption performance of different gas–solid pairs employing physical, chemical, and composite adsorbents for adsorption ice makers on fishing boats and observed that the adsorption performance of the composite adsorbent prepared using $CaCl_2$ and activated carbon is the best among all. The cooling power of the composite adsorption ice maker is observed to be 10 times that of the physical adsorption ice maker and 1.38 times that of the chemical adsorption ice maker [60]. It is explained that the problem of agglomeration during ammonia adsorption of a chemical adsorbent ($CaCl_2$) is solved by using the composite of the chemical adsorbent ($CaCl_2$ + activated carbon) and the solidified composite is proven to have better performance and more stability [61, 62]. The investigations on the cooling performances, namely, coefficient of performance (COP) and specific cooling power (SCP), of the adsorption refrigeration system have been done extensively under different operating temperature conditions using various salts among which the most commonly used salt adsorbent is $CaCl_2$. Table 5.2 represents the performance values and operating conditions of the adsorption cooling system for different applications as reported in the literature. Rivera et al. [74] established that the ammoniated salt-based thermochemical system can operate at a very low temperature of the heat source, as low as 53 °C, by examining the sorption between $BaCl_2$ and NH_3. The adsorption system is found to be suitable for solar-powered applications in which the refrigeration or cooling effect is produced during night time, whereas solar energy available during daytime is utilized to release ammonia gas from halide salt to the evaporator [67]. It is also observed that the adsorption system is capable enough to provide deep freezing by attaining the evaporator temperature as low as −35 °C using $MnCl_2$ + ENG as an adsorbent where the COP and SCP of the system are reported as 200 W/kg and 0.33, respectively [71]. A hybrid system of a compressor-driven refrigeration unit and thermochemical refrigeration unit has been proposed by Ferrucci et al. [75] in which the compressor-driven refrigeration unit works during the day either to produce refrigeration or to dissociate $BaCl_2$–NH_3, whereas the thermochemical unit produces the cooling effect during the evening.

Meanwhile, researchers have found out that certain halide salts can desorb ammonia below room temperatures and can be used in place of an ammonia evaporator/condenser to produce the cooling effect [76, 77]. In this type of gas–solid sorption system, ammonia is exchanged between two halide salt adsorbers and is termed the resorption system which is explained in Section 5.3.

5.3 Resorption Refrigeration System

The resorption refrigeration system consists of two reactors individually filled with two different halide salts between which ammonia vapor is exchanged. The principle of resorption refrigeration system is shown in Figure 5.3. The enthalpy required to desorb ammonia from one of the salts is used for the production of refrigeration effect where the respective salt is termed the refrigerating salt. The desorbed ammonia vapor is then simultaneously adsorbed by another salt adsorbent which thereby releases adsorption heat to the atmosphere. The ammonia vapor is

Table 5.2 Performance values of the adsorption system for different applications.

Adsorbent	Application	Evaporator temperature °C	Ambient temperature °C	Specific cooling power (SCP) W/kg	Coefficient of performance (COP)	Work
$CaCl_2$ + AC	Icemaker for fishing boat	−15	25	339.5	—	[60]
	Icemaker for fishing boat	−20	30	770.4	0.39	[62]
	Solar icemaker	−15	30	161.2	0.12	[62]
	Refrigeration system	−5	25	205.2	0.197	[63]
$CaCl_2$ + ENG	Icemaker	−11	30	900	0.36	[64]
	Refrigeration system	−2.7	30	415	0.28–0.46	[65]
		−18.3		255		
$CaCl_2$	Icemaker for fishing boat	−15	25	246.0	—	[60]
	Refrigeration system	−5	37–40	265	0.28	[66]
$BaCl_2$	Solar cooling system	−30	15–30	—	0.031	[67]
	Air-conditioning	15	35	—	0.6	[41]
$BaCl_2$ + ENG	Solar cooling system	0–15	25, 30	50–500	0.50–0.53	[68]
NaBr + ENG	Air-conditioning	15	30	129–137	0.43–0.46	[69]
	Air-conditioning	10	25	231–575	0.28–0.48	[70]
$MnCl_2$ + ENG	Deep freezing	−35 to 0	25	200–700	0.33–0.37	[71]
	Refrigeration system	10	25	—	0.05–0.25	[72]
LiCl + ENG	Solar icemaker	−10 to −5	25–30	91–117	0.47	[73]

ENG, expanded natural graphite [65, 68–73]; AC, activated carbon [60, 62–64].

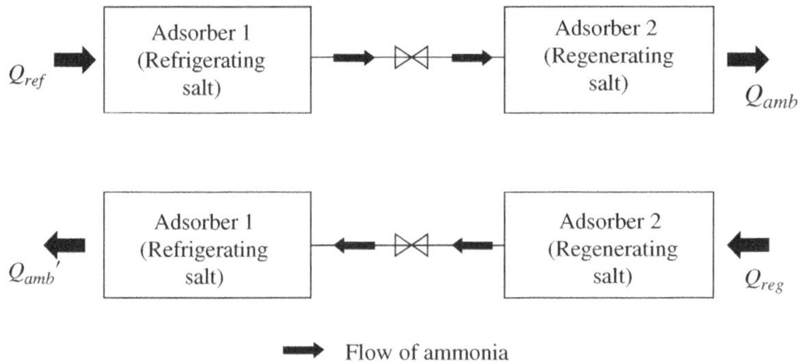

Figure 5.3 Principle of resorption refrigeration system.

desorbed back from the second salt to the refrigerating salt to complete the cycle. The second salt is termed the regenerating salt as it desorbs ammonia back to the refrigerating salt which produces the cooling effect in a further cycle. In contrast to the adsorption refrigeration system, no change in the phase of ammonia occurs during the cyclic process of resorption refrigeration system, and ammonia is cycled in vapor form which allows the usage of resorption system at places where the positioning of liquid ammonia is a problem and continuous occurrence of vibration.

The halide salts are chosen as either refrigerating or regenerating salt by their ammonia sorption characteristics which are unique to each salt [76]. Figure 5.4 shows the pressure–temperature (*P–T*) equilibrium lines (Van't Hoff plots) of ammonia sorption of different halide salts [75]. The different halide salts are categorized as either LTS, MTS, or HTS depending upon their equilibrium states of pressure and temperature during ammonia sorption. Jiang et al. [21] suggested that the salts which have equilibrium temperatures higher than 150 °C at an equilibrium pressure of 1.5 MPa come under the category of HTS, while those having temperature between 90 and 150 °C and below 90 °C at 1.5 MPa are categorized as MTS and LTS, respectively [21]. The choice of refrigerating salt for the resorption system is done from LTS which can desorb ammonia at low temperatures, while the regenerating salt is chosen from MTS or HTS so that to ensure a sufficient pressure difference during the refrigeration process [77].

So far, very few working pairs have been practiced in the resorption system to achieve a cooling or heating effect. Among which the most commonly used LTS to produce cooling or refrigeration effects are $BaCl_2$, $PbCl_2$, NaBr, and NH_4Cl, while $MnCl_2$ and $NiCl_2$ are the two HTSs used for regeneration. The pair of $BaCl_2$–$MnCl_2$, each mixed with graphite to improve heat transfer, is firstly used by Lepinasse et al. [76] to analyze a thermochemical resorption transformer for the refrigeration effect and later by Goetz et al. [77] to produce refrigeration and air-conditioning. Gradually, researchers have started to investigate more about

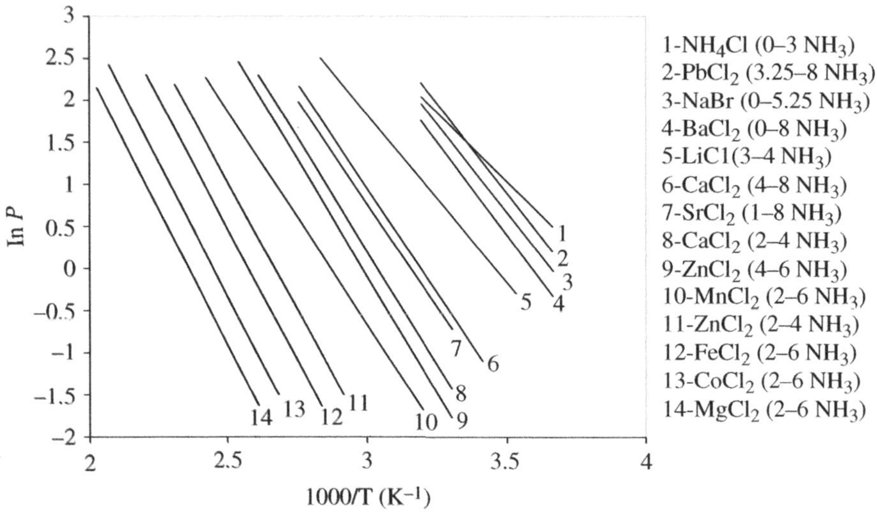

Figure 5.4 *P*–*T* equilibrium lines of ammonia sorption of halide salt. Source: Adapted from Goetz et al. [77].

the resorption system and proposed modifications. Simultaneous heating and cooling are achieved by Vasiliev et al. by using the pair of $BaCl_2$–$NiCl_2$ in the two reactors' resorption heat pump which is modified to the $BaCl_2$–$MnCl_2$-based four reactors' resorption heat pump to produce continuous thermal effects where Busofit is used to make a composite of each halide salt [78]. Lepinasse et al. investigated the prospect of resorption system for cold storage and successfully maintained a box at 5 °C for three hours using the pair $PbCl_2$–$MnCl_2$ [79].

As the cooling effect is produced by the desorption enthalpy of halide salt in the resorption cooling system which is higher than the latent heat of evaporation of ammonia, the COP of the resorption system is higher than the adsorption system. Wang et al. [66] and Li et al. [72] investigated the cooling performance of the resorption system using the $BaCl_2$–$MnCl_2$ pair and compared with that of the adsorption system. It is observed that the resorption system operates at lower pressure than the adsorption system which critically affects the mass transfer. The resorption system can also be used to produce heat at high temperature. Van Der Pal et al. experimentally demonstrated that a temperature lift of 50 °C could be achieved by using a rarely studied pair of LiCl and $MgCl_2$ in the resorption heat pump [80].

Subsequently, researchers have performed experiments with other LTSs like NaBr and NH_4Cl and compared with the performance of the $BaCl_2$-based resorption system. Bao et al. conducted an experiment to compare the performance of the resorption system based on different pairs of NH_4Cl–$MnCl_2$, NaBr–$MnCl_2$, and $BaCl_2$–$MnCl_2$ to produce the cooling effect at 0 °C or below [81, 82]. It has been observed by the study that the resorption system based on NH_4Cl–$MnCl_2$ has the highest SCP among the three pairs, while $BaCl_2$–$MnCl_2$ has the lowest. The high SCP is due to the high degree of adsorption and desorption of ammonia by NH_4Cl in comparison to the other two LTSs. Accordingly, Bao et al. reported that NH_4Cl–$MnCl_2$ is best suitable for the resorption system to produce the cooling effect at 0 °C or below [81, 82].

A prototype of the NH_4Cl–$MnCl_2$ resorption system is designed and tested by Xu et al. [83] to obtain simultaneous cooling and heating effects. The cooling effect is observed at a temperature as low as −5 °C and a heating effect at 80 °C which can be further utilized to drive a silica gel–water adsorption chiller [83]. The pair of NH_4Cl–$MnCl_2$ has been further utilized in the resorption refrigeration system for cold storage and long-distance refrigeration [84, 85]. Van Der Pal et al. have used the concept of resorption system to increase the temperature of available waste heat by combining it with the compression system [86].

Some researchers show that the resorption system can also be applied in applications other than producing useful thermal effects. Wang et al. have coupled the resorption refrigeration system with a turbine to produce power along with cooling production [87]. A resorption power generation (RPG) cycle has been proposed by Bao et al. to produce power through the resorption system utilizing low-grade heat [88, 89]. Lu et al. have placed two expansion devices termed the scroll expander between LTS and HTS to produce power each time during the transfer of ammonia between salt beds one after the other [90, 91].

The operation of the adsorption refrigeration system based on one adsorber salt bed (Figure 5.1) involves sensible heating or cooling of the adsorber in between refrigeration or regeneration processes which results in the discontinuous production of the useful cooling effect. Two adsorber salt beds are employed which work opposite to each other in such a way that if one adsorber undergoes adsorption by coupling with the evaporator, the other desorbs ammonia to the condenser and vice versa to produce the cooling effect continuously, and the sorption system is termed a continuous adsorption refrigeration system. Similarly, in the resorption system based on a single pair of LTS and HTS (Figure 5.3), the cooling effect is produced intermittently. Two pairs of LTS and HTS are used in which each pair works out of phase to another pair to achieve the continuous cooling effect, and the system is called the continuous resorption refrigeration system. Furthermore, work has been done in the meantime to improve the performance of the basic adsorption and resorption refrigeration system by developing a gas–solid advanced sorption system.

5.4 Advanced Sorption Systems

To overcome the problem of low COP associated with ammoniated salt-based adsorption or resorption systems, advanced sorption systems have been proposed in the literature [6, 7]. The working principles of different advanced sorption systems are explained below.

5.4.1 Thermal Wave Cycle

Thermal or heat wave cycle is based on the principle of thermal coupling of the two adsorber beds of the continuous adsorption refrigeration system by a single common heat transfer fluid circuit [92]. A heat transfer fluid is circulated through the two adsorber beds of the adsorption refrigeration system to utilize the heat released by the adsorber bed undergoing adsorption which is supplied to facilitate desorption from another adsorber bed which consequently decreases the amount of heat required by the heat transfer fluid from the external heat source. Figure 5.5 shows the schematic of a thermal wave cycle which consists of two adsorber beds, an

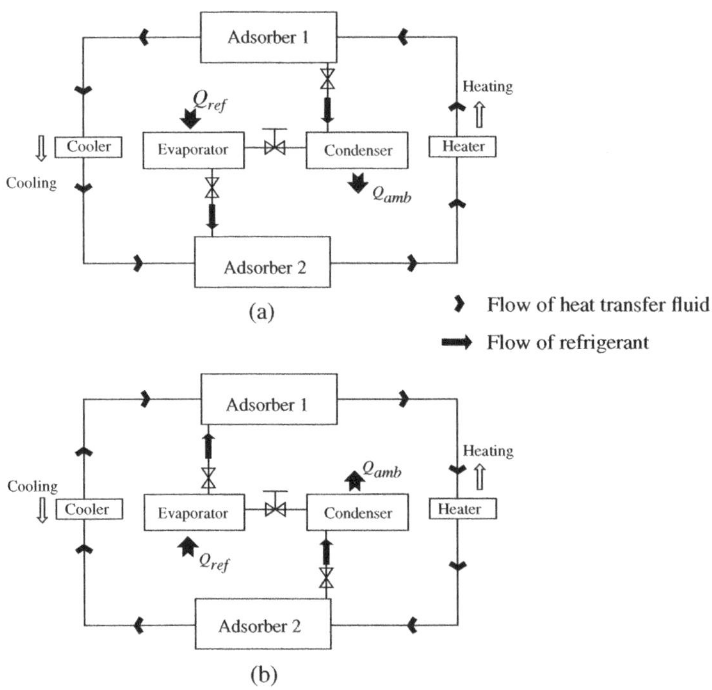

Figure 5.5 Schematic of the thermal wave cycle. Source: Shelton [92]/U.S. Department of Energy/Public Domain.

evaporator, and a condenser. The two adsorber beds are thermally coupled along with a cooler and heater by a heat transfer fluid circuit.

Figure 5.5a shows the initial operation of the thermal wave cycle in which adsorber 1 is ready for desorption and connected with a condenser, while adsorber 2 is coupled with the evaporator for the adsorption process. The heat transfer fluid is cooled by the cooler and allowed to pass through adsorber 2. The adsorption process begins with the cooling of adsorber 2, and the released adsorption heat is taken up by the heat transfer fluid. After attaining heat from adsorber 2, the heat transfer fluid is further heated in the heater to achieve the temperature required to enable desorption from adsorber 1. The temperature of the heat transfer fluid becomes low at the end of adsorber 1 and is further lowered in the cooler before passing through adsorber 2. In this way, the adsorption heat released by adsorber 2 is transferred to adsorber 1 which decreases the external heat input. This process continues till adsorber 2 reaches the saturation adsorption state and desorption in adsorber 1 is completed. Figure 5.5b represents the next operation in which the heat transfer fluid flows in the opposite direction, and the two adsorber beds switch their processes, i.e., adsorber 1 undergoes adsorption and adsorber 2 undergoes desorption.

Goetz and Llobet [93] carried out an experiment on the thermal wave cycle in which various salts are filled in a single reactor and the heat transfer fluid is allowed to pass through it in such a manner that it exchanges heat with each salt one after another during refrigeration or regeneration processes.

5.4.2 Mass and Internal Heat Recovery Cycles

Another method employed to improve the performance of advanced sorption systems is the coupling of reactors to cause either mass or heat exchange or both between them [94]. The adsorption system in which the exchange of the ammonia refrigerant is carried out between the adsorbent beds is termed the mass-recovery adsorption system [95].

In the mass-recovery adsorption system, the two adsorber beds are connected with each other for the transfer of ammonia during switch time. Before the switch time, the adsorber which adsorbs ammonia from the evaporator is at low pressure, while the adsorber which desorbs ammonia to the condenser is at high pressure. At this condition, the high-pressure adsorber is connected with the low-pressure adsorber which increases the desorption rate of the high-pressure adsorber and thus improves its adsorption capacity in the next cycle. Figure 5.6 illustrates the schematic of the mass-recovery adsorption system [95]. Figure 5.6a represents the cooling and regeneration processes, whereas Figure 5.6b shows the mass-recovery process during switch time in which ammonia flows from adsorber 1 to adsorber 2. The adsorbers interchange their function in Figure 5.6c, while the

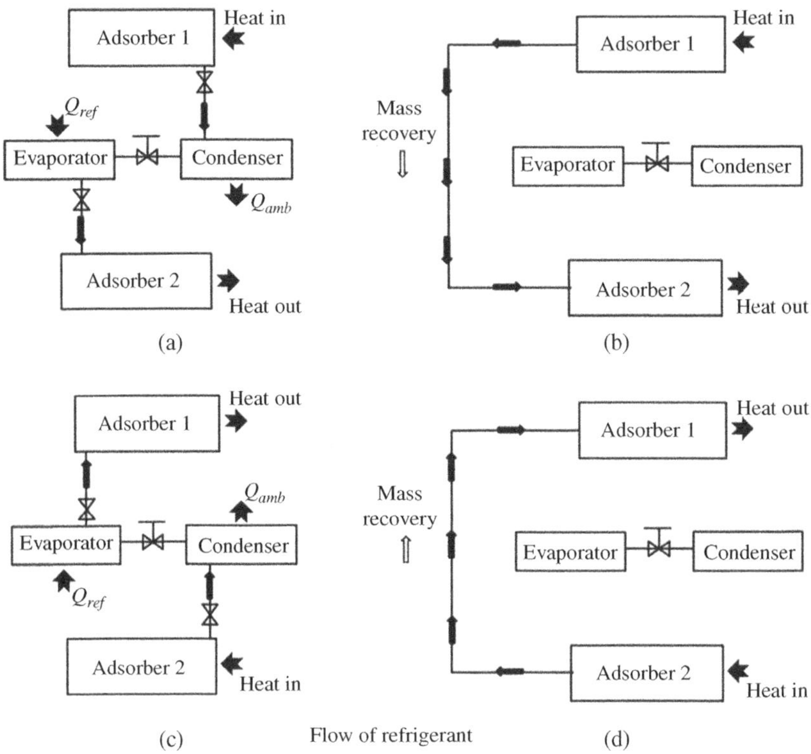

Figure 5.6 Schematic of the mass-recovery adsorption system. Source: Wang et al. [95]/John Wiley & Sons.

mass recovery is observed by transferring ammonia from adsorber 2 to adsorber 1 in Figure 5.6d which reinstates the initial condition to perform the next cycle.

Neveu and Castaing described that internal heat recovery could be possible by using two different salts in the continuous adsorption refrigeration system in which the adsorption heat of one salt adsorber is utilized to drive the desorption process from another salt adsorber, and the system is termed the internal heat-recovery adsorption system [7]. As ammonia is chemisorbed by a wide variety of halide salts at different temperatures and pressures, it is entirely possible to use the adsorption heat of one salt to desorb another in the ammoniated salt-based sorption system. Figure 5.7 shows the schematic of the internal heat recovery adsorption system.

In the first half cycle, as depicted in Figure 5.7a, the first cooling effect is produced by the evaporation of ammonia which is adsorbed by salt B in adsorber 2, while salt A in adsorber 1 desorbs ammonia by consuming heat from the external heat sources which is condensed in the condenser. After the completion of the

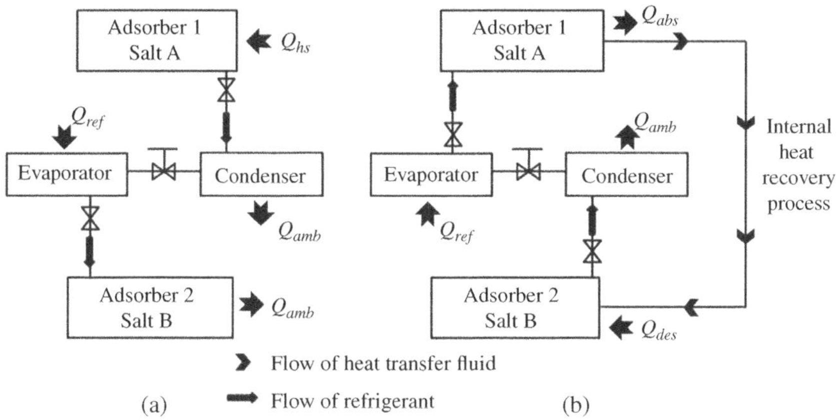

Figure 5.7 Schematic of the internal heat-recovery adsorption system. Source: Adapted from Neveu and Castaing [7].

first half cycle, the working operation of both adsorbers interchange, i.e., adsorber 1 (salt A) is coupled with the evaporator and adsorber 2 (salt B) with the condenser. In the second half cycle (Figure 5.7b), the cooling effect is produced in the evaporator where ammonia is adsorbed by salt A in adsorber 1. The adsorption heat released by salt A is collected by the heat transfer fluid and supplied to salt B in adsorber 2 to desorb ammonia which condenses in the condenser. The process of utilization of adsorption heat of one salt adsorber (salt A) by another salt adsorber (salt B) for the desorption process is termed an internal heat-recovery process. In this type of adsorption system, the cooling effect is produced twice at the expense of a single heat input from the external heat source to the system, resulting in the performance improvement of the system. The two salts A and B are selected in such a manner that the adsorption heat of one of the salts is sufficient enough to drive the desorption of ammonia from another.

Wang et al. conducted an experiment on the prototype of the adsorption refrigeration system and found that the COP and SCP are improved by 48.6% and 54.5%, respectively, by employing the concept of mass recovery [96]. Later, the mass-recovery adsorption cycles are tested for ice making by Oliveira et al. [97] and reported the highest COP of the system as 0.1. The experimental setup of novel heat-pipe type adsorption refrigeration system has been developed in Shanghai Jiao Tong University, China, in which both mass and heat recoveries are carried out [98–102]. Li et al. [99, 100] have investigated the performance of the heat-pipe type adsorption system for different cyclic operations, which is observed by executing the mass and heat recovery in various sequences like mass and heat recovery with heating and cooling, with and without mass recovery, heat and mass recovery, and mass and heat recovery. The significant

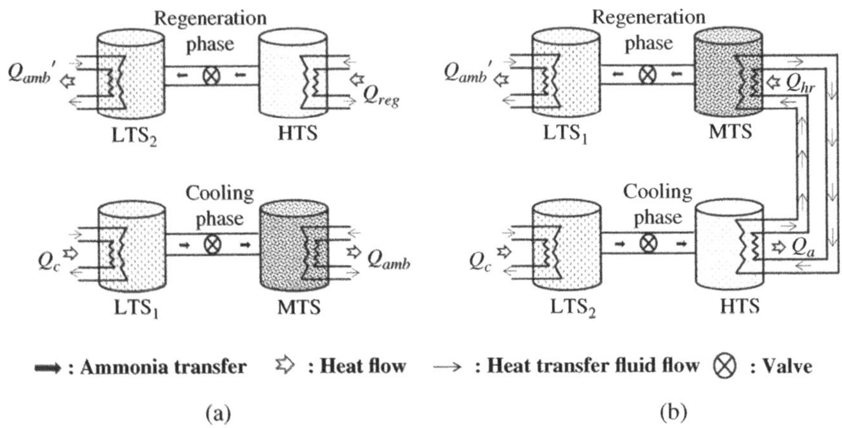

Figure 5.8 Heat-recovery resorption cooling system. (a) First half cycle. (b) Second half cycle. Source: Sharma and Anil Kumar [106]/with permission of Elsevier.

decrease in the heat source temperature is achieved by operating the heat-pipe type sorption system in a two-stage thermodynamic cycle based on two heat recovery processes [101]. A heat-recovery sorption refrigeration system has been thermodynamically analyzed by Rao et al. using different pairs of halide salts, and the performance is compared with the sorption system without heat recovery [103]. Recently, the concept of mass and heat recovery has been applied in the resorption cogeneration system to produce the refrigeration effect and power simultaneously which suggests significant improvement in the COP (38%) and energy efficiency of electricity (8–12%) [104]. Furthermore, Li et al. have proposed a triple-effect adsorption refrigeration system by employing two internal heat recovery processes between three different salts coupled with an evaporator and a condenser and observed the COP value in the range between 0.75 and 0.97 for different mass ratio ranges from 4 to 15 [105].

In another work, two different regenerating salts, one each from medium- and high-temperature salt, are included in heat-recovery resorption cooling system [106]. Two distinct refrigerating salts are individually coupled with two different regenerating salts (forming pairs LTS_1–MTS and LTS_2–HTS) which operate in cooling and regeneration phases in a succession. The schematic diagram of the heat-recovery resorption cooling system is illustrated in Figure 5.8 in which one complete cycle involves two half cycles where each half cycle produces a cooling effect. Thermal coupling of HTS and MTS is performed in the second half cycle by a heat transfer fluid circuit to achieve the heat-recovery process. Thus, the cooling effect is obtained twice in one complete cycle by supplying heat input to HTS only once from the external heat source. Since the individual pairs of LTS_1–MTS and LTS_2–HTS operate distinctly, two separate LTSs are taken to pair

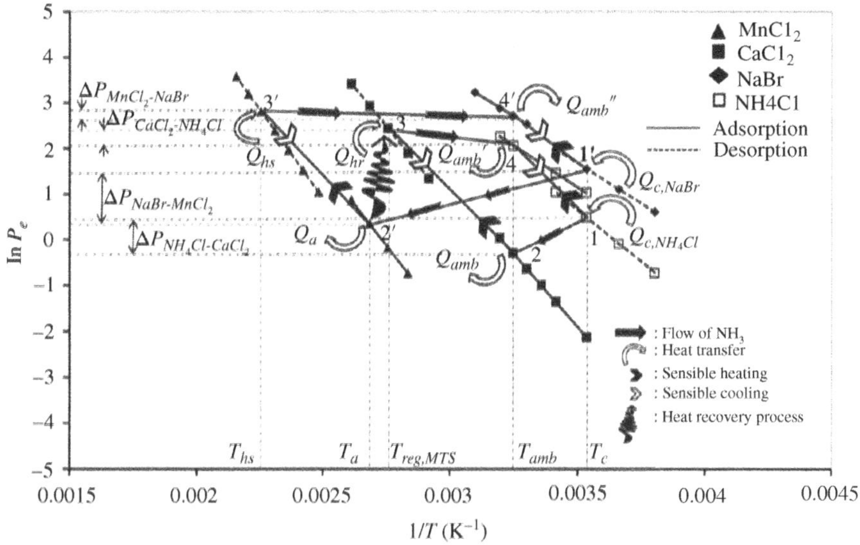

Figure 5.9 Thermodynamic cycle for the combination of $NH_4Cl-CaCl_2$: $NaBr-MnCl_2$. Source: Sharma and Anil Kumar [106]/with permission of Elsevier.

with MTS and HTS. Figure 5.9 demonstrates the thermodynamic cycle involving two different pairs as $NH_4Cl-CaCl_2$: $NaBr-MnCl_2$. The cycle 1–2–3–4 indicates the working operation of the $NH_4Cl-CaCl_2$ pair, and 1′–2′–3–4′ represents that of $NaBr-MnCl_2$. Cooling effects are achieved by the desorption enthalpy of NaBr and NH_4Cl, and the theoretical thermodynamic efficiency is observed to be 0.64 in the $NH_4Cl-CaCl_2$: $NaBr-MnCl_2$ heat-recovery resorption cooling system [106].

5.4.3 Combined Cycles

An innovative gas–solid sorption system which is a combination of evaporation and resorption processes has been suggested by Li et al. [107]. In this system, the cooling effect is obtained by the evaporation of ammonia in the evaporator as well as by the desorption of ammonia from halide salt. To further improve the performance of the above system and to increase the cooling capacity, the heat-recovery process is proposed along with the evaporation and resorption processes [107]. The heat recovery is realized between HTS and MTS, while cooling effects are observed by the evaporation of ammonia and by resorption processes between LTS and MTS as well as LTS and HTS, and the COP is observed in the range of 0.91–1.8 for different salts [107]. Oliveira et al. [108] studied the operation of three different types of cycles, namely, conventional adsorption cycle with and without internal heat recovery, resorption cycle with heat recovery, and combined adsorption and resorption cycle with heat recovery. It has been observed that the highest COP of 1.1 is obtained in the case of combined adsorption and resorption cycle with heat recovery but with a low cooling power density of 21.5 kW/m^3, while the maximum cooling power density (499 kW/m^3) is reported for the conventional adsorption cycle without heat recovery but with lowest COP value (0.16) [108]. It has been suggested by Oliveira et al. [108] that an optimum value of COP as 0.72 with the cooling power density of 47.2 kW/m^3 can be realized by using the resorption system with heat recovery. Later, Li et al. [109, 110] tested the feasibility of combined adsorption and resorption processes in a sorption cycle by constructing an experimental prototype using $MnCl_2$ and $BaCl_2$ as halide salt adsorbents. It is observed that the theoretical COP has been improved by 167% and 60% in comparison to the conventional adsorption cycle and resorption cycle, respectively [109, 110]. Further investigation on the performance of the combined sorption cycle involving adsorption and resorption processes has been performed by Xu et al. [111] who has later studied the performance of the combined system with the heat-recovery process [112]. A two-stage desorption process is proposed in which the regeneration temperature is decreased by desorbing ammonia from the first adsorber salt to second adsorber salt which is later desorbed into the condenser [113–115]. As the equilibrium pressure of reactive salt is lower than the saturation pressure of ammonia at the same temperature, the decrease in regeneration temperature is observed. Gao et al. employed the concept of two-stage desorption process to develop a freezing system for refrigerated trucks [116, 117].

Thus, the performance of ammoniated salt-based gas–solid sorption can be greatly improved by the development of advanced sorption systems. Also, different advanced sorption systems provide a suitable option for functioning with variable solar insolation [118].

5.5 Thermal Energy Storage System

The thermal energy storage based on thermochemical sorption has advantages of high energy storage density with a more extended storage period over other two heat storage systems, i.e., sensible and latent heat [119–123].

The working of thermochemical sorption energy storage is based on the principle of reversible adsorption and desorption processes between the adsorbent and adsorbate (Figure 5.10). The storage of thermal energy is achieved by desorbing the adsorbate from the adsorbent, whereas the stored thermal energy is released during adsorption. Hence, thermal energy can be stored in the thermochemical sorption system for the time until the interaction between the adsorbate and adsorbent is carried out. The thermochemical sorption heat storage method is more suitable for solar applications due to its fluctuating energy

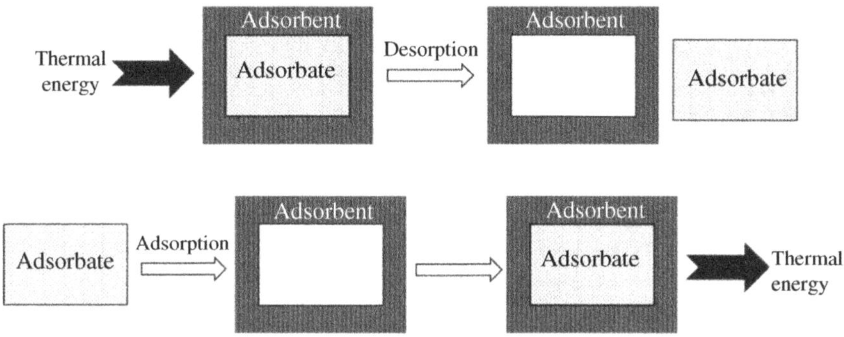

Figure 5.10 Principle of thermochemical sorption energy storage.

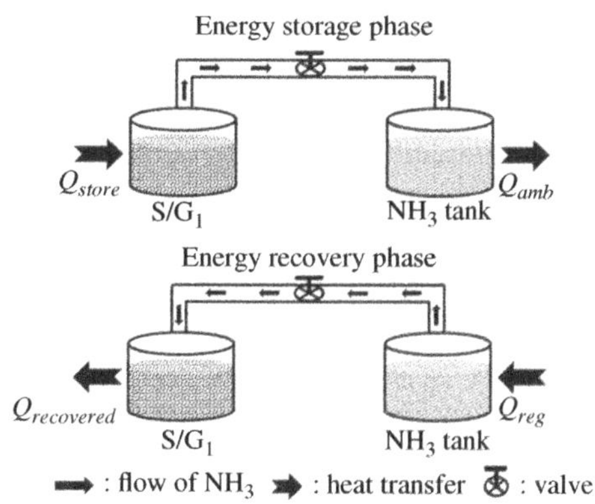

Figure 5.11 Thermochemical sorption energy storage based on the adsorption process.

storage requirements. The thermal energy storage device based on sorption between ammonia and halide salts can be technically realized using the adsorption process (Figure 5.11) or resorption process (Figure 5.12). In the adsorption process, halide salt adsorber is coupled with ammonia tank [124] in which ammonia is evaporated or condensed according to the adsorption or desorption process respectively. Whereas in the resorption process, two different halide salt adsorbers are coupled together between which ammonia vapor is exchanged by simultaneously undergoing adsorption and desorption process [125].

Significant investigations have been conducted using the ammoniated salt-based sorption system for the storage of solar thermal energy for the short and long span of time, for the upgradation of stored thermal energy, and to analyze sorption thermal batteries. During the short-term storage, the storage of energy is observed during sunshine hours and is recovered during night or at the time of cloudy day. Similarly, in long-term storage which is also termed the seasonal storage, solar thermal energy is stored in summer and recovered during

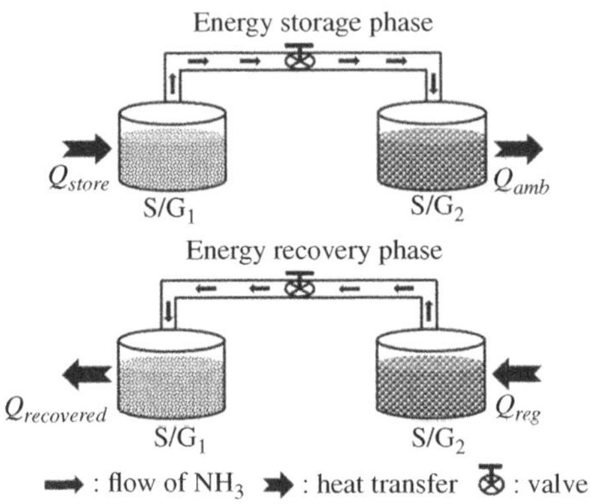

Figure 5.12 Thermochemical sorption energy storage based on the resorption process.

winter. The investigation of a solar air-conditioning plant has been done for two years to store solar energy during daytime and to achieve the cooling effect during the night by using sorption between $BaCl_2$ and NH_3 [126]. Li et al. [127] have studied an innovative dual-mode thermochemical sorption system for the seasonal storage of solar energy, which can supply useful heat at high ambient temperature and at low ambient temperatures in winter, using the NH_3–$CaCl_2$ pair and observed the system COP and energy density as 0.6 and 1000 kJ/kg of salt, respectively. The upgradation of stored thermal energy to the desired temperature can also be achieved in the gas–solid sorption technique due to monovariant characteristics according to which by increasing the pressure, the temperature of the energy released can be elevated [128, 129]. A thermochemical transformer has been developed by Li et al. [128] in which several ammoniated salt pairs are tested for the energy storage and upgradation and observed the temperature lift from 87 to 171 °C using the $MnCl_2$–$CaCl_2$–NH_3 pair. In another work, the low-grade heat is upgraded from 96 to 161 °C using the $MnCl_2$–$SrCl_2$–NH_3 sorption pair where the exergy and energy efficiencies of the system are reported as 0.75 and 0.43, respectively [129]. Based on the resorption process between $MnCl_2$ and $CaCl_2$, a resorption thermal energy storage (RTES) system has been developed for the seasonal storage of thermal energy [130], heat and refrigeration cogeneration [131], and energy storage and upgradation [132]. The effective heat storage density and COP of the RTES system are observed to be in the range of 398–1047 kJ/kg and 0.21–0.58, respectively [130].

The gas–solid sorption energy storage has supplementary advantage over sensible and latent heat storage by the development of sorption thermal cells [133] and thermal batteries [134] which work in a similar fashion of charging and discharging like electrical batteries, but the output is obtained as a temperature difference. The sorption thermal cell or battery consists of a sorption reactor and a liquid refrigerant tank enclosed in a tank. The reactor and refrigerant tank are connected by a tube and valve arrangement to facilitate mass transfer. The thermal effects are exchanged to the end user through the heat transfer fluid. A multilevel sorption thermal battery has been developed by filling several halide salts operating at various temperatures in a reactor to store the varying solar thermal energy at different cascaded temperatures [135]. The sorption thermal cell and battery are also identified as a useful method for the simultaneous heating, cooling, and energy storage and for low-grade waste heat recovery [136, 137]. The performance analysis of multilevel sorption thermal energy storage (STES) has been conducted by developing the bi-salt and tri-salt composite adsorbents and observed the highest energy storage density of 1949 kJ/kg for the bi-salt adsorbent and 1823.3 kJ/kg for the tri-salt adsorbent [138]. The thermodynamic study of the RTES system using several pairs of reactive salts is carried out, and significant heat upgradation is achieved [139]. Recently, the model of a cascade thermal energy storage system is proposed using four salts and analysis is performed using measured sorption thermodynamic properties [46]. A simultaneous electric and thermal energy storage cycle based on ammoniated salt chemisorption has been proposed by Bao et al. [140] in which the system stores the electrical power and low-grade heat in the form of chemical energy. The stored energy is recovered as electric power, which is generated by the expansion of high-pressure desorbing ammonia, simultaneously with upgraded thermal energy and cooling effect [140].

5.6 Concluding Remarks

The gas–solid sorption pair of ammonia and metal halides finds its place as a promising candidate to develop sorption devices not only for cooling and heating but also for power generation and energy storage. The low cost and abundant availability of ammonia and metal halides encourage the progress of thermochemical devices. However, the ammonia–salt-based cooling and heating device reaches the advanced stage of prototype testing; the study is still going on in the progress of energy storage and power generation systems. The information on thermodynamic and thermophysical properties are also important for the better designing of the thermochemical system. The use of the ammoniated salt system provides a suitable alternative as a clean energy technology and can resolve the problem of energy scarcity in the near future. The better understanding of ammonia–salt sorption and optimal designing of the thermochemical system can deliver the outstanding success in the use of the ammoniated salt system.

References

1 Li, T.X., Wang, R.Z., and Li, H. (2014). Progress in the development of solid–gas sorption refrigeration thermodynamic cycle driven by low-grade thermal energy. *Progress in Energy and Combustion Science* 40: 1–58.

2 Lepinasse, E. and Spinner, B. (1994). Production de froid par couplage de reacteurs solide–gas I: analyse des performance de perormances de tells systemes. *International Journal of Refrigeration* 17: 309–322.
3 Wongsuwan, W., Kumar, S., Neveu, P., and Meunier, F. (2001). A review of chemical heat pump technology and applications. *Applied Thermal Engineering* 21: 1489–1519.
4 Wang, R.Z. and Oliveira, R.G. (2006). Adsorption refrigeration—an efficient way to make good use of waste heat and solar energy. *Progress in Energy and Combustion Science* 32: 424–458.
5 Meunier, F. (2013). Adsorption heat powered heat pumps. *Applied Thermal Engineering* 61: 830–836.
6 Spinner, B. (1993). Ammonia-based thermochemical transformers. *Heat Recovery Systems & CHP* 13: 301–307.
7 Neveu, P. and Castaing, J. (1993). Solid–gas chemical heat pumps: field of application and performance of the internal heat of reaction recovery process. *Heat Recovery Systems & CHP* 13: 233–251.
8 Rockenfeller, U. (1988). System for low temperature refrigeration and chill storage using ammoniated complex compounds. U.S. Patent 4848994.
9 Mauran, S., Bodiot, D., and Crozat, G. (1983). Optimisation des densites energetiques de systemes de stockage chimique bases sur des reactions solide–gaz renversables. *Revue de Physique Appliquee* 18: 107–112.
10 Mauran, S., Crozat, G., Bodiot, D. et al. (1985). Chemical home heat pumps: acquisition of thermodynamic and kinetic data. *J. Chim. Phys.* 82: 791–802.
11 Marty, A. (1991). Etude par microcalorimetrie de la reactmte de deux ammoniacates de chlorure de manganese. *Journal of Thermal Analysis* 37: 479–498.
12 Goetz, V. and Marty, A. (1992). A model for reversible solid–gas reactions submitted to temperature and pressure constraints: simulation of the rate of reaction in solid-gas reactor used as chemical heat pump. *Chemical Engineering Science* 47: 4445–4454.
13 Coste, C., Crozat, G., and Mauran, S. (1986). Gaseous–solid reaction. U.S Patent 4,595,774.
14 Mauran, S., Prades, P., and L'Haridon, F. (1993). Heat and mass Transfer in consolidated reacting beds for thermochemical systems. *Heat Recovery Systems & CHP* 13: 315–319.
15 Mauran, S., Lebrun, M., Prades, P. et al. (1994). Active composite and its use as reactive medium. US Patent 5,283,219.
16 Han, J., Lee, K., and Kim, H. (1999). Effective thermal conductivity of graphite-metallic salt complex for chemical heat pumps. *Journal of Thermophysics and Heat Transfer* 13: 481–488.
17 Han, J. and Lee, K. (2001). Gas permeability of expanded graphite-metallic salt composite. *Applied Thermal Engineering* 21: 453–463.

18 Wang, K., Wu, J., Wang, R.Z., and Wang, L.W. (2006). Effective thermal conductivity of expanded graphite-$CaCl_2$ composite adsorbent for chemical adsorption chillers. *Energy Conversion and Management* 47: 1902–1912.

19 Jiang, L., Wang, L.W., Jin, Z.Q. et al. (2012). Permeability and thermal conductivity of compact adsorbent of salts for sorption refrigeration. *ASME-Journal of Heat Transfer* 134: 1045031–1045036.

20 Tian, B., Jin, Z.Q., Wang, L.W., and Wang, R.Z. (2012). Permeability and thermal conductivity of compact chemical and physical adsorbents with expanded natural graphite as host matrix. *International Journal of Heat and Mass Transfer* 55: 4453–4459.

21 Jiang, L., Wang, L.W., Jin, Z.Q. et al. (2013). Effective thermal conductivity and permeability of compact compound ammoniated salts in the adsorption/desorption process. *International Journal of Thermal Sciences* 71: 103–110.

22 Jiang, L., Wang, L.W., and Wang, R.Z. (2014). Investigation on thermal conductive consolidated compound $CaCl_2$ for adsorption refrigeration. *International Journal of Thermal Sciences* 81: 68–75.

23 Dellero, T., Sarmeo, D., and Touzain, P. (1999). A chemical heat pump using carbon fibers as additive. Part I: enhancement of thermal conduction. *Applied Thermal Engineering* 19: 991–1000.

24 Dellero, T. and Touzain, P. (1999). A chemical heat pump using carbon fibres as additive. Part II: study of constraint parameters. *Applied Thermal Engineering* 19: 1001–1011.

25 Aidoun, Z. and Ternan, M. (2002). Salt impregnated carbon fibres as the reactive medium in a chemical heat pump: the NH_3–$CoCl_2$ system. *Applied Thermal Engineering* 22: 1167–1173.

26 Aidoun, Z. and Ternan, M. (2002). The synthesis reaction in a chemical heat pump reactor filled with chloride salt impregnated carbon fibres: the NH_3–$CoCl_2$ system. *Applied Thermal Engineering* 22: 1943–1954.

27 Goetz, V., Elie, F., and Spinner, B. (1993). The structure and performance of single effect solid–gas chemical heat pumps. *Heat Recovery Systems & CHP* 13: 79–96.

28 Vasiliev, L.L., Kanonchik, L.E., Antuh, A.A., and Kulakov, A.G. (1996). NaX zeolite, carbon fibre and $CaCl_2$ ammonia reactors for heat pumps and refrigerators. *Adsorption* 2: 311–316.

29 Vasiliev, L.L., Mishkinis, D.A., Antukh, A.A., and Vasiliev, L.L. (2001). Solar-gas solid sorption refrigerator. *Adsorption* 7: 149–161.

30 Mbaye, M., Aidoun, Z., Valkov, V., and Legault, A. (1998). Analysis of chemical heat pumps (CHPS): basic concepts and numerical model description. *Applied Thermal Engineering* 18: 131–146.

31 Castaing-Lasvignottes, J. and Neveu, P. (1998). Equivalent Carnot cycle concept applied to a thermochemical solid/gas resorption system. *Applied Thermal Engineering* 18: 745–754.

32 Enibe, S.O. and Iloeje, O.C. (2000). Heat and mass transfer in porous spherical pellets of $CaCl_2$ for solar refrigeration. *Renewable Energy* 20: 305–324.

33 Meunier, F. (1998). Solid sorption heat powered cycles for cooling and heat pumping applications. *Applied Thermal Engineering* 18: 715–729.

34 Pons, M., Meunier, F., Cacciola, G. et al. (1999). Thermodynamic based comparison of sorption systems for cooling and heat pumping. *International Journal of Refrigeration* 22: 5–17.

35 Trudel, J., Hosatte, S., and Ternan, M. (1999). Solid–gas equilibrium in chemical heat pumps: the NH_3-$CoCl_2$ system. *Applied Thermal Engineering* 19: 495–511.

36 Aidoun, Z. and Ternan, M. (2001). Pseudo-stable transitions and instability in chemical heat pumps: the NH_3-$CoCl_2$ system. *Applied Thermal Engineering* 21: 1019–1034.

37 Anjorin, S.A. (2011). Measurement of adsorption capacity of ammonia on calcium chloride composites. *Journal of Industrial Research and Technology* 3: 39–48.

38 Veselovskaya, J.V., Critoph, R.E., Thorpe, R.N. et al. (2010). Novel ammonia sorbents "porous matrix modified by active salt" for adsorptive heat transformation: 3. Testing of "$BaCl_2$/vermiculite" composite in a lab-scale adsorption chiller. *Applied Thermal Engineering* 30: 1188–1192.

39 Veselovskaya, J. and Tokarev, M. (2011). Novel ammonia sorbents "porous matrix modified by active salt" for adsorptive heat transformation: 4. Dynamics of quasi-isobaric ammonia sorption and desorption on $BaCl_2$/vermiculite. *Applied Thermal Engineering* 31: 566–572.

40 Yu, Y., Wang, L.W., and An, G.L. (2018). Experimental study on sorption and heat transfer performance of NaBr-NH_3 for solid sorption heat pipe. *International Journal of Heat and Mass Transfer* 117: 125–131.

41 Zhong, Y., Critoph, R., Thorpe, R. et al. (2007). Isothermal sorption characteristics of the $BaCl_2$-NH_3 pair in a vermiculite host matrix. *Applied Thermal Engineering* 27: 2455–2462.

42 Zhou, Z.S., Wang, L.W., Jiang, L. et al. (2016). Non-equilibrium sorption performances for composite sorbents of chlorides–ammonia working pairs for refrigeration. *International Journal of Refrigeration* 65: 60–68.

43 Jiang, L., Wang, L.W., Zhou, Z.S. et al. (2016). Investigation on non-equilibrium performance of composite adsorbent for resorption refrigeration. *Energy Conversion and Management* 119: 67–74.

44 Sharma, R. and Anil Kumar, E. (2016). Measurement of thermodynamic properties of ammoniated salts and thermodynamic simulation of resorption cooling system. *International Journal of Refrigeration* 67: 54–68.

45 Sharma, R. and Anil Kumar, E. (2017). Performance evaluation of simple and heat recovery adsorption cooling system using measured NH_3 sorption characteristics of halide salts. *Applied Thermal Engineering* 119: 459–471.

46 Wu, S., Li, T.X., and Wang, R.Z. (2018). Experimental identification and thermodynamic analysis of ammonia sorption equilibrium characteristics on halide salts. *Energy* 161: 955–962.

47 Sharonov, V.E. and Aristov, Y.I. (2005). Ammonia adsorption by $MgCl_2$, $CaCl_2$ and $BaCl_2$ confined to porous alumina: the fixed bed adsorber. *React. Kinet. Catal. Lett.* 85: 183–188.

48 Bevers, E.R.T., Van Ekeren, P.J., Haije, W.G., and Oonk, H.A.J. (2006). Thermodynamic properties of lithium chloride ammonia complexes for application in a high-lift high temperature chemical heat pump. *Journal of Thermal Analysis and Calorimetry* 86: 825–832.

49 Zhong, Y., Critoph, R.E., Thorpe, R.N., and Tamainot-Telto, Z. (2009). Dynamics of $BaCl_2$–NH_3 adsorption pair. *Applied Thermal Engineering* 29: 1180–1186.

50 Iwata, R., Yamauchi, T., Hirota, Y. et al. (2014). Reaction kinetics of ammonia absorption/desorption of metal salts. *Applied Thermal Engineering* 72: 244–249.

51 Liu, C. and Aika, K. (2004). Ammonia absorption into alkaline earth metal halide mixtures as an ammonia storage material. *Ind. Eng. Chem. Res.* 43: 7484–7491.

52 Grekova, A., Veselovskaya, J., Tokarev, M. et al. (2014). Ammonia sorption on the composites "($BaCl_2$+$BaBr_2$) inside vermiculite pores". *Colloids and Surfaces A: Physicochemical and Engineering Aspects* 448: 169–174.

53 Sharonov, V.E., Veselovskaya, J.V., and Aristov, Y.I. (2006). Ammonia sorption on composites '$CaCl_2$ in inorganic host matrix': isosteric chart and its performance. *International Journal of Low Carbon Technologies* 1: 191–200.

54 Wang, L.W., Chen, L., Wang, H., and Liao, D. (2009). The adsorption refrigeration characteristics of alkaline-earth metal chlorides and its composite adsorbents. *Renewable Energy* 34: 1016–1023.

55 Veselovskaya, J., Tokarev, M., and Aristov, Y. (2010). Novel ammonia sorbents "porous matrix modified by active salt" for adsorptive heat transformation 1. Barium chloride in various matrices. *Applied Thermal Engineering* 30: 584–589.

56 Tokarev, M., Veselovskaya, J., Yanagi, H., and Aristov, Y. (2010). Novel ammonia sorbents "porous matrix modified by active salts" for adsorptive heat transformation: 2. $CaCl_2$ in ACF felt. *Applied Thermal Engineering* 30: 845–849.

57 Yan, T., Li, T.X., Li, H., and Wang, R.Z. (2014). Experimental study of the ammonia adsorption characteristics on the composite sorbent of $CaCl_2$ and multi-walled carbon nanotubes. *International Journal of Refrigeration* 46: 165–172.

58 Gao, J., Wang, L.W., Wang, R.Z., and Zhou, Z.S. (2017). Solution to the sorption hysteresis by novel compact composite multi-salt sorbents. *Applied Thermal Engineering* 111: 580–585.

59 Gao, J., Wang, L.W., An, G.L. et al. (2018). Performance analysis of multi-salt sorbents without sorption hysteresis for low-grade heat recovery. *Renewable Energy* 118: 718–726.

60 Wang, L.W., Wang, R.Z., Wu, J.Y. et al. (2004). Adsorption ice makers for fishing boats driven by the exhaust heat from diesel engine: choice of adsorption pair. *Energy Conversion and Management* 45: 2043–2057.

61 Wang, L.W., Wang, R.Z., Wu, J.Y., and Wang, K. (2004). Compound adsorbent for adsorption ice maker on fishing boats. *International Journal of Refrigeration* 27: 401–408.

62 Lu, Z., Wang, R.Z., Wang, L.W., and Chen, C. (2006). Performance analysis of an adsorption refrigerator using activated carbon in a compound adsorbent. *Carbon* 44: 747–752.

63 Pan, Q., Wang, R.Z., Lu, J., and Wang, L.W. (2014). Experimental investigation of an adsorption refrigeration prototype with the working pair of composite adsorbent-ammonia. *Applied Thermal Engineering* 72: 275–282.

64 Oliveira, R.G., Wang, R.Z., and Wang, C. (2007). Evaluation of the cooling performance of a consolidated expanded graphite-calcium chloride reactive bed for chemisorption icemaker. *International Journal of Refrigeration* 30: 103–112.

65 Oliveira, R.G. and Wang, R.Z. (2007). A consolidated calcium chloride-expanded graphite compound for use in sorption refrigeration systems. *Carbon* 45: 390–396.

66 Wang, L.W., Bao, H.S., and Wang, R.Z. (2009). A comparison of the performances of adsorption and resorption refrigeration systems powered by the low grade heat. *Renewable Energy* 34: 2373–2379.

67 Neveu, P., Mazet, N., and Stitou, D. (2007). Experimental results of a solar powered cooling system at low temperature. *International Journal of Refrigeration* 30: 1050–1058.

68 Li, T.X., Wang, R.Z., Wang, L.W., and Kiplagat, J.K. (2009). Study on the heat transfer and sorption characteristics of a consolidated composite sorbent for solar-powered thermochemical cooling systems. *Solar Energy* 83: 1742–1755.

69 Oliveira, R.G., Wang, R.Z., Kiplagat, J.K., and Wang, C. (2009). Novel composite sorbent for resorption systems and for chemisorption air conditioners driven by low generation temperature. *Renewable Energy* 2757–2764.

70 Kiplagat, J.K., Wang, R.Z., Oliveira, R.G. et al. (2013). Experimental study on the effects of the operation on the performance of a chemisorption air conditioner powered by low grade heat. *Applied Energy* 103: 571–580.

71 Li, T.X., Wang, R.Z., Kiplagat, J.K., and Wang, L.W. (2009). Performance study of a consolidated manganese chloride-expanded graphite compound for sorption deep-freezing processes. *Applied Energy* 86: 1201–1209.

72 Li, T.X., Wang, R.Z., Kiplagat, J.K., and Chen, H. (2010). Experimental study and comparison of thermochemical resorption refrigeration cycle and adsorption refrigeration cycle. *Chemical Engineering Science* 65: 4222–4230.

73 Kiplagat, J.K., Wang, R.Z., Oliveira, R.G., and Li, T.X. (2010). Lithium chloride-expanded graphite composite sorbent for solar powered ice maker. *Solar Energy* 84: 1587–1594.

74 Rivera, C., Pilatowsky, I., Mendez, E., and Rivera, W. (2007). Experimental study of a thermo-chemical refrigerator using the barium chloride-ammonia reaction. *International Journal of Hydrogen Energy* 32: 3154–3158.

75 Ferrucci, F., Stitou, D., Ortegaa, P., and Lucas, F. (2018). Mechanical compressor-driven thermochemical storage for cooling applications in tropical insular regions. Concept and efficiency analysis. *Applied Energy* 219: 240–255.

76 Lepinasse, E., Goetz, V., and Crozat, G. (1994). Modelling and experimental investigation of a new type of thermochemical transformer based on the coupling of two solid–gas reactions. *Chemical Engineering and Processing* 33: 125–134.

77 Goetz, V., Spinner, B., and Lepinasse, E. (1997). A solid–gas thermochemical cooling system using $BaCl_2$ and $NiCl_2$. *Energy* 22: 49–58.

78 Vasiliev, L., Mishkinis, D., Antukh, A. et al. (2004). Resorption heat pump. *Applied Thermal Engineering* 24: 1893–1903.

79 Lepinasse, E., Marion, M., and Goetz, V. (2001). Cooling Storage with a resorption process. Application to a box temperature control. *Applied Thermal Engineering* 21: 1251–1263.

80 Van Der Pal, M., De Boer, R., Veldhuis, J., and Smeding, S. (2009). Thermally driven ammonia-salt type II heat pump: development and test of a prototype. In: *Heat Powered Cycles Conference*. Berlin.

81 Bao, H.S., Oliveira, R.G., Wang, R.Z., and Wang, L.W. (2010). Choice of low temperature salt for a resorption refrigerator. *Ind. Eng. Chem. Res.* 49: 4897–4903.

82 Bao, H.S., Oliveira, R.G., Wang, R.Z. et al. (2011). Working pairs for resorption refrigerator. *Applied Thermal Engineering* 31: 3015–3021.

83 Xu, J., Oliveira, R.G., and Wang, R.Z. (2011). Resorption system with simultaneous heat and cold production. *International Journal of Refrigeration* 34: 1262–1267.

84 Bao, H.S., Wang, R.Z., and Wang, L.W. (2011). A resorption refrigerator driven by low grade thermal energy. *Energy Conversion and Management* 52: 2339–2344.

85 Bao, H.S., Wang, R.Z., Oliveira, R.G., and Li, T.X. (2012). Resorption system for cold storage and long-distance refrigeration. *Applied Energy* 93: 479–487.

86 Van Der Pal, M., Wemmers, A., Smeding, S., and Van Den Heuvel, K. (2011). Study on the performance of hybrid adsorption-compression type II heat pumps based on ammonia salt adsorption. *International Journal of Low-Carbon Technologies* 6: 207–211.

87 Wang, L.W., Ziegler, F., Roskilly, A.P. et al. (2013). A resorption cycle for the cogeneration of electricity and refrigeration. *Applied Energy* 106: 56–64.

88 Bao, H.S., Ma, Z., and Roskilly, A.P. (2017). An optimised chemisorption cycle for power generation using low grade heat. *Applied Energy* 186: 251–261.

89 Bao, H.S., Ma, Z., and Roskilly, A.P. (2017). A chemisorption power generation cycle with multi-stage expansion driven by low grade heat. *Energy Conversion and Management* 150: 956–965.

90 Lu, Y., Roskilly, A.P., Tang, K. et al. (2017). Investigation and performance study of a dual-source chemisorption power generation cycle using scroll expander. *Applied Energy* 204: 979–993.

91 Lu, Y., Roskilly, A.P., Wang, Y., and Wang, L.W. (2017). Study of a novel dual-source chemisorption power generation system using scroll expander. *Energy Procedia* 105: 921–926.

92 Shelton, S. (1986). Solid adsorbent heat pump system. US Patent 4,610,148.

93 Goetz, V. and Llobet, J. (2000). Testing and modelling of a temperature front solid–gas reactor applied to thermochemical transformer. *Applied Thermal Engineering* 20: 155–177.

94 Wang, R.Z. (2001). Performance improvement of adsorption cooling by heat and mass recovery operation. *International Journal of Refrigeration* 24: 602–611.

95 Wang, R.Z., Wang, L.W., and Wu, J. (2014). *Adsorption Refrigeration Technology: Theory and Application*. China: Shanghai Jiao Tong University: Wiley.

96 Wang, L.W., Wang, R.Z., Lu, Z. et al. (2006). Comparison of the adsorption performance of compound adsorbent in a refrigeration cycle with and without mass recovery. *Chemical Engineering Science* 61: 3761–3770.

97 Oliveira, R.G., Silveira, V. Jr., and Wang, R.Z. (2006). Experimental study of mass recovery adsorption cycles for ice making at low generation temperature. *Applied Thermal Engineering* 26: 303–311.

98 Lu, Z., Wang, R.Z., Li, T.X. et al. (2007). Experimental investigation of a novel multifunction heat pipe solid sorption icemaker for fishing boats using $CaCl_2$/activated carbon compound-ammonia. *International Journal of Refrigeration* 30: 76–85.

99 Li, T.X., Wang, R.Z., Wang, L.W. et al. (2007). Performance study of a high efficient multifunction heat pipe type adsorption ice making system with novel mass and heat recovery processes. *International Journal of Thermal Sciences* 46: 1267–1274.

100 Li, T.X., Wang, R.Z., Wang, L.W. et al. (2008). Influence of mass recovery on the performance of a heat pipe type ammonia sorption refrigeration system using $CaCl_2$/activated carbon as compound adsorbent. *Applied Thermal Engineering* 28: 1638–1646.

101 Li, T.X., Wang, R.Z., Wang, L.W., and Lu, Z. (2008). Experimental study on an innovative multifunction heat pipe type heat recovery two-stage refrigeration system. *Energy Conversion and Management* 49: 2505–2512.

102 Li, T.X., Wang, R.Z., Wang, L.W., and Lu, Z. (2008). Experimental investigation of an innovative dual-mode chemisorption refrigeration system based on multifunction heat pipes. *International Journal of Refrigeration* 31: 1104–1112.

103 Rao, K.M., Ramgopal, M., and Bhattacharyya, S. (2013). Thermodynamic analysis of a sorption refrigeration system with and without internal heat recovery. *International Journal of Low-Carbon Technologies* 8: 124–133.

104 Lu, Y., Wang, Y., Bao, H.S. et al. (2015). Analysis of an optimal resorption cogeneration using mass and heat recovery processes. *Applied Energy* 160: 892–901.

105 Li, T.X., Wang, R.Z., Kiplagat, J.K. et al. (2009). A conceptual design and performance analysis of a triple-effect solid–gas thermochemical sorption refrigeration system with internal heat recovery. *Chemical Engineering Science* 64: 3376–3384.

106 Sharma, R. and Anil Kumar, E. (2020). Assessment of different combinations of ammoniated halide salt for resorption cooling system with and without heat-recovery using sorption thermodynamic properties. *Thermal Science and Engineering Progress* 19: 100652.

107 Li, T.X., Wang, R.Z., Oliveira, R.G., and Wang, L.W. (2007). Performance analysis of an innovative multimode, multi-salt and multi-effect chemisorption refrigeration system. *AIChE Journal* 53: 3222–3230.

108 Oliveira, R.G., Wang, R.Z., and Li, T.X. (2008). Transient analysis of a chemisorption air conditioning system operating under different kinds of cycle. *Ind. Eng. Chem. Res.* 47: 1102–1110.

109 Li, T.X., Wang, R.Z., Oliveira, R.G. et al. (2009). A combined double-way chemisorption refrigeration cycle based on adsorption and resorption processes. *International Journal of Refrigeration* 32: 47–57.

110 Li, T.X., Wang, R.Z., Kiplagat, J.K. et al. (2009). Thermodynamic study of a combined double-way solid–gas thermochemical sorption refrigeration cycle. *International Journal of Refrigeration* 32: 1570–1578.

111 Xu, L., Wang, R.Z., Li, T.X., and Wang, L.W. (2011). Experimental study on a combined double-way chemisorption refrigeration system. *International Journal of Refrigeration* 34: 914–921.

112 Xu, L., Wang, R.Z., Li, T.X., and Wang, L.W. (2011). Experimental study on the performance of double-effect and double-way thermochemical sorption refrigeration cycle. *Applied Thermal Engineering* 31: 3658–3663.

113 Hu, Y., Wang, L.W., Xu, L. et al. (2011). A two-stage deep freezing chemisorption cycle driven by low-temperature heat source. *Front. Energy* 5: 263–269.

114 Li, T.X., Wang, R.Z., Kiplagat, J.K. et al. (2011). A new target-oriented methodology of decreasing the regeneration temperature of solid-gas thermochemical sorption refrigeration system driven by low-grade thermal energy. *International Journal of Heat and Mass Transfer* 54: 4719–4729.

115 Jiang, L., Wang, L.W., Luo, W., and Wang, R.Z. (2015). Experimental study on working pairs for two-stage chemisorption freezing cycle. *Renewable Energy* 74: 287–297.

116 Gao, P., Zhang, X., Wang, L.W. et al. (2016). Study on $MnCl_2/CaCl_2$-NH_3 two stage solid sorption freezing cycle for refrigerated trucks at low engine load in summer. *Energy Conversion and Management* 109: 1–9.

117 Gao, P., Wang, L.W., Wang, R.Z. et al. (2016). Experimental investigation of a $MnCl_2/CaCl_2$-NH_3 two stage solid sorption freezing system for a refrigerated truck. *Energy* 103: 16–26.

118 Li, T.X., Wang, R.Z., Kiplagat, J.K., and Ma, L. (2012). Performance analysis of a multi-mode thermochemical sorption refrigeration system for solar-powered cooling. *International Journal of Refrigeration* 35: 532–542.

119 Abedin, A. and Rosen, M. (2012). Closed and open thermochemical energy storage: energy- and exergy- based comparisons. *Energy* 41: 83–92.

120 Yu, N., Wang, R.Z., and Wang, L.W. (2013). Sorption thermal storage for solar energy. *Progress in Energy and Combustion Science* 39: 489–514.

121 Gordeeva, L., Solovyeva, M., and Aristov, Y. (2016). NH_2-MIL-125 as a promising material for adsorptive heat transformation and storage. *Energy* 100: 18–24.

122 Zhao, Y., Wang, R.Z., Li, T.X., and Nomura, Y. (2016). Investigation of a 10 kWh sorption heat storage device for effective utilization of low-grade thermal energy. *Energy* 113: 739–747.

123 Hamdan, M., Rossides, S., and Haj, K.R. (2013). Thermal energy storage using thermo-chemical heat pump. *Energy Conversion and Management* 65: 721–724.

124 Li, T.X., Wang, R.Z., Yan, T., and Ishugah, T. (2014). Integrated energy storage and energy upgrade, combined cooling and heating supply, and waste heat recovery with solid-gas thermochemical sorption heat transformer. *International Journal of Heat and Mass Transfer* 76: 237–246.

125 Jiang, L., Zhu, F., Wang, L.W. et al. (2016). Experimental investigation on a $MnCl_2$-$CaCl_2$-NH_3 thermal energy storage system. *Renewable Energy* 91: 130–136.

126 Stitou, D., Mazet, N., and Mauran, S. (2012). Experimental investigation of a solid/gas thermochemical storage process for solar air-conditioning. *Energy* 41: 261–270.

127 Li, T.X., Wang, R.Z., Kiplagat, J.K., and Kang, Y. (2013). Performance analysis of an integrated energy storage and energy upgrade thermochemical solid-gas sorption system for seasonal storage of solar thermal energy. *Energy* 50: 454–467.

128 Li, T.X., Wang, R.Z., and Kiplagat, J.K. (2013). A target-oriented solid-gas thermochemical sorption heat transformer for integrated energy storage and energy upgrade. *AICHE Journal* 59: 1334–1347.

129 Wua, S., Li, T.X., Yan, T., and Wang, R.Z. (2017). Experimental investigation on a novel solid-gas thermochemical sorption heat transformer for energy upgrade with a large temperature lift. *Energy Conversion and Management* 148: 330–338.

130 Jiang, L., Wang, R.Z., Wang, L.W., and Roskilly, A.P. (2017). Investigation on an innovative resorption system for seasonal thermal energy storage. *Energy Conversion and Management* 149: 129–139.

131 Zhu, F.Q., Jiang, L., Wang, L.W., and Wang, R.Z. (2016). Experimental investigation on a $MnCl_2$-$CaCl_2$-NH_3 resorption system for heat and refrigeration cogeneration. *Applied Energy* 181: 29–37.

132 Jiang, L., Wang, L.W., Wang, R.Z. et al. (2017). Experimental investigation on an innovative resorption system for energy storage and upgrade. *Energy Conversion and Management* 138: 651–658.

133 Jiang, L., Wang, R.Z., Lu, Y. et al. (2017). Investigation on novel sorption thermal cell with improved energy charging and discharging performance. *Energy Conversion and Management* 148: 110–119.

134 Li, T.X., Wang, R.Z., and Yan, T. (2015). Solid–gas thermochemical sorption thermal battery for solar cooling and heating energy storage and heat transformer. *Energy* 84: 745–758.

135 Li, T.X., Wu, S., Yan, T. et al. (2016). A novel solid-gas thermochemical multilevel sorption thermal battery for cascaded solar thermal energy storage. *Applied Energy* 161: 1–10.

136 Li, T.X., Xu, J., Yan, T., and Wang, R.Z. (2016). Development of sorption thermal battery for low-grade waste heat recovery and combined cold and heat energy storage. *Energy* 107: 347–359.

137 Jiang, L., Roskilly, A.P., Wang, R.Z. et al. (2017). Analysis on innovative modular sorption and resorption thermal cell for cold and heat cogeneration. *Applied Energy* 204: 767–779.

138 Jiang, L., Gao, J., Wang, L.W. et al. (2017). Investigation on performance of multi-salt composite sorbents for multilevel sorption thermal energy storage. *Applied Energy* 190: 1029–1038.

139 Sharma, R. and Anil-Kumar, E. (2017). Study of ammoniated salts based thermochemical energy storage system with heat up-gradation: a thermodynamic approach. *Energy* 141: 1705–1716.

140 Bao, H.S., Ma, Z., and Roskilly, A.P. (2016). Integrated chemisorption cycles for ultra-low grade heat recovery and thermo-electric energy storage and exploitation. *Applied Energy* 164: 228–236.

6

Thermochemical Energy Storage Systems

Kartik Jain[1], Akshay Chate[1], Susmita Dash[1], and Pradip Dutta[1,2]

[1] *Department of Mechanical Engineering, Indian Institute of Science, Bangalore, India*
[2] *Interdisciplinary Centre for Energy Research, Indian Institute of Science, Bangalore, India*

6.1 Introduction/Background

In recent years, there has been a surge in energy demand worldwide due to economic progress and increase of energy-intensive activities [1]. However, the availability of conventional fossil fuels is decreasing, accompanied by a significant increase in global greenhouse gas emissions. As a result, there is an urgent need to widely adopt non-conventional renewable energy sources to ensure sustainable development for future generations. The importance of this transition is highlighted by one of the objectives of the Paris Agreement, which aims to completely shift away from fossil fuels toward renewable energy. However, the full potential of renewable energy on a global scale has not been fully harnessed. This highlights the pressing need for advancements and innovations in the technologies that facilitate the utilization of renewable energy sources.

There are several renewable energy resources, including solar, biomass, wind, geothermal, and so on. All these resources show ample promise in meeting the renewable energy demand for mitigating the adverse effects of fossil fuels on the environment. Among these options, solar energy offers distinct advantages such as abundant availability, ease of implementation, and reliability. Solar energy can be utilized in a variety of ways, including electricity generation using photovoltaic panels or concentrating solar power, and heat generation by solar thermal energy means for process heating, cooking, drying, and more. Energy usage in the form of heat constitutes a substantial share of the total energy generated. In this regard, the International Energy Agency [2] reported that in 2020, buildings consumed approximately 47% of the total heat produced, specifically for space and water

Advancements in Non-Conventional Cooling and Thermal Storage Strategies, First Edition.
Edited by Bidyut Baran Saha and Dibakar Rakshit.

heating purposes. This highlights the significance of heating applications in buildings. In regions with cold climates, providing adequate heating becomes an even more challenging task. According to recent estimates, more than 80% of heat required in cold regions of the United States is supplied by burning fossil fuels [3]. In recent times, the demand for heating has also been extended to automobiles, particularly electric vehicles (EVs) which require cabin heating and battery preheating in cold weather conditions. By utilizing solar thermal energy in combination with locally available waste heat sources for these heating applications, heavy dependence on conventional fossil fuel-based energy resources can be significantly reduced, leading to a decrease in greenhouse gas emissions. However, the intermittent and fluctuating nature of solar energy presents a significant challenge to its efficient utilization for the aforementioned applications. In this context, the development of high-density thermal energy storage (TES) systems becomes crucial. TES systems play a vital role in storing heat from solar thermal energy or waste heat sources when they are available and releasing it later at a desired time for heating applications. This helps bridge the gap between the variable thermal energy supply and fluctuating user demand, ensuring optimal utilization of renewable energy sources.

Based on the mechanism of heat storage, there can be three categories of TES systems: sensible heat storage, latent heat storage, and thermochemical energy storage. Sensible storage involves storing heat based on temperature differences in the storage medium. These systems have some clear advantages in terms of cost, simplicity, and applicability over a wide range of temperatures. However, there are certain disadvantages including lower heat capacities, heat leakage to the surrounding, and a progressive decrease in heat recovery temperature during the discharging process [4]. Latent heat-based systems, on the other hand, are characterized by higher storage density as a large amount of heat can be stored during phase transition. At a constant pressure, the phase transition can be isothermal (for pure substances) or over a range of temperatures (for mixtures and alloys), thus offering adequate flexibility for the heat recovery process. However, there are challenges associated with the cyclic operation of phase change materials (PCMs), which include volume change upon phase change, undercooling during solidification, and phase segregation [5]. Additionally, both sensible and latent storage systems are not suitable for long-term heat storage as they tend to lose heat to the surroundings. In contrast, thermochemical energy storage systems (TESSs) work on the principle of storing energy in the form of heat of reaction. During the endothermic reaction, surplus or waste heat is added at a specific temperature to separate the reactants, and the chemical products formed can be stored separately. Heat can be recovered whenever desired by bringing the products back together to undergo the backward exothermic reaction. Thermochemical storage materials (TSMs) can be stored at ambient temperatures once the heat is stored,

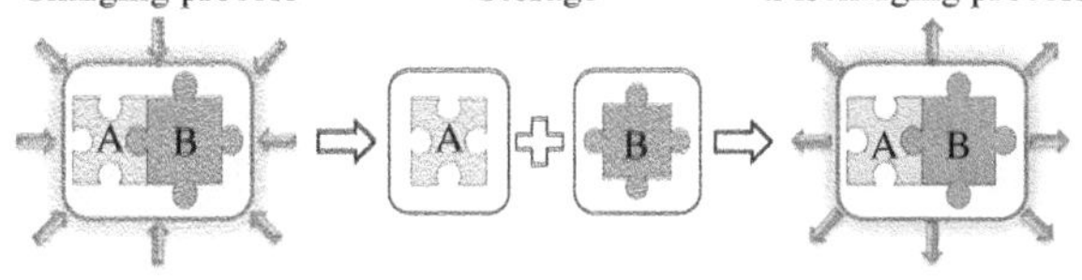

Figure 6.1 Operation of a typical thermochemical energy storage system (TESS).

eliminating storage losses, and enabling transportation over long distances. TESSs also offer higher energy storage density, being approximately an order of magnitude higher compared to sensible storage and a few times more than phase change storage [6]. However, TSMs usually come with issues such as low thermal conductivity, agglomeration, and change in thermophysical properties during cyclic operation. The operation of a typical TESS involves two key processes, as shown in Figure 6.1. The heat "charging" process involves the supply of an external source of heat to make the TSM (usually a solid–gas pair) undergo an endothermic reaction to store heat. The separated solid and gas are stored in different containers. In the heat "discharging" process, the gas and solid are combined to undergo an exothermic reaction to release the stored heat. During this process, typically, a heat transfer fluid (HTF) is passed through the reactor to absorb the released heat which is subsequently delivered for specific applications such as space heating and domestic hot water heating [7]. While TESS is a relatively newer and less explored form of energy storage compared to conventional sensible and latent heat storage systems, it holds great potential for advancement. This chapter aims to shed light on various intriguing aspects and parameters that significantly influence the performance of TESSs.

Depending on the desired application of the TESS, a suitable TSM needs to be selected. Therefore, the subsequent section explores various classes of TSMs available, specifically in the context of solid–gas pairs. Additionally, an overview of the equilibrium curve and reaction kinetics associated with TSMs is provided. Then, a case study on performing the thermodynamic analysis of the operating cycle of the TESS is explained. This proposed TESS is envisioned to function as a heat source for various heating activities in cold climates. The thermodynamic analysis of a TESS allows for a preliminary understanding of the performance of the TESS under different operating conditions and parameters. The chapter also covers the classification of TESSs based on their mode of operation. For instance, an open type of TESS allows for mass and energy exchange with the environment, while a closed TESS only permits energy interactions with the surroundings. Moreover, the different aspects and configurations of reactors used in practical TESS setups are highlighted. Finally, the chapter concludes with a discussion of some notable practical applications of TESS, showcasing their real-world relevance and potential.

6.2 Types of Thermochemical Storage Materials

The TSMs can be broadly classified into two categories: sorption processes and chemical reactions, as depicted in Figure 6.2. Sorption processes involve energy storage through surface interactions between a solid adsorbent and a gaseous adsorbate. These processes are suitable for storing low-grade (<100 °C) and medium-grade (100–400 °C) heat. The enthalpy associated with sorption processes typically ranges between 20 and 70 kJ/mol of the adsorbate. On the other hand, chemical reactions facilitate energy storage through the breaking and forming of chemical bonds between an adsorbate and an adsorbent. Due to their energy-intensive nature, chemical reactions offer higher reaction enthalpy, typically ranging from 80 to 180 kJ/mol of the adsorbate. These reactions find applications in high-grade TESs (>400 °C).

The solid–gas pairs of storage materials in the sorption processes are further classified into three major categories, namely, ammoniates, hydrates, and metal hydrides (MHs).

6.2.1 Ammoniates

A typical ammonia–metal halide salt sorption reaction can be expressed as

$$MX_x \cdot n_1NH_{3\,(s)} + \Delta H \Longleftrightarrow MX_x \cdot (n_1 - n_2)NH_{3\,(s)} + n_2NH_{3\,(g)}. \tag{6.1}$$

In the reversible reaction shown in Eq. (6.1), the left-hand side contains a metal halide salt, MX_x, which is bonded with ammonia molecules in the form of an ammoniated salt $MX_x \cdot n_1NH_3$. In the forward reaction, external heat is supplied

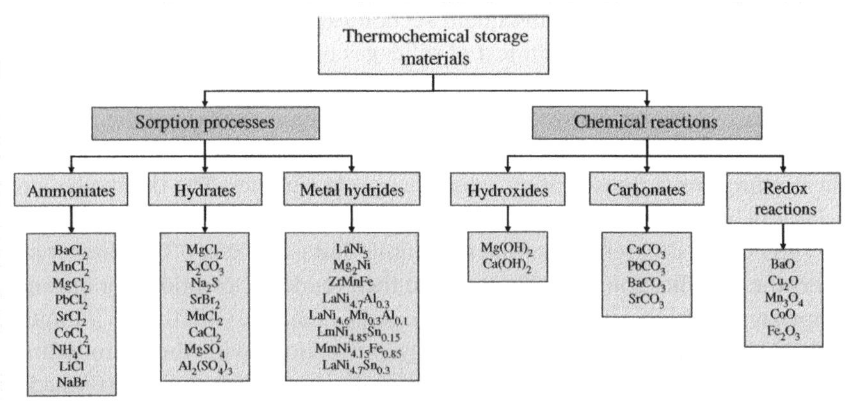

Figure 6.2 Categorization of various thermochemical storage materials (TSMs). Source: Adapted from Refs. [8–11].

to decompose the ammoniated salt partially or completely. Heat is released during the backward reaction when ammonia recombines with the salt [3]. As ammonia is in the vapor form at ambient conditions (normal boiling point is −33 °C), and the gas can combine with various halide salts, ammonia–halide salt pairs can operate over a wide range of temperatures, typically from −50 to 350 °C. Another advantage of this gas–solid pair is that the storage of ammonia after the charging process can be done in a secondary reactor by a sorption process using another suitable metal halide salt which can dissociate in the room temperature range. During the discharging process, the stored ammonia in the secondary reactor can be released using ambient heat so that the gas recombines with the metal halide salt in the heat storage reactor to undergo an exothermic reaction [12, 13]. Ammoniated salts have been extensively studied for various energy storage applications, including short-term (daily) and long-term (seasonal) TESs. In daily storage, solar heat is captured during daytime and then recovered during the night or cloudy days for applications requiring heat or cold. Seasonal storage involves storing solar thermal energy collected during the summer season and using it in the winter season as needed. Furthermore, the ammoniated halide salt sorption process has been adapted for the development of multifunctional thermal batteries capable of delivering heat at multiple temperatures and simultaneously providing a cooling effect when heat is absorbed by the secondary reactor during the discharging process. The concept of multifunctional thermal batteries has been elaborated subsequently in Section 6.8.3 in the context of MH-based sorption systems.

6.2.2 Salt Hydrates

Like ammoniates, a typical salt hydrate-based sorption process can be depicted as follows:

$$M \cdot n_1 H_2O_{(s)} + \Delta H \Longleftrightarrow M \cdot (n_1 - n_2) H_2O_{(s)} + n_2 H_2O_{(g)}. \tag{6.2}$$

In this case, $M \cdot n_1 H_2O$ is a solid salt hydrate complex produced by an interaction between generic salt (M) and water vapor. When external heat is applied to the salt hydrate complex, it dissociates, resulting in the desorption of water vapor, as shown in Eq. (6.2). This stored heat can be recovered when water vapor is reabsorbed at a suitable temperature and pressure to form the original salt hydrate complex [14]. Salt hydrates consist of two types of water molecules: adsorbed water and crystallized water. Adsorbed water comes from the surrounding environment and attaches to the surface of the salt hydrate through physical adsorption, which can be easily removed by heating. Crystallized water, on the other hand, is held in the salt hydrate through ion–dipole interactions or hydrogen bonds. The salt hydrates can store/release heat through physisorption as well as chemisorption processes. The water molecules adsorbed on the surface of salt hydrate are primarily responsible for physisorption processes, whereas crystallized water molecules in the salt hydrate participate in the chemisorption processes.

Similar to ammoniates, salt hydrates have the potential to be used for short-term as well as long-term (seasonal) heat storage applications. However, unlike ammoniates which can operate even at subzero temperatures, salt hydrates cannot handle temperatures near or below the freezing point of water.

6.2.3 Metal Hydrides

In MH-based sorption systems, the metal (or the alloy) bonds with hydrogen to form a solid–gas pair. Because of their hydrogen storage capability, MHs have been extensively studied in the context of storage of hydrogen under moderate pressures. However, lately, because of the high energy storage density of these materials, there have been several new studies on the application of MHs in the context of thermal storage. Like the metal halide salts, MHs can also offer a wide range of operating temperatures because of the availability of a wide variety of alloys. The high reaction enthalpies and diverse range of available materials and compositions make them suitable for use in various heating and cooling systems. The formation and decomposition of MHs involve a combination of physisorption and chemisorption processes. Most of the heat storage occurs in the form of chemisorption (as heat of reaction), while a relatively small amount is also stored as part of the physisorption process. In addition to hydrogen storage and TES applications, MHs find use in a wide range of other applications including thermal compression of hydrogen, adsorption-based refrigeration and heat pump systems, fuel cells, and more [15]. The general stoichiometric reaction between metals/alloys (M) and hydrogen involved during the energy storage application can be written as follows:

$$\mathrm{MH}_{x\,(s)} + \Delta H \Longleftrightarrow \mathrm{M}_{(s)} + \frac{x}{2}\mathrm{H}_{2\,(g)}, \tag{6.3}$$

where the forward reaction represents the dissociation of the MH (MH_x) into metal/alloy (M) and hydrogen. This is an endothermic reaction and can be utilized for energy storage or cooling applications The backward reaction, which depicts the formation of MH (MH_x), is an exothermic reaction, and the released heat can be utilized for heating applications.

6.2.4 Hydroxides

The general reaction scheme for the hydroxide-based TES system is given by

$$\mathrm{M(OH)}_{2\,(s)} + \Delta H \Longleftrightarrow \mathrm{MO}_{(s)} + \mathrm{H_2O}_{(g)}. \tag{6.4}$$

Here, $M(OH)_2$ represents the alkaline earth metal-based hydroxide which dissociates into metal oxide (MO) and steam upon heating. This forms the energy storage process, and when the stored energy is to be retrieved, steam at a suitable temperature and pressure is reacted with the MO, which results in an exothermic reaction. Two promising hydroxides studied in this category of TSMs are $Mg(OH)_2$ and $Ca(OH)_2$ [10, 11]. The charging temperatures for $Mg(OH)_2$ are around 300–350 °C, and for $Ca(OH)_2$, the charging temperature range is 400–600 °C. $Mg(OH)_2$ offers high energy storage density, low cost, and superior cyclic stability up to 500 cycles (compared to 200 cycles with $Ca(OH)_2$). Based on the discharging temperature range, $Mg(OH)_2$ has found novel applications in temperature upgradation of waste heat through the use of a sorption-based heat pump. On the other hand, $Ca(OH)_2$, which operates at relatively higher temperatures, has the potential to be used for high-grade heat storage for use in process heating and concentrated solar plants.

6.2.5 Carbonates

Carbonates are an appealing class of TSMs due to their high-energy storage density, compatibility with working temperatures required for coupling with concentrated solar power plants, cost-effectiveness, abundance, and low toxicity. These TSMs typically operate within a charging temperature range of 800–1200 °C, and the performance of the system is influenced by CO_2 partial pressure and temperature. Among the carbonate TSMs, $CaCO_3$ stands out as the most promising candidate, particularly for critical applications like post-combustion CO_2 capture. This process aims to sequester CO_2 to mitigate greenhouse gas emissions through the use of the Ca-looping (CaL) cycle. $CaCO_3$ can also be utilized in limestone production. A typical carbonate system reaction, where $M(CO)_3$ represents carbonate and MO represents metal oxide, is expressed as

$$M(CO)_{3\,(s)} + \Delta H \Longleftrightarrow MO_{(s)} + CO_{2\,(g)}. \tag{6.5}$$

6.2.6 Redox (Reduction–Oxidation) Reactions

The general reaction equation for this class of TSMs can be shown as follows:

$$MO_{x\,(s)} + \Delta H \Longleftrightarrow MO_{x-y\,(s)} + \frac{y}{2}O_{2\,(g)}. \tag{6.6}$$

In Eq. (6.6), the forward reaction represents a reduction reaction, where the oxidation state of the metal oxide (MO_x) is reduced by heating it, resulting in the generation of oxygen gas. These reactions typically occur at high temperatures, exceeding 800 °C. As the process involves the release of oxygen, there is an option to store it separately for other uses. For the reverse exothermic reaction involving reoxidation, ambient air can be used instead. In the latter case, however, the purity of air can become an important factor affecting the system performance.

6.3 Key Challenges Associated with TSMs

The classes of TSMs mentioned earlier often share common drawbacks, one of which is their inherent low thermal conductivity. This limitation negatively affects the heat and mass transfer performance of the reactor of the storage system. During the discharging process, when an exothermic reaction releases heat in the reactor, it must be rapidly dissipated to the HTF for the discharging reaction to proceed at an adequate rate. However, due to the low thermal conductivity of the TSM within the reactor, there is a high likelihood of an adverse situation occurring, where the rate of heat dissipation from the reactor to the HTF is lower than the rate of heat generation within the reactor. This situation hinders the progress of the discharging reaction and results in poor performance of the storage system. To overcome this challenge, various approaches can be implemented. One solution is to incorporate metallic fins in the TSM domain, which enhances the overall thermal conductivity of the reactor. By doing so, the rate of heat dissipation can be improved, ensuring efficient discharging reactions. Another approach is to use composite materials instead of pure TSMs. A common method involves impregnating the TSM onto an expanded graphite matrix to form a composite. This results in a significant increase in both thermal conductivity and the overall surface area available for solid–gas interactions. Consequently, such composite materials exhibit improved heat transfer properties, addressing the issues related to low thermal conductivity and facilitating better performance of the storage system.

The variation in the volume of a TSM during continuous charging and discharging processes presents another challenge. During the discharging process, when gas adsorbs or reacts with the solid TSM, its volume increases, and the reverse occurs during the charging process. These volume variations lead to changes in particle size and distribution within the TSM domain. Over multiple operating cycles, the probability of local particle agglomerations or breakages increases, which can have a negative impact on important system performance parameters such as reactor porosity and permeability. Moreover, as the volume variations continue over several cycles, the thermophysical properties of the TSMs tend to degrade, resulting in reduced system performance over time. To address these issues, the use of composites and the addition of suitable doping materials can play a significant role in mitigating the rapid degradation of system performance. Composites and doping materials can help stabilize the TSM structure during cyclic processes, reducing the negative effects of volume variations. This can lead to improved system performance and longevity, enhancing the overall efficiency and reliability of the TESS.

Apart from these, a major challenge with TESSs is concerning the storage of gas released during the charging process. Subsequently, this released gas must be reintroduced into the reactor at specific temperature and pressure levels during the discharging process. As a result, gas storage becomes a critical task as it directly influences the overall volume requirement for the storage system. One potential solution for this issue involves designing the gas storage tank (GST) in a manner that allows the released gas to be stored in a condensed state by transferring its heat to the ambient surroundings. Subsequently, the gas can be converted back to its gaseous form using ambient heat. Another possible solution is the development of resorption systems, wherein two distinct TSMs are interconnected in such a way that when gas is released from the reactor of one TSM, it gets adsorbed or reacts in the reactor containing the other TSM, and vice versa. This arrangement enables gas storage within the TSM materials themselves, eliminating the need for a separate GST. Although these solutions reduce the system volume requirement, they introduce additional complexity to the overall system design. It is also important to note that these solutions may not be universally applicable for all classes of TSMs, as their feasibility depends on the specific characteristics and behavior of the storage materials.

The discussion presented in this section aims to provide an overview of the various classes of TSMs that are currently available or under investigation. However, for more in-depth and comprehensive information, readers are encouraged to go through the relevant references [3, 6, 8–11, 14, 15] as well as other additional sources.

6.4 Reaction Kinetics Model

In chemical reactions, the rate of reaction depends on several factors including pressure, temperature, extent of reaction, and concentration of species. This study is referred to as reaction kinetics. In the context of chemisorption based on a solid–gas pair, the main practical objective of studying chemical reaction rates is to determine how quickly the reaction proceeds in a particular direction and approaches the equilibrium condition. However, a complete understanding of every step of the reaction mechanism is often challenging due to the complexities of the underlying phenomena [16, 17]. Therefore, modeling based on experimental data is commonly employed to determine reaction rates.

A common approach to model kinetics is by quantifying the rate of advancement of reaction from the amount of reacted solid present relative to the total solid reactants initially present. Over the past decades, several models have been developed to describe sorption kinetics, such as the linear drive force (LDF) and equilibrium models. The equilibrium model is particularly suitable when pore diffusion is adequate such that the mass transfer rate is not the limiting factor for reaction. On the other hand, the solid diffusion model is more cumbersome but considerably more accurate and has been widely used. In contrast, the LDF model simplifies the solid diffusion model through a mathematical assumption, but it tends to underestimate the system performance [18].

In the design of TESSs, the rate of thermal energy absorbed or released by the thermochemical materials (TCMs) depends on the chemical reaction kinetics, which also provides the necessary solid/gas equilibrium condition for the chemical reaction to occur. Therefore, this section briefly discusses the solid/gas reaction equilibrium first, followed by some commonly used models for the reaction kinetics of TCMs.

6.4.1 Solid/Gas Reaction Equilibrium

The reversible solid/gas absorption–desorption reactions used in TESSs are based on a monovariant equilibrium of the form given by Eq. (6.7),

$$\underset{\text{(Solid)}}{S_0} + \underset{\text{(Gas)}}{\nu G} \underset{\text{Endothermic}}{\overset{\text{Exothermic}}{\rightleftarrows}} \underset{\text{(Solid)}}{S_1} + \nu \Delta h_r^0. \tag{6.7}$$

The equilibrium conditions (p_{eq}, T_{eq}) of the solid/gas reaction can be described using the Clausius–Clapeyron relation [19]. This relation is derived from the principle that the free Gibbs energy of the transformation reaches equilibrium when it equals zero,

$$\Delta G_r = \Delta G_r^0 + \overline{R}T_{eq} \ln(K_{eq}) = \nu\Delta h_r^0 - \nu T_{eq}\Delta s_r^0 + \overline{R}T_{eq} \ln(K_{eq}) = 0, \tag{6.8}$$

where K_{eq} is the equilibrium constant for the solid/gas reaction, $\overline{R}$ is the universal gas constant, and Δh_r^0 and Δs_r^0 are the standard enthalpy and entropy changes, respectively. The direction of the reversible reaction is determined by the equilibrium constant. For a perfect gas, K_{eq} becomes

$$K_{eq} = \left(\frac{p_{eq}}{p_{ref}}\right)^{\nu}, \tag{6.9}$$

where p_{ref} is the reference pressure, set at 1 bar.

The thermodynamic equilibrium condition is finally defined by p_{eq} and T_{eq} according to the following equation:

$$\ln\left(\frac{p_{eq}}{p_{ref}}\right) = -\frac{\Delta h_r^0}{\overline{R}T_{eq}} + \frac{\Delta s_r^0}{\overline{R}}. \tag{6.10}$$

Eq. (6.10) is called the Van't Hoff equation and decides the direction of the reversible reaction based on the local temperature and pressure. The reaction described in Eq. (6.7) proceeds in the forward direction (exothermic reaction) when the supplied gas pressure is more than the equilibrium pressure at the

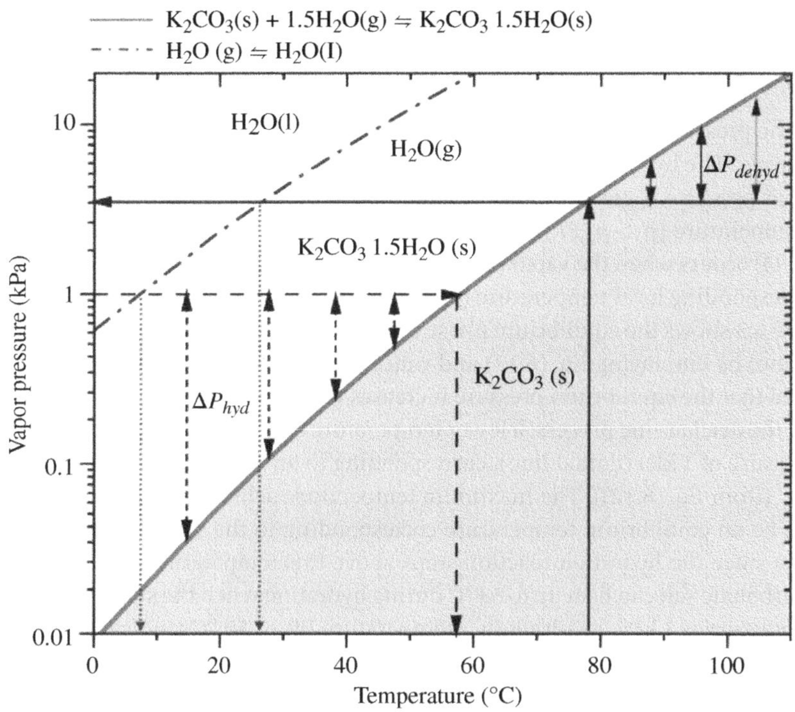

Figure 6.3 Equilibrium lines of the solid/gas reaction of K_2CO_3/H_2O (solid line) and the liquid/vapor phase change of water (dash dotted line). The pressure difference Δp between the equilibrium condition and the water vapor drives the hydration or dehydration reaction. Source: Adapted from Houben et al. [20].

corresponding local temperature ($p_g > p_{eq}(T)$). The endothermic reaction occurs when the gas pressure is less than the equilibrium pressure at the corresponding local temperature ($p_g < p_{eq}(T)$).

For example, the equilibrium phase diagram is plotted in Figure 6.3 for potassium carbonate (K_2CO_3) salt hydrate and water vapor. The gas–solid reaction of potassium carbonate is a simple single-step reversible reaction, represented by Eq. (6.11) [21],

$$K_2CO_3(S_0) + 1.5\,H_2O(G) \underset{\text{Dehydration}}{\overset{\text{Hydration}}{\rightleftarrows}} K_2CO_3 \cdot 1.5H_2O\,(S_1) + 1.5\Delta h_r^0 \quad (6.11)$$

Potassium carbonate/H_2O has been identified as a suitable reactive solid–gas pair for house heating applications [22, 23]. In Eq. (6.11), K_2CO_3 and $K_2CO_3 \cdot 1.5H_2O$ are dehydrated (S_0) and hydrated (S_1) salts and reactive gas (G) is water vapor. The equilibrium relation for the K_2CO_3/H_2O pair established experimentally by Gaeini et al. [21] is given by Eq. (6.12),

$$p_{eq} = 4.228 * 10^{12} \exp\left(\frac{-7337}{T}\right), \tag{6.12}$$

where the pressure (p_{eq}) and temperature (T) are measured in Pascal and Kelvin, respectively. The hydration reaction occurs (forward reaction in Eq. (6.11)) when the vapor pressure is greater than the equilibrium pressure at the corresponding local temperature ($p_v > p_{eq}(T)$). The dehydration reaction (backward reaction in Eq. (6.11)) occurs when the vapor pressure is less than the equilibrium pressure at the corresponding local temperature ($p_v < p_{eq}(T)$).

Figure 6.3 shows the equilibrium phase diagram of the potassium carbonate salt (solid line) by employing Eq. (6.12) and water vapor (dash dotted line). It can be observed that the equilibrium pressure increases with an increase in temperature. During the discharging process, a lower temperature of 7 °C generates water vapor at a pressure of 1 kPa (dotted line), corresponding to an equilibrium temperature of 58 °C (from Eq. (6.12)). The maximum temperature achieved during discharging can be an equilibrium temperature corresponding to the source water vapor pressure since the hydration reaction stops above this temperature. Thus, potassium carbonate salt can heat up to 58 °C during hydration when the supplied water vapor pressure is 1 kPa, resulting in a temperature lift of 50 °C (indicated by the horizontal black dashed line in Figure 6.3). Similarly, during the charging process, a vapor pressure of 4 kPa (solid black horizontal line in Figure 6.3) corresponds to a solid/gas equilibrium temperature of 80 °C. The dehydration reaction of potassium carbonate salt is possible at temperatures above 80 °C. The water vapor generated must be condensed by a source of a temperature below the saturation temperature at 4 kPa (28 °C) to continue the charging process.

6.4.2 Reaction Rate Expression

Reaction kinetics can be experimentally determined by measuring several parameters such as temperature (T), pressure (p), and the extent of reaction (α). Based on the systemic studies performed for kinetic rates in heterogeneous reactions, the global form of reaction rate in terms of time-dependence conversion rate is given as follows [21, 24]:

$$\frac{\partial \alpha}{\partial t} = k(T) \cdot f(\alpha) \cdot h(p). \tag{6.13}$$

In Eq. (6.13), $k(T)$ represents the temperature dependence of the reaction rate and typically parametrized by the Arrhenius-type function, which is given as follows:

$$k(T) = A_f \exp\left(-\frac{E_a}{\overline{R}T}\right), \tag{6.14}$$

where A_f is the pre-exponential factor and E_a is the activation energy. This function shows that the rate of reaction always increases with an increase in temperature. The pressure dependence is represented by $h(p)$. It is often neglected in most kinetic computational methods used in thermal analysis. However, pressure can significantly influence the kinetics of processes involving gases as reactants and/or products. Many reversible solid–gas reactions follow simple stoichiometry, such as $A_{solid} \leftrightarrow B_{solid} + C_{gas}$, and the pressure dependence of their rates can be represented as follows [25]:

$$h(p) = 1 - \frac{p_g}{p_{eq}}, \tag{6.15}$$

where p_g is the partial pressure of the gas and p_{eq} is the equilibrium pressure.

The function $f(\alpha)$ depicts the dependence of the reaction rate on reactant conversion and is represented by the reaction model. The extent of conversion (α) typically represents the progress of the overall transformation of a reactant to final products. The value of α can be experimentally determined as a fraction of the overall change in a physical property during the chemical reaction process. When a reaction involves mass loss, α may represent a fraction of the total mass lost during the process. Similarly, in processes accompanied by heat release or absorption, α is a fraction of the total heat released or absorbed during the process. As the process advances from initiation to completion, α increases from 0 to 1 [25]. Numerous reaction models can be used to represent the dependence on the extent of conversion, $f(\alpha)$, some of which are presented in Table 6.1 [25]. $g(\alpha)$ is the integral form of the reaction model given by Eq. (6.16) as follows (Table 6.1):

$$g(\alpha) \equiv \int_0^{\alpha} \frac{d\alpha}{f(\alpha)}. \tag{6.16}$$

Although there are several reaction models, they can be mainly categorized into three types: accelerating, decelerating, and sigmoidal (autocatalytic). Accelerating models represent reaction processes with continuously increasing reaction rates as the extent of conversion increases, reaching the maximum reaction rate at the end of the process. The power-law model is used for accelerating reaction, given as follows:

$$f(\alpha) = n\alpha^{(n-1)/n}, \tag{6.17}$$

where n is a constant. Decelerating models represent processes with maximum reaction rates at the beginning of the process, decreasing continuously as the extent of conversion increases. The reaction-order model is the most common example of a decelerating model, which is given as follows:

$$f(\alpha) = (1-\alpha)^n, \tag{6.18}$$

Table 6.1 Some of the reaction models used in kinetics of solid-state reactions.

Reaction model	Code	$f(\alpha)$	$g(\alpha)$
Power law	P4	$4\alpha^{3/4}$	$\alpha^{1/4}$
Power law	P3	$3\alpha^{2/3}$	$\alpha^{1/3}$
Power law	P2	$2\alpha^{1/2}$	$\alpha^{1/2}$
Power law	P2/3	$2/3\,\alpha^{-1/2}$	$\alpha^{3/2}$
One-dimensional diffusion	D1	$1/2\,\alpha^{-1}$	α^2
Two-dimensional diffusion	D2	$[-\ln(1-\alpha)]^{-1}$	$(1-\alpha)\ln(1-\alpha)+\alpha$
Three-dimensional diffusion	D3	$3/2\,(1-\alpha)^{2/3}\,[1-(1-\alpha)^{1/3}]^{-1}$	$[1-(1-\alpha)^{1/3}]^2$
Mampel (first order)	F1	$1-\alpha$	$-\ln(1-\alpha)$
Avrami–Erofeev	A4	$4(1-\alpha)\,[-\ln(1-\alpha)]^{3/4}$	$[-\ln(1-\alpha)]^{1/4}$
Avrami–Erofeev	A3	$3(1-\alpha)\,[-\ln(1-\alpha)]^{2/3}$	$[-\ln(1-\alpha)]^{1/3}$
Avrami–Erofeev	A2	$2(1-\alpha)\,[-\ln(1-\alpha)]^{1/2}$	$[-\ln(1-\alpha)]^{1/2}$
Contracting sphere	R3	$3(1-\alpha)^{2/3}$	$1-(1-\alpha)^{1/3}$
Contracting cylinder	R2	$2(1-\alpha)^{1/2}$	$1-(1-\alpha)^{1/2}$

Source: Vyazovkin et al. [25]/with permission of Elsevier.

where n is the reaction order. Sigmoidal models represent processes with initial accelerating behavior and final decelerating behavior, with the process rate reaching its maximum at intermediate values of the extent of conversion. The Avrami–Erofeev model is an example of the sigmoidal model, which is given as follows:

$$f(\alpha) = n(1-\alpha)[-\ln(1-\alpha)]^{(n-1)/n}. \tag{6.19}$$

Reliable methods should be capable of treating all three types of conversion dependencies. Sestak and Berggren [26] proposed an empirical model represented by Eq. (6.20), which can encompass different reaction models based on the values of m, n, and p,

$$f(\alpha) = \alpha^m(1-\alpha)^n[-\ln(1-\alpha)]^p. \tag{6.20}$$

As mentioned earlier, salt hydrates are very promising TCMs for low- and medium-temperature applications. Fopah Lele et al. [27] derived the following equation to describe the reaction kinetics of salt hydrate, incorporating a series of experimental results and theoretical analyses:

$$\frac{\partial\alpha}{\partial t} = A_f \exp\left(-\frac{E_a}{\overline{R}T}\right)\alpha^m(1-\alpha)^n\left(1-\frac{p_v}{p_{eq}}\right). \tag{6.21}$$

During simulation, one can substitute in Eq. (6.21) with $m = 1, n = 0$ for dehydration and $m = 0, n = 1$ for hydration [28]. Based on experimental data, Mamani et al. found that the cylindrical particle contraction model and the first-order reaction model are suitable for representing the dehydration reaction. Based on experimental characterizations, Gaeini et al. [21] formulated the kinetics models for the hydration (Eq. 6.22a) and dehydration (Eq. 6.22b) reactions of potassium carbonate (K_2CO_3) salt hydrate,

$$\frac{\partial \alpha}{\partial t} = A_{f,h} \exp\left(\frac{-E_{a,h}}{\overline{R}T}\right) \cdot (1-\alpha)^{\frac{2}{3}} \cdot \left(\frac{p_v}{p_{eq}}\right), \tag{6.22a}$$

$$\frac{\partial \alpha}{\partial t} = A_{f,d} \exp\left(\frac{-E_{a,d}}{\overline{R}T}\right) \cdot (1-\alpha)^{\frac{2}{3}} \cdot \left(\frac{p_{eq}}{p_v}\right). \tag{6.22b}$$

The deviation between water vapor pressure and equilibrium pressure (Δp_{hyd}) drives the hydration reaction. The rate of hydration reaction is an essential factor in releasing heat at a high rate, and this rate increases with increasing vapor pressure or decreasing temperature (Figure 6.3). Conversely, the hydration reaction releases exothermic heat and absorbs vapor, leading to an increase in temperature and a reduction in vapor pressure. On the other hand, the rate of the dehydration reaction is essential in storing heat at a high rate, which increases with decreasing vapor pressure or increasing temperature (Figure 6.3). However, the dehydration reaction absorbs endothermic heat and desorbs vapor, resulting in a reduction in temperature and an increase in vapor pressure. To efficiently design the reactor for the TESS, a heat-exchanging surface must remove/supply the heat, and vapor must be supplied/removed continuously during the hydration/dehydration reaction. Therefore, adequate heat and mass transfer rates are essential for the reactor design to ensure the efficient performance of the TESS.

6.5 Different Operating Modes of TESS

According to the operating modes of the thermochemical reactor during the charging and discharging processes, TESSs can be categorized into closed and open systems [29, 30]. The choice of the system is primarily determined by the adsorbent material [31]. The two operating modes entail different designs and dimensions regarding the system configuration. A schematic representation of open and closed systems is presented in Figures 6.4 and 6.5, respectively, and a summary of their main advantages and drawbacks is provided in Table 6.2.

6.5.1 Closed Sorption System

Closed sorption systems, extensively studied for refrigeration [32], heat pumps [33], and energy storage applications [34], provide attractive options

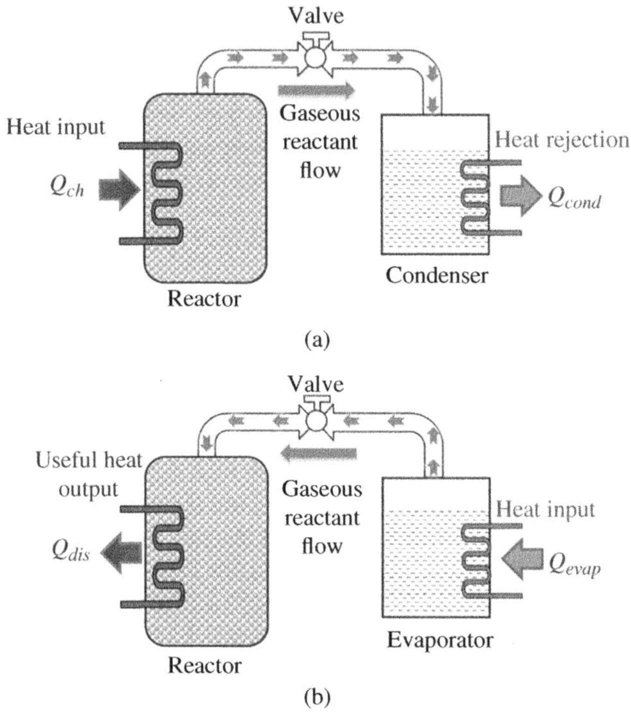

Figure 6.4 Schematic representation of the operation of the closed TESS during (a) charging and (b) discharging processes.

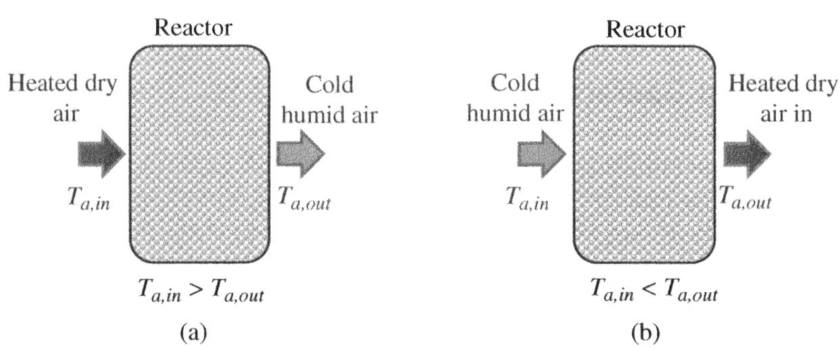

Figure 6.5 Schematic representation of the operation of the open TESS during (a) charging and (b) discharging process.

Table 6.2 Summary of advantages and drawbacks of open and closed systems.

Type of system	Advantages	Drawbacks
Open system	• Operates at atmospheric pressure. • Simple and technologically less complex than closed systems. • Less number of components than in a closed system. • Heat transfer can be enhanced by forced circulation. • Less expensive • Higher energy storage density	• External air flow and partial humidification required. • High-pressure drop due to high sorbate flow rate. • Need to avoid undesirable side effects.
Closed system	• No restriction on gaseous reactants; both NH_3 and H_2O can be used. • For the same vapor pressure, the equilibrium discharging temperature is higher than for an open system. • Can be used as an adsorption heat pump to provide both heating and cooling.	• Condenser/evaporator unit is required. • Complicated operation due to the requirement of vacuum conditions. • Complex system design • Periodic evacuation to avoid non-condensable gases (NCGs) • Higher cost • Heat source is required for evaporation.

for compact and highly efficient devices in small-scale applications. In a closed operating mode, the pure gaseous reactant (sorbate) is supplied at a vacuum pressure to the reactor bed to react with the solid reactant (sorbent) [35]. The heat is transferred to/from the reactive bed during the charging/discharging process through a separate heat exchanger. The applications of closed sorption systems include refrigeration [36], and these closed systems have the advantage of not releasing toxic gases to the environment, such as NH_3 [37, 38].

The closed system primarily consists of a reactor vessel for storing the solid reactant (or sorbent) and a separate vessel for storing the gaseous reactant (or sorbate) in liquid form. The gaseous reactants are condensed, stored in liquid form, and later evaporated in this separate vessel, so it is essentially a condenser/evaporator unit. The two vessels are connected by a conduit or pipe to allow the passage of the gaseous reactant. Figure 6.4 illustrates the operating principle of a closed TESS during the charging and discharging processes. The charging process involves a desorption reaction in the reactor and a gas–liquid phase change reaction in the condenser (Figure 6.4a). During the charging process, solar/waste heat (Qch) from the micro combined heat and power (micro-CHP) is added to the reactor using suitable HTF, leading to the detachment of the gaseous reactant (sorbate), which is bonded to the solid sorbent. The released gas (or sorbate) flows toward the condenser/evaporator unit through the pipe. The condenser is maintained at a low-temperature level, where the gaseous reactant rejects the heat of condensation (Q_{cond}) and condenses into liquid form. Once the charging process is complete, the reactor and the condenser/evaporator unit are separated by closing the connecting valve.

Reconnection of the reactor and condenser/evaporator unit is performed when heating or cooling demands arise. Discharging is performed by supplying sufficient heat (Q_{evap}) to the evaporator/condenser unit such that the gas/vapor is passed on to the reactor end to perform the exothermic reaction. The heat released (Q_{dis}) during the exothermic reaction is then transferred to the heat load using HTF (Figure 6.4b). Depending on the specific requirements, the evaporator can produce a cooling effect, or the reactor can generate a heating effect. This flexibility allows closed sorption thermal storage systems to offer both "cold storage" in summer and "heat storage" in winter [1, 39].

The discharging process requires heat to evaporate the liquid, and the ambient is often used as the heat source [39]. Air is often used due to its availability, ease of designing heat exchangers, and lack of location restrictions [40]. However, in situations where the ambient temperature is too low to facilitate evaporation, the discharging process may face significant challenges. Alternatively, the temperature of the ground is relatively higher and more stable. Some sorption storage projects have employed ground-source heat exchangers to extract heat from the earth [37, 41]. However, the construction of ground-source heat exchangers can be time consuming and expensive.

Compared to open systems (discussed next), closed systems are capable of delivering higher output temperatures for heating applications and a small amount of energy is required for vapor transport [31]. Nevertheless, closed TESSs have high costs and are complex in design. Moreover, the low operating pressure imposes significant technological constraints on the reactor and evaporator/condenser design and manufacturing, including poor heat and mass transfer coefficients [42]. Additionally, the presence of the evaporator/condenser and heat exchanger in the reactive bed increases the volume of the system, reducing the apparent energy density of the storage system relative to its volume [19]. Another major disadvantage of a closed system includes a rapid decrease in performance when non-condensable gases (NSGs) are introduced, inhibiting the mass diffusion of sorbent in a reactant [43, 44].

6.5.2 Open Sorption System

Open sorption systems are designed to operate at atmospheric pressure and are primarily used for nonhazardous solid reactant materials. In an open operating mode, the gaseous reactant is supplied with the ambient air as the carrier through the reactive solid bed at atmospheric pressure. In the case of an open system, water vapor typically serves as the gaseous reactant, and the thermodynamic equilibrium conditions are related to the partial pressure of water vapor in moist air [45].

Figure 6.5 demonstrates the operating principle of an open TESS during the charging and discharging processes. It utilizes a single vessel (reactor) containing the solid reactant and relies on atmospheric air as the carrier of heat and gaseous reactant (usually water vapor) [46]. During the charging process, a dry air stream is sufficiently heated by an external heat source, such as solar/waste heat from the micro-CHP. The transformed hot and dry air stream is then supplied to the reactor filled with reactant material (sorbent) to facilitate the endothermic reaction (Figure 6.5a), becoming cooler in the process. The cooled air leaving the reactor carries the water vapor desorbed by the sorbent. Finally, the cooled and wetter air is then exhausted into the surroundings.

In the discharging process, a humid and cool air stream is supplied to the reactor containing the previously desorbed solid reactant to meet the water vapor requirements for the exothermic reaction (Figure 6.5b). Some of the water vapor in the air is drawn in by the sorbent, releasing heat of sorption that heats up the air, making it suitable for heating purposes. The heated air leaving the reactor is directed toward the heat load.

The condition of the reactive salt in an open TESS is actually determined by the temperature and humidity of the ambient air provided to the bed during the charging and discharging processes. Depending on the geographical location and local weather conditions, the ambient air can be highly humid. Excessive humidity in the ambient air leads to the formation of an aqueous solution of the reactive salt within the reactive bed. This significantly affects the performance of the reactive bed during charging and discharging and also reduces the cyclability of the reactive salt. Nevertheless, open sorption systems offer practical advantages, such as lower investment costs and improved heat and mass transfer conditions compared to closed systems, making them an attractive choice for TES projects [19, 47]. However, it is essential to consider strategies to minimize pressure drop in the air stream so as to avoid excessive power consumption by the fan/blower. Following are the advantages and disadvantages of the open system in comparison to the closed system:

Advantages:

- The open systems function at atmospheric pressure, eliminating the need for an evaporator/condenser unit and a water storage container. As a result, it offers the potential for a simpler and more cost-effective reactor design.
- In an open system, the air (with reactive vapor) also acts as the HTF; thus, a separate heat exchanger is not required.

Disadvantages:

- In an open system, water vapor is obtained from moist air coming from the surrounding environment. However, during winter, the partial pressure of water vapor in the ambient air may be too low, necessitating humidification before entering the reactor. In this case, a humidifier may be required [19].
- Other drawbacks of the open system concept are the limited temperature lift it can achieve, and it requires power to drive the forced airflow [20]. Also, a limited choice of working pairs is available for reactant materials.

A detailed first and second law analysis was carried out by Abedin and Rosen [48] for both open and closed systems. It was revealed that, for the cases studied, both energy and exergy efficiencies were significantly higher for an open system. However, the analysis was heavily influenced by fundamental system parameters such as adsorbate flow rate, pressure, and temperature of heat recovery. Consequently, it is not possible to draw a definitive conclusion regarding the relative performance of the two types of systems.

6.6 Thermodynamic Analysis of a Thermochemical Energy Storage System

There are multiple ways to study the performance of a TESS designed for a certain heating application (in increasing order of their complexity), which can be summarized as follows: thermodynamic analysis, numerical modeling, and experimental studies. Thermodynamic analysis is a preliminary-level analysis methodology where the end states of various processes are considered for analysis. The energy balance calculations, along with estimations of mass and volume for the TESS, are utilized to analyze its performance under varying operating conditions. The numerical modeling approach involves the application of governing equations to define and numerically solve the multi-physics phenomena occurring during the operating cycle of the TESS under different conditions. This modeling can be further divided into component-level modeling and complete system-level modeling. Then, experimental studies can also be performed to observe the real performance of TESS in practical conditions. These studies commence at the laboratory scale. Based on the outcomes of small-scale experiments, the TESS design is refined, leading to the construction of commercial-level experimental systems. In this section, readers are introduced to the thermodynamic analysis methodology for a TESS designed to act as a heat source for various heating activities in cold climatic conditions. In cold climates, performing essential heating activities for buildings poses significant challenges. Additionally, there is an increasing demand for heating in automobile applications, particularly in EVs and hybrid EVs [49–52]. The conventional heating systems predominantly rely on electrical resistance heaters or fossil fuel-based sources of heat. In this section, a TESS is proposed to potentially fulfill these heating activities. Considering the intended applications, a closed type of TESS is deemed to be the appropriate choice.

If the TESS is to be designed for heating requirements in cold climate, it implies that the heat discharge is to occur at above-ambient temperature. The minimum discharge temperature needs to be fixed depending on the application. For instance, if the application is primarily space heating, one may set the minimum discharge temperature at about 25 °C. Similarly, the upper temperature limit (corresponding to the highest charging temperature) is determined from the available external heat source. For instance, if non-concentrating solar thermal energy is planned to be used as the resource, the charging temperature is limited to about 90 °C. Both the energy storage material and the HTF must operate within these specified temperature limits to enable the efficient and reliable performance of the proposed TESS. The ethylene glycol–water mixture (50 : 50) is observed to be the suitable choice for the HTF within these temperature limits. Also, within these operating temperature limits, classes of TSMs falling under the category of sorption processes look promising for storage materials. If the TESS is intended for heating activities in automotive applications, certain TSM classes, such as ammoniates and MHs, may not be suitable due to the involvement of ammonia and hydrogen gas, respectively. Thus, salt hydrates such as based on K_2CO_3 emerge as a viable option. One also needs to consider other factors such as cost-effectiveness, chemical stability, reaction kinetics, and operational safety [9, 23] before finalizing the TCM.

Figure 6.6 provides a schematic representation of the TESS during these charging and discharging processes. During the charging process, thermal energy obtained from a suitable external energy source is stored as reaction enthalpy resulting from an endothermic reaction occurring within the energy storage bed (ESB). The ESB serves as a reactor where these chemical reactions take place. The water vapor generated during the charging process is collected and stored within a dedicated GST. The choice between storing the released gas in gaseous form or condensed liquid form depends on the selected storage material for the TESS. Figure 6.6a depicts a system in which water vapor is stored in the GST in condensed liquid form, with its excess heat being rejected to the ambient. For a given bed temperature in the ESB, the driving pressure difference for vapor flow from ESB to GST is the pressure difference between the equilibrium pressure

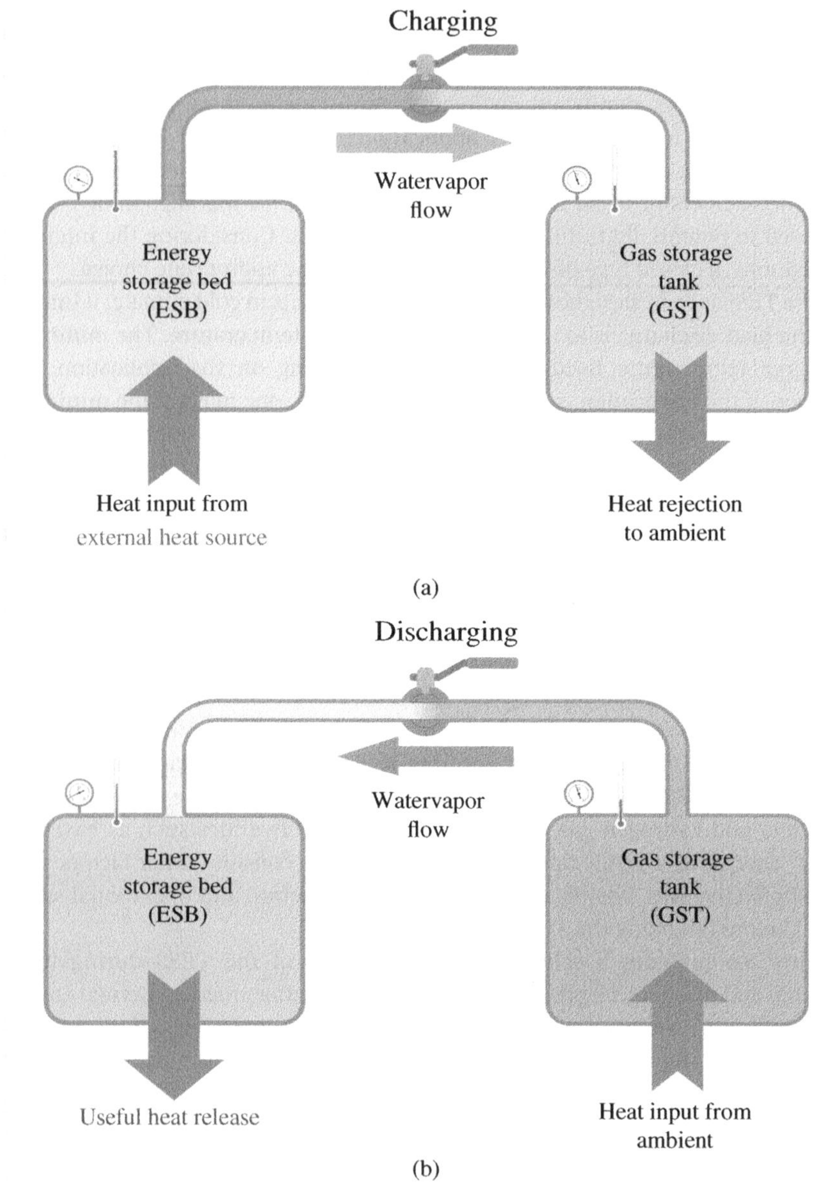

Figure 6.6 Operating principle of the TESS. Source: Chate et al. [2]/with permission of Elsevier.

of the salt hydrate (corresponding to the bed temperature) and the vapor/liquid equilibrium pressure in the GST. If ambient air is used to cool the condenser, the corresponding equilibrium (condensation) pressure in the GST is the saturation pressure at the ambient temperature. Subsequently, during the discharging process, the condensed liquid water in the GST is evaporated with the assistance of ambient heat. This process results in the release of useful heat due to the exothermic nature of the interaction between anhydrous K_2CO_3 salt and water vapor, as illustrated in Figure 6.6b. Water evaporates in the GST at the saturation pressure corresponding to the evaporation of water vapor that occurs at the saturation pressure corresponding to the ambient temperature. To maintain the flow of vapor from GST to ESB during the discharging process, the corresponding equilibrium pressure in the salt hydrate bed (ESB) should be slightly lower than that in GST. Even for low ambient temperature conditions, the corresponding equilibrium temperature for heat recovery in the ESB by hydration is expected to be significantly higher than the ambient temperature, ensuring the supply of heat at a desired temperature for the heating application.

The charging process typically occurs at higher temperatures than the discharging process. Consequently, additional processes, such as sensible heating and cooling of the ESB, become integral components of the operating cycle of the TESS. Figure 6.7 shows various processes involved in the operating cycle of the TESS on

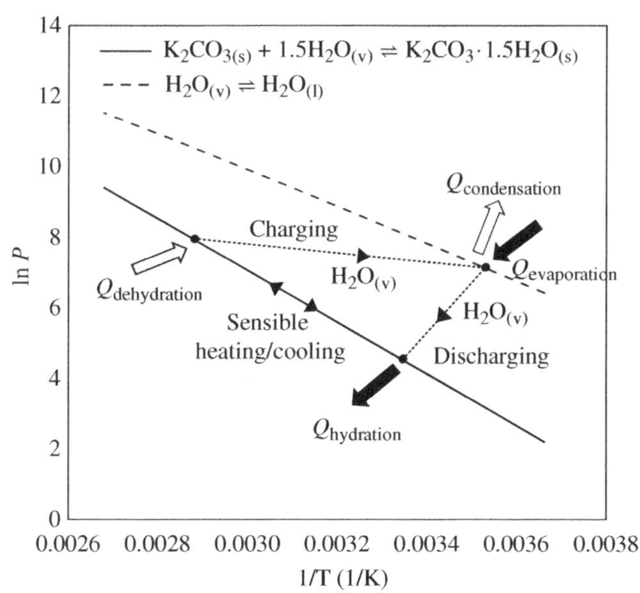

Figure 6.7 Operating cycle of the TESS on the Van't Hoff plot. Source: Chate et al. [2]/ with permission of Elsevier.

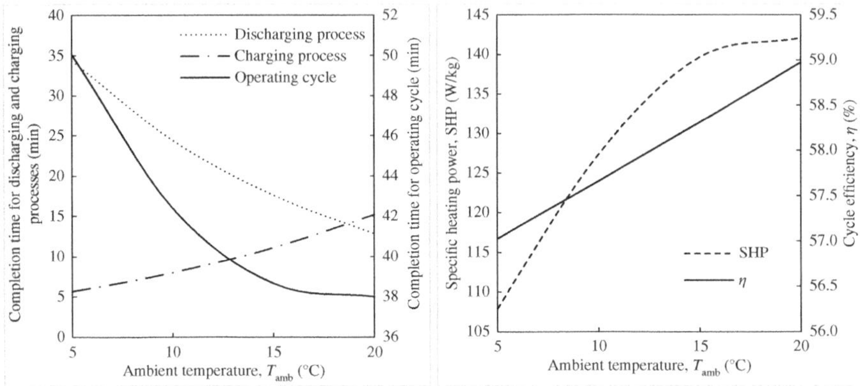

(a) (b)

Figure 6.8 (a) Effect of ambient temperature on the half-cycle completion time during discharging and charging processes. (b) Effect of ambient temperature on specific heating power and cycle efficiency. Source: Chate et al. [2]/with permission of Elsevier.

Table 6.3 Various processes involved in the operating cycle of the TESS.

Process in operating cycle	Changes take place in ESB	Changes take place in GST
Sensible heating of ESB	Temperature of ESB rises from ambient temperature to suitable charging temperature.	GST remains at ambient temperature.
Charging process	Endothermic dissociation reaction in ESB.	Water vapor released from ESB is stored in GST as condensed liquid water
Sensible/Natural cooling of ESB	Temperature of ESB drops to suitable discharging temperature.	GST remains at ambient temperature.
Discharging process	Exothermic recombination reaction in ESB.	Water vapor is evaporated in GST using ambient heat, and it flows back to the ESB.

the Van't Hoff plot. A charging temperature of 75 °C, a discharging temperature of 25 °C, and an ambient temperature of 10 °C are considered for plotting Figure 6.8. For a TESS that starts in a fully discharged state at ambient temperature, the various processes involved in its operating cycle are detailed in Table 6.3.

6.6.1 Methodology of Analysis

As thermodynamic analysis serves as a preliminary technique for evaluating the performance of the TESS, it necessitates incorporating various assumptions, as referenced in [2]. However, the primary objective of this discussion is to present the analysis methodology applicable to diverse storage systems, considering different classes of TSMs and heating applications. To reduce deviations between the actual TESS performance and the outcomes of thermodynamic analysis, accurate reaction rate expressions and realistic thermophysical properties of the TSM must be integrated into the analysis. The methodology for thermophysical analysis can be summarized as follows:

- Determine the required storage capacity of the TESS to meet the demands of the heating applications.
- Calculate the mass and volume requirements for the selected TSM based on the determined storage capacity.
- Assume realistic initial values for charging, discharging, and ambient temperatures, along with other operating parameters such as reactor porosity and the ratio of solid container mass of the reactor to the salt mass inside the reactor, etc. These initial values are subject to revision to achieve optimal TESS performance.
- Utilize the calculated mass and volume requirements of the TSM, relevant thermophysical properties of the TSM, and initial values of various operating parameters to calculate the energy interactions for each process in the operating cycle of the TESS.
- Modify the reaction rate expressions to obtain an approximate estimation of the completion time for the various processes in the operating cycle, as detailed in [2].
- Define evaluation parameters, such as cycle efficiency, volumetric energy storage density, specific heating power (SHP), etc., to quantitatively assess the system performance under different operating conditions.
- Perform parametric studies to understand the influence of specific operating parameters on the overall performance of the TESS while keeping other parameters constant. This helps in determining the optimal values for each operating parameter.
- Repeat the optimization process for all significant operating parameters to establish the set of optimized parameters that yield the best performance for the TESS.

This methodology of analysis is applied to the present K_2CO_3 salt hydrate-based TESS, considering a representative energy storage capacity of 1 MJ. Among various factors influencing the TESS performance, the ambient temperature plays a critical role as it affects the supply and outlet pressure of water vapor during the charging and discharging processes. The effects of ambient temperature variation on various performance parameters of the TESS are analyzed, and the results are depicted in Figure 6.8.

As the ambient temperature increases, the completion time for the charging process extends, while the completion time for the discharging process reduces. This behavior is attributed to the higher saturation pressure for water vapor generated and condensed in the GST at higher ambient temperatures. The increased outlet pressure of water vapor during the charging process adversely impacts the dehydration reaction rate in the ESB, leading to an elongated charging process. Conversely, the elevated supply pressure of water vapor during discharging results in a faster hydration reaction in the ESB, resulting in a shortened discharging process. However, the rate of decrease in the discharging process completion time exceeds the rate of increase in the charging process completion time. Consequently, the overall completion time of the operating cycle decreases with higher ambient temperatures, as indicated in Figure 6.8a. Cycle efficiency, defined as the ratio of net useful energy output to the total energy input of the complete TESS (including ESB and GST), increases with rising ambient temperatures. This rise in cycle efficiency is attributed to the decrease in energy required from the external source for sensible heating of the ESB, resulting in a reduced net energy input to the operating cycle. Therefore, higher ambient temperatures lead to an increase in SHP, as depicted in Figure 6.8b.

In a similar way, parametric studies can be carried for various other parameters to understand their effects on the performance of the TESS using the abovementioned methodology of thermodynamic analysis. A detailed discussion of these parametric studies is mentioned in Ref. [2].

6.7 Different Reactor Configurations

6.7.1 Fixed/Packed Bed Reactor Configuration

In a fixed/packed bed reactor configuration, the reactor vessel contains stationary solid reactant particles that remain immobile when exposed to the flow of gaseous reactant. These solid reactive materials are in the form of small particles, pellets, granules, or powder. They are either randomly packed or arranged in a specific structure within the reactor vessel. The gaseous reactant, on the other hand, passes through the empty spaces or voids between the solid particles. Fixed bed reactors have a simple design and are compact and easy to operate [53, 54]. This particular reactor type offers excellent reaction conversion per unit of mass conversion and low operating costs [55]. Additionally, fixed bed reactors demand a lower power input compared to fluidized bed reactors (discussed next) and have a

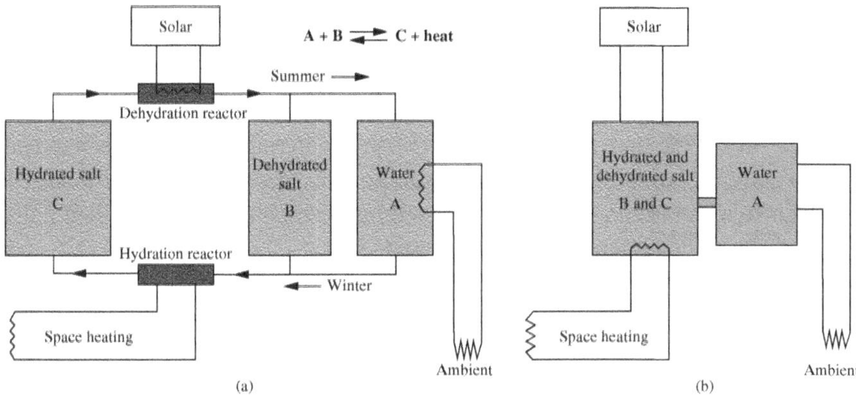

Figure 6.9 Schematic of (a) separated and (b) integrated reactor configurations. Source: Yu et al. [39]/with permission of Elsevier.

reduced risk of reactor erosion. However, some drawbacks exist, such as the possibility of undesired thermal gradients within the reactor, poor temperature control, and challenges associated with the maintenance and cleaning of the units. The fixed bed reactor configurations can be further divided into the following types:

(1) **Integrated reactor**. An integrated reactor contains the entire solid reactive material, and therefore, the reactive material need not be transported once it is filled in the reactor vessel. However, sensible heat losses increase with the reactor size. On the other hand, integrated systems are advantageous due to their simple construction.
(2) **Separated reactor**. In a separated reactor configuration, the sorbent is fed back to the storage vessel once the reaction is complete. Figure 6.9 provides a schematic representation of the thermochemical seasonal storage system with a separate bed reactor configuration. The separated reactor design involves the transfer of solid reactive material to and from the reactor vessel as needed. As a result, only a small quantity of reactive material is present in the reactor vessel at any given time, leading to a lower thermal mass of reactive material that needs to be heated during the reaction [56]. Separate bed reactor offers advantages such as optimized mass transport capability and reduced heat requirements for inert metallic components. Separate reactors are more suitable for large-scale storage applications compared to integrated reactors [57]. However, the complexity of the reactor design arises from the requirement to transfer the reactive material between the reactor and the storage vessel. Also, a separate reactor system requires additional electrical energy for powder transport and handling within the reactor. The concept of separate reactors has been implemented in a collaborative project called CWS-NT.

(3) **Modular reactor.**
Modular reactors are a type of integrated reactor where the solid reactive material is divided and placed into separate small modules [54]. Each module functions as an independent reactor with a smaller quantity of reactive material. As a result, the thermal mass of the reactive material per module is reduced compared to an integrated reactor. Additionally, the use of smaller modules leads to lower pressure drops [58], resulting in less input power. Utilizing the modular design offers several advantages [39]:

- Modular reactors allow for easy extension or reduction of the power and storage capacity based on practical requirements with minimal technical efforts. However, full system design and control strategies can become complicated for large-scale systems.
- The performance of a reactor relies on the ratio of sorbent volume to reactor volume. The heating and cooling processes occurring in a cyclic manner result in performance degradation due to thermal mass. By discreetly selecting the size of modular reactors, the heat transfer distance and thermal capacity of metallic components can be minimized. However, achieving an optimal thermal capacity ratio can be challenging for smaller modules due to a poorer surface-to-volume ratio and nonlinear scalability of certain parts. Therefore, an optimization design approach is necessary to obtain the best parameters.
- The design and manufacturing strategies of modular reactors can be standardized, thus offering distinct cost advantages and scalability.

6.7.2 Fluidized Bed Reactor Configuration

In a fluidized bed, the small particles of the solid reactant are suspended within a high-velocity flow of the gaseous reactant. The gaseous reactant enters the reactor at a high velocity through a perforated plate, known as the distributor, on which the solid particles rest when they are not in motion. At low gas velocities, the solid particles settle on the distributor, allowing the gas to flow through the gaps between them, resembling a fixed bed reactor. Fluidized bed reactors provide a large surface area contact between the solid particles and the gaseous reactant, promoting efficient heat and mass transfer between them. However, the high-velocity flow of gas in the reactor requires substantial input power, and the movement of solid materials can lead to erosion of the reactor material [55].

In contrast, fixed bed reactors demand a lower power input compared to fluidized bed reactors and have a lower probability of reactor erosion. The solid particles remain stationary within the fixed bed, providing a more stable and less erosive environment for the reactor.

6.8 Specific Applications of TESS

6.8.1 Building Heating Application

For seasonal heat storage, TESSs can be charged using solar thermal energy during the summer. Moreover, surplus electricity or industrial waste heat can also be utilized to charge the system, as shown in Figure 6.10. The stored heat can later be utilized for domestic applications, such as space heating or providing domestic hot water during the winter or whenever needed.

While closed TESSs have been studied [60, 61], open TESSs are popular for building/space heating applications. This popularity is due to their operation at atmospheric pressure conditions, using ambient air, resulting in a simpler design

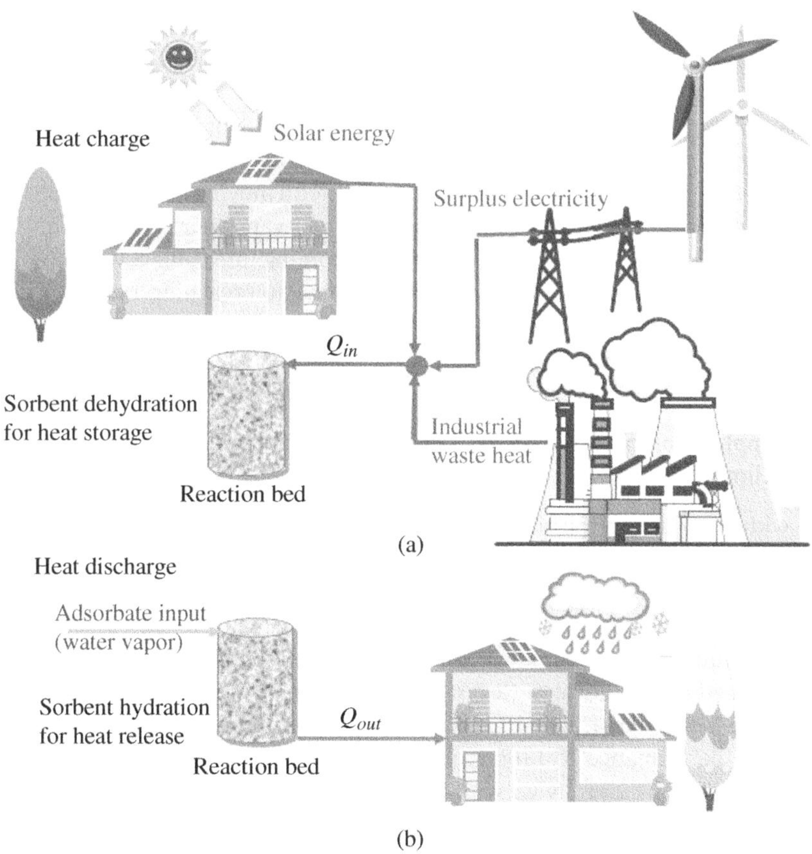

Figure 6.10 Schematic representation of the operation of the TESS employed for domestic applications. Source: Li et al. [59]/with permission of Elsevier.

compared to closed TESSs operating under vacuum conditions. Based on screening criteria, salt hydrates are identified as the best choice for such applications [9, 23]. Since the system is intended for seasonal operation in residential buildings for space heating, the chosen salt hydrate should be nontoxic. Also, as the gaseous reactant is water vapor [62, 63], salt hydrates are considered safe. Moreover, it should be capable of dehydrating below 100 °C and exhibiting reversibility during the cycles of dehydration and hydration. Some salt hydrates, such as the potassium carbonate/water vapor and strontium bromide/water vapor reactive pair, meet these criteria and have been identified as suitable choices for the system.

Successful implementations of the strontium bromide salt/water vapor reactive pair have been reported for various low-temperature applications, such as building floor heating [40] and residential space heating [63, 64]. The strontium bromide–water vapor reactive pair demonstrates a suitably high energy density of 2.49 GJ/m^3, making it ideal for compact heat storage and ensuring space efficiency in residential dwellings.

Moreover, the fixed bed reactor configuration is preferred for residential or space heating applications compared to fluidized bed reactors due to their advantages mentioned in Section 1.7. Farcot et al. [65] constructed a moving bed open TESS for residential space heating, employing the $SrBr_2$–water vapor reaction pair. In this system, the solid reactive salt falls due to gravity, coming into contact with humid air flowing horizontally in a cross-flow arrangement. After reaching a steady state, the reactor achieved a homogeneous temperature of 41 °C which is suitable for space heating applications.

6.8.2 Application of the TESS for Battery Preheating of EVs in Cold Climates

Along with reduction in the air pollution and the greenhouse gas emissions, EVs exhibit higher energy efficiency relative to vehicles reliant on fossil fuels. Among the cellular chemistries feasible for batteries of EVs, lithium-ion batteries (LIBs) are commonly acknowledged as the most fitting choice owing to their discernible benefits, including increased power density, protracted operational lifespan, and diminished self-discharge rates. Nevertheless, it is imperative to highlight that the efficiency of LIBs is intricately linked to the operational temperature. Attaining the optimal performance of LIBs requires a specified operational temperature span ranging from 15 to 35 °C, with a maximal temperature variance of 5 °C between individual battery cells. In instances where the temperature of the EV battery decreases beneath this preferred range, there is a notable escalation in the internal resistance of the battery. Hence, to avert battery harm and guarantee its sustained durability, it becomes necessary to decrease the charging current, thereby resulting in an extended charging duration for the EV. Furthermore, in colder temperatures, it is essential to curtail the current drawn from the battery, leading to restricted initial acceleration capabilities for the EV. Moreover, there exist numerous other unfavorable outcomes, encompassing reduced driving range and battery longevity, diminished regenerative braking efficiency, and higher demands on heating, ventilation, and air conditioning systems. Thus, to address these challenges and optimize the performance of EVs in cold climates, battery preheating becomes essential [7].

Commonly, the favored energy source for the preheating of EV batteries is the grid power supplied at charging stations. However, in scenarios involving cold climates where an EV remains parked for extended durations without convenient access to a nearby charging station, the preheating process relies on utilizing a substantial portion of the battery's stored charge. Among the array of techniques for preheating batteries, the prevalent choice in EVs is the employment of an indirect heating method employing a liquid HTF. This preference stems from the demonstrated safety, dependability, and established technical sophistication associated with this method. Currently, the liquid HTF employed for battery preheating undergoes heating through electric heaters, which source energy from either the grid power present at the charging station or the EV battery itself. In this section, we explore the application of a TESS to facilitate the battery preheating process in cold climates. The K_2CO_3 salt hydrate-based TESS, previously discussed in Section 1.6, can also be utilized for this purpose. This TESS is engineered to operate in tandem with the current heating system infrastructure within EVs, augmenting adaptability by functioning as an onboard heat source for battery preheating in situations where conventional electric heating sources are not accessible.

Figure 6.11 illustrates the existing architecture employed for managing the battery temperature. Battery cooling involves transferring heat to the battery evaporator using the HTF. The battery evaporator is typically a cooling plate with multiple channels for HTF flow. Subsequently, this thermal energy is released to the surroundings through a condenser, following a conventional vapor compression cycle. Conversely, the process of battery heating is achieved through the utilization of electric heaters, wherein the designated HTF facilitates the transfer of thermal energy from the heater to the battery pack. Figure 6.12 illustrates the battery heating circuit featuring the integration of the suggested TESS. When the TESS is getting charged, water vapor getting released from various ESB modules is condensed and stored in a common GST, as shown in Figure 6.12a. In addition, the electric heaters are bypassed during the discharging of TESS as heating is performed by the ESB modules, as shown in Figure 6.12b.

The duty cycle of the EV under consideration determines the charging frequency for the TESS. If the EV is anticipated to frequently endure prolonged periods in

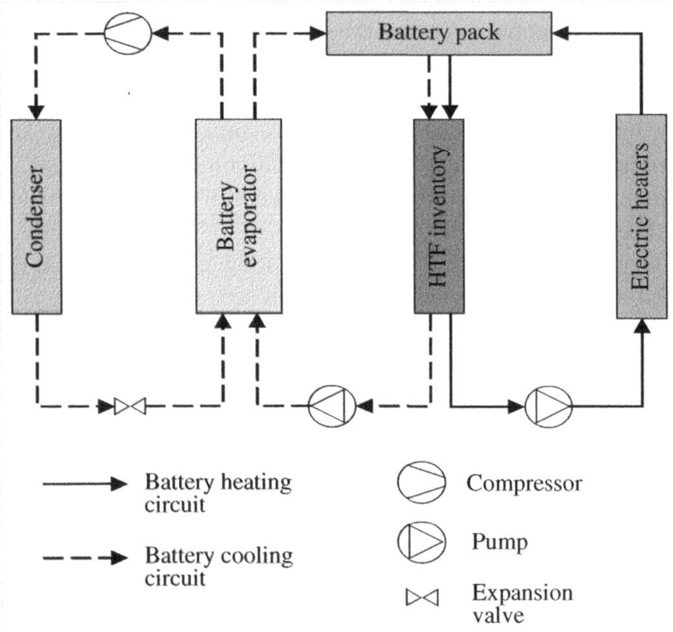

Figure 6.11 Existing arrangements for battery thermal management in EVs. Source: Chate et al. [7]/with permission of Elsevier.

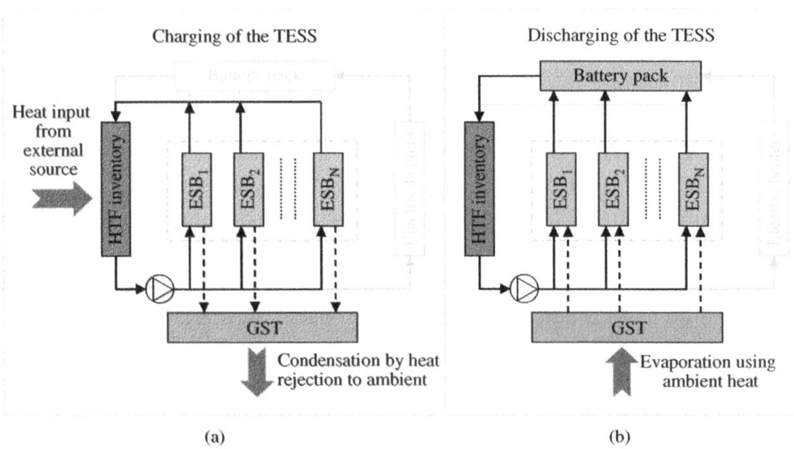

Figure 6.12 Arrangement for the battery preheating with proposed TESS. Source: Chate et al. [7]/with permission of Elsevier.

cold environmental conditions, such as overnight parking at remote sites, the TESS necessitates recurring charging prior to the commencement of the operational cycle. For other cases, the TESS serves as an onboard emergency preheating system with lesser requirement of the charging frequency. The setup essential for thermally recharging the TESS must be tailored to the specific characteristics of the EV in question. For instance, in the case of a passenger electric car, the TESS can be charged within a home garage or the parking facility of a residential complex, utilizing available resources like residential waste heat, hot water, or electricity. Conversely, for an electric bus, a viable approach would involve conducting the charging process at the central depot, where the bus is typically stationed after fulfilling its scheduled journeys for the day. In addition to the proposed heat sources such as electrical energy or industrial/residential waste heat, the potential integration of solar thermal energy could also be examined for large-scale charging stations located at bus depots.

6.8.3 Metal Hydride-Based Multifunctional Thermal Batteries

MHs are attractive materials for use in various heating and cooling systems mainly because of their high reaction enthalpies, diverse range of available materials and compositions, and flexible operating conditions. Referring to the general stoichiometric reaction (Eq. 6.3) forward endothermic reaction can be utilized for energy storage or cooling applications, while the backward exothermic reaction can be utilized for energy release or heating applications. These multiple functions can be realized in closed cycle metal hydride-based TES systems (MH-TESSs), in which a secondary vessel is required for parking the hydrogen released from the high-temperature reactor. As in any other closed cycle sorption system, the couple (MH-TESS) needs a one-time charge of the gas which circulates between the two reactors. As the gaseous mode of storage requires a very high volume, a low-temperature hydride bed (which can desorb at nearly ambient temperature) is typically used for hydrogen storage. The selection of a secondary hydride for hydrogen storage is crucial in the design of closed MH systems. The hydride materials chosen for the two beds depend on the desired operating parameters. The high temperature hydride is selected based on the available heat source temperature range and operating pressure. Similarly, the material for the hydrogen storage bed is selected based on the ambient temperature range and the given operating pressure. It is important to choose a material that can release hydrogen at near-ambient temperature in the operating pressure range. All these compatibility factors will make the two materials thermodynamically compatible for operation in a cycle. MH systems, in this regard, can offer a wide variety of alloys to satisfy the needs for various applications. Figure 6.13 shows the Van't Hoff plots for some selected materials, covering three broad ranges of dehydrogenation temperature, namely,

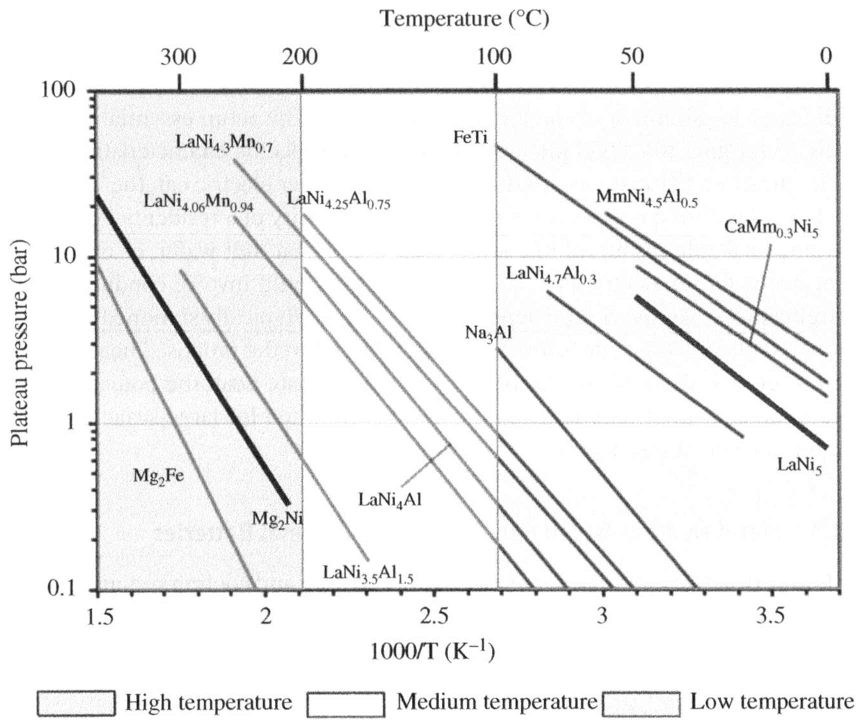

Figure 6.13 Van't Hoff plot for some of the selected hydrides. Source: Malleswararao et al. [66]/with permission of Elsevier.

high temperature (~300 °C), medium temperature (~150 °C), and low temperature (near ambient). Typically, materials for the hydrogen storage bed belong to the low-temperature class of MHs so that near-ambient or low-grade waste heat can be used for heat recovery from a thermal battery.

Figure 6.14 shows an example of cyclic operation of a MH-based multifunctional thermal battery. In the first stage, the thermal battery (MH1) is charged using external heat (Q_w) available at about 150 °C, and this endothermic process releases hydrogen which is subsequently adsorbed by MH2 containing a low-temperature MH. MH2 operates at a near-ambient temperature of 30 °C, and the corresponding heat of reaction (Q_h) released due to the exothermic reaction becomes an available resource for low-temperature heating application such as space heating. During the discharging process, the low-temperature reactor MH2 absorbs heat to take cooling load (Q_c) at 30 °C and releases hydrogen. The desorbed hydrogen flows to the high-temperature reactor MH1 and releases stored heat through an exothermic reaction. The released heat (Q_d) can be delivered to a desired application using a HTF with an inlet temperature of 50 °C. This completes the cyclic operation of the

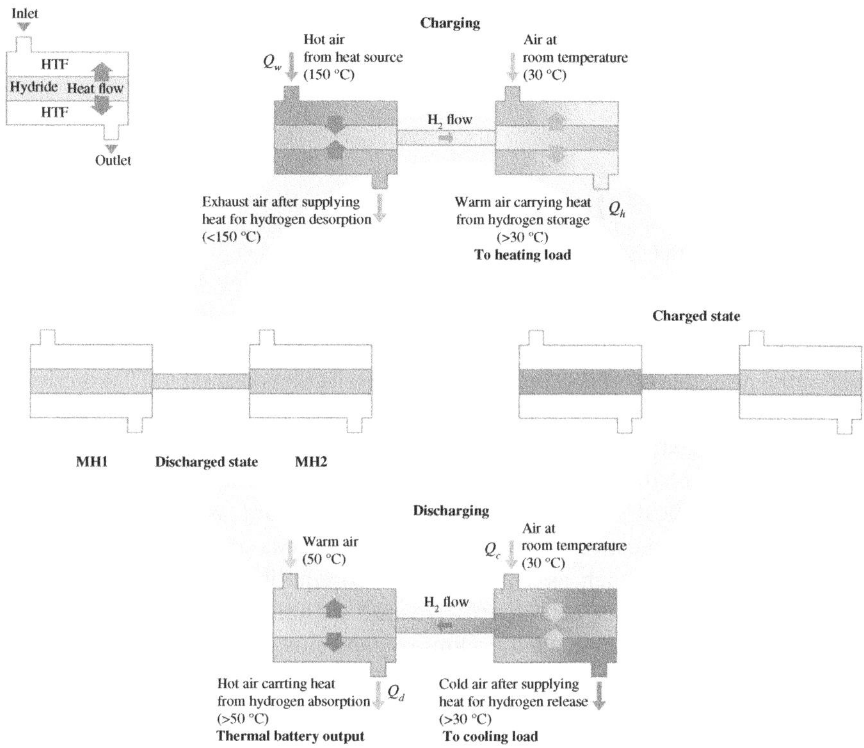

Figure 6.14 Operating cycle of a metal hydride-based multifunctional thermal battery. Source: Malleswararao et al. [67]/with permission of Elsevier.

Figure 6.15 Van't Hoff plot for cyclic operation of a coupled thermal battery. Source: Malleswararao et al. [67]/with permission of Elsevier.

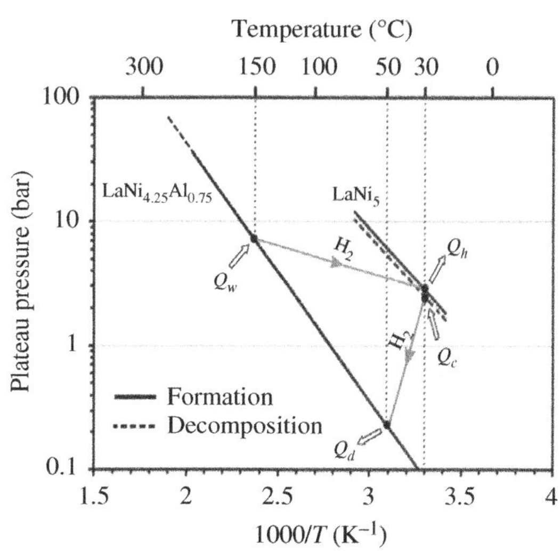

metal hydride based thermal energy storage system (MH-TES). In summary, the system performs heating as well as cooling. Figure 6.15 shows the thermodynamic operating cycle on a Van't Hoff plot.

6.9 Summary

The present chapter focuses on various aspects of the gas–solid pair-based TESSs. These energy storage systems offer a compelling solution for addressing the challenge posed by the disparity between the availability of thermal energy from renewable sources and fluctuating demand from users. Initially, the fundamental operating principle of a TESS is explained. The selection of suitable TSM is critical for optimal performance of the TESS. Diverse categories of available TSMs are discussed, elucidating their inherent characteristics and potential applications. The significant challenges that are inherent in these TSMs are also discussed, along with practical solutions. Furthermore, important aspects affecting the performance of respective chemical reactions of TSMs are highlighted, which encompass concepts such as solid–gas reaction equilibrium, reaction kinetics, reaction rate expressions, etc. Thermodynamic analysis of the operating cycle of the TESS provides much needed preliminary insights about the performance of TESS under various operating conditions. The methodology of analysis for the thermodynamic analysis is discussed with the help of a case study of a TESS designed for performing numerous heating applications in cold climates. Apart from the selection of TSM, the performance of the TESS is significantly influenced by the configuration of the reactor employed within the TESS. Advantages and challenges related to different reactor configurations used in the practical TESS are also presented. The chapter is concluded with discussion on some of the intriguing and innovative applications of the TESS, such as building heating, battery preheating in EVs, multifunctional MH-based thermal batteries.

Nomenclature

ESB	energy storage bed
EV	electric vehicle
GST	gas storage tank
HTF	heat transfer fluid
LIB	lithium-ion battery
MHs	metal hydrides
TESS	thermochemical energy storage system
TSM	thermochemical storage material

References

1 Lin, J., Zhao, Q., Huang, H. et al. (2021). Applications of low-temperature thermochemical energy storage systems for salt hydrates based on material classification: a review. *Solar Energy* 214: 149–178. https://doi.org/10.1016/j.solener.2020.11.055.

2 Chate, A., Sharma, R., Srinivasa Murthy, S., and Dutta, P. (2022). Studies on a potassium carbonate salt hydrate based thermochemical energy storage system. *Energy* 258: 124873. https://doi.org/10.1016/j.energy.2022.124873.

3 Yang, Z., Qu, M., and Gluesenkamp, K.R. (2020). Ammonia-based chemisorption heat pumps for cold-climate heating applications: a comprehensive review. *Applied Thermal Engineering* 179: 115674. https://doi.org/10.1016/j.applthermaleng.2020.115674.

4 Vasta, S., Brancato, V., La Rosa, D. et al. (2018). Adsorption heat storage: state-of-the-art and future perspectives. *Nanomaterials* 8: https://doi.org/10.3390/NANO8070522.

5 Sarbu, I. and Dorca, A. (2019). Review on heat transfer analysis in thermal energy storage using latent heat storage systems and phase change materials. *International Journal of Energy Research* 43: 29–64. https://doi.org/10.1002/er.4196.

6 Aydin, D., Casey, S.P., and Riffat, S. (2015). The latest advancements on thermochemical heat storage systems. *Renewable and Sustainable Energy Reviews* 41: 356–367. https://doi.org/10.1016/j.rser.2014.08.054.

7 Chate, A., Dutta, P., and Murthy S, S. (2023). Performance analysis of a thermochemical energy storage system for battery preheating in electric vehicles. *Applied Thermal Engineering* 219: 119439. https://doi.org/10.1016/j.applthermaleng.2022.119439.

8 Cot-Gores, J., Castell, A., and Cabeza, L.F. (2012). Thermochemical energy storage and conversion: a-state-of-the-art review of the experimental research under practical conditions. *Renewable and Sustainable Energy Reviews* 16: 5207–5224. https://doi.org/10.1016/j.rser.2012.04.007.

9 Donkers, P.A.J., Sögütoglu, L.C., Huinink, H.P. et al. (2017). A review of salt hydrates for seasonal heat storage in domestic applications. *Applied Energy* 199: 45–68. https://doi.org/10.1016/j.apenergy.2017.04.080.

10 Pardo, P., Deydier, A., Anxionnaz-Minvielle, Z. et al. (2014). A review on high temperature thermochemical heat energy storage. *Renewable and Sustainable Energy Reviews* 32: 591–610. https://doi.org/10.1016/j.rser.2013.12.014.

11 Carrillo, A.J., González-Aguilar, J., Romero, M., and Coronado, J.M. (2019). Solar energy on demand: a review on high temperature thermochemical heat storage systems and materials. *Chemical Reviews* 119: 4777–4816. https://doi.org/10.1021/acs.chemrev.8b00315.

12 Sharma, R., Anil Kumar, E., Dutta, P. et al. (2021). Ammoniated salt based solid sorption thermal batteries: a comparative study. *Applied Thermal Engineering* 191: https://doi.org/10.1016/j.applthermaleng.2021.116875.

13 Li, T.X., Wu, S., Yan, T. et al. (2016). A novel solid-gas thermochemical multilevel sorption thermal battery for cascaded solar thermal energy storage. *Applied Energy* 161: 1–10. https://doi.org/10.1016/j.apenergy.2015.09.084.

14 Hua, W., Yan, H., Zhang, X. et al. (2022). Review of salt hydrates-based thermochemical adsorption thermal storage technologies. *Journal of Energy Storage* 56: 106158. https://doi.org/10.1016/j.est.2022.106158.

15 Malleswararao, K., Dutta, P., and Murthy S, S. (2022). Applications of metal hydride based thermal systems: a review. *Applied Thermal Engineering* 215: 118816. https://doi.org/10.1016/j.applthermaleng.2022.118816.

16 Ašperger, S. (2003). *Chemical Kinetics and Inorganic Reaction Mechanisms.* Springer https://doi.org/10.1007/978-1-4419-9276-5.

17 Daniels, F. and Alberty, R.A. (2004). Kinetics and mechanism, second edition. *The Journal of Physical Chemistry* 65: 384–384. https://doi.org/10.1021/J100820A601.

18 Fopah Lele, A. (2016). State-of-art of thermochemical heat storage systems. In: *A Thermochemical Heat Storage System for Households*, Springer Theses, 15–58. Cham: Springer https://doi.org/10.1007/978-3-319-41228-3_2.

19 Michel, B., Neveu, P., and Mazet, N. (2014). Comparison of closed and open thermochemical processes, for long-term thermal energy storage applications. *Energy* 72: 702–716. https://doi.org/10.1016/j.energy.2014.05.097i.

20 Houben, J., Sögütoglu, L., Donkers, P. et al. (2020). K_2CO_3 in closed heat storage systems. *Renewable Energy* 166: 35–44. https://doi.org/10.1016/J.RENENE.2020.11.119.

21 Gaeini, M., Shaik, S.A., and Rindt, C.C.M. (2019). Characterization of potassium carbonate salt hydrate for thermochemical energy storage in buildings. *Energy and Buildings* 196: 178–193. https://doi.org/10.1016/J.ENBUILD.2019.05.029.

22 Donkers, P.A.J., Pel, L., and Adan, O.C.G. (2016). Experimental studies for the cyclability of salt hydrates for thermochemical heat storage. *Journal of Energy Storage* 5: 25–32. https://doi.org/10.1016/J.EST.2015.11.005.

23 Sögütoglu, L.C., Donkers, P.A.J., Fischer, H.R. et al. (2018). In-depth investigation of thermochemical performance in a heat battery: cyclic analysis of K_2CO_3, $MgCl_2$ and Na_2S. *Applied Energy* 215: 159–173. https://doi.org/10.1016/J.APENERGY.2018.01.083.

24 Fopah Lele, A. (2016). Thermal management modeling in thermo-chemical heat storage systems. In: *A Thermochemical Heat Storage System for Households*, Springer Theses, 59–94. Cham: Springer. https://doi.org/10.1007/978-3-319-41228-3_3.

25 Vyazovkin, S., Burnham, A.K., Criado, J.M. et al. (2011). ICTAC kinetics committee recommendations for performing kinetic computations on thermal analysis data. *Thermochimica Acta* 520: 1–19. https://doi.org/10.1016/J.TCA.2011.03.034.

26 Šesták, J. and Berggren, G. (1971). Study of the kinetics of the mechanism of solid-state reactions at increasing temperatures. *Thermochimica Acta* 3: 1–12. https://doi.org/10.1016/0040-6031(71)85051-7.

27 Fopah Lele, A., Kuznik, F., Rammelberg, H.U. et al. (2015). Thermal decomposition kinetic of salt hydrates for heat storage systems. *Applied Energy* 154: 447–458. https://doi.org/10.1016/J.APENERGY.2015.02.011.

28 Fopah Lele, A., Kuznik, F., Opel, O., and Ruck, W.K.L. (2015). Performance analysis of a thermochemical based heat storage as an addition to cogeneration systems. *Energy Conversion and Management* 106: 1327–1344. https://doi.org/10.1016/J.ENCONMAN.2015.10.068.

29 Hauer, A. (2007). Evaluation of adsorbent materials for heat pump and thermal energy storage applications in open systems. *Adsorption* 13: 399–405. https://doi.org/10.1007/S10450-007-9054-0/METRICS.

30 Deshmukh, H., Maiya, M.P., and Murthy S, S. (2017). Study of sorption based energy storage system with silica gel for heating application. *Applied Thermal Engineering* 111: 1640–1646. https://doi.org/10.1016/J.APPLTHERMALENG.2016.07.069.

31 Fumey, B., Weber, R., and Baldini, L. (2019). Sorption based long-term thermal energy storage – process classification and analysis of performance limitations: a review. *Renewable and Sustainable Energy Reviews* 111: 57–74. https://doi.org/10.1016/J.RSER.2019.05.006.

32 Wongsuwan, W., Kumar, S., Neveu, P., and Meunier, F. (2001). A review of chemical heat pump technology and applications. *Applied Thermal Engineering* 21: 1489–1519. https://doi.org/10.1016/S1359-4311(01)00022-9.

33 Ziegler, F. (1999). Recent developments and future prospects of sorption heat pump systems. *International Journal of Thermal Sciences* 38: 191–208. https://doi.org/10.1016/S1290-0729(99)80083-0.

34 Füldner, G., Henning, H.-M., Schossig, P., and Schmidt, F.P. (2011). General considerations on thermal energy storage with closed adsorption systems. https://doi.org/10.24406/PUBLICA-FHG-375129 (accessed 26 April 2024).

35 De Jong, A.J., Van Vliet, L., Hoegaerts, C. et al. (2016). Thermochemical heat storage – from reaction storage density to system storage density. *Energy Procedia* 91: 128–137. https://doi.org/10.1016/J.EGYPRO.2016.06.187.

36 Li, T.X., Wang, R.Z., and Li, H. (2014). Progress in the development of solid–gas sorption refrigeration thermodynamic cycle driven by low-grade thermal energy. *Progress in Energy and Combustion Science* 40: 1–58. https://doi.org/10.1016/J.PECS.2013.09.002.

37 Stitou, D., Mazet, N., and Mauran, S. (2012). Experimental investigation of a solid/gas thermochemical storage process for solar air-conditioning. *Energy* 41: 261–270. https://doi.org/10.1016/J.ENERGY.2011.07.029.

38 Li, T., Wang, R., and Kiplagat, J.K. (2013). A target-oriented solid-gas thermochemical sorption heat transformer for integrated energy storage and energy upgrade. *AIChE Journal* 59: 1334–1347. https://doi.org/10.1002/AIC.13899.

39 Yu, N., Wang, R.Z., and Wang, L.W. (2013). Sorption thermal storage for solar energy. *Progress in Energy and Combustion Science* 39: 489–514. https://doi.org/10.1016/J.PECS.2013.05.004.

40 Lahmidi, H., Mauran, S., and Goetz, V. (2006). Definition, test and simulation of a thermochemical storage process adapted to solar thermal systems. *Solar Energy* 80: 883–893. https://doi.org/10.1016/J.SOLENER.2005.01.014.

41 N'Tsoukpoe, K.E., Le Pierrès, N., and Luo, L. (2012). Numerical dynamic simulation and analysis of a lithium bromide/water long-term solar heat storage system. *Energy* 37: 346–358. https://doi.org/10.1016/J.ENERGY.2011.11.020.

42 Zhang, Y. and Wang, R. (2020). Sorption thermal energy storage: concept, process, applications and perspectives. *Energy Storage Materials* 27: 352–369. https://doi.org/10.1016/J.ENSM.2020.02.024.

43 Donkers, P., Gao, K., Houben, J. et al. (2020). Effect of non-condensable gasses on the performance of a vacuum thermochemical reactor. *Energies* 13: 362. https://doi.org/10.3390/EN13020362.

44 Yan, T., Wang, R.Z., Li, T.X. et al. (2015). A review of promising candidate reactions for chemical heat storage. *Renewable and Sustainable Energy Reviews* 43: 13–31. https://doi.org/10.1016/J.RSER.2014.11.015.

45 Michel, B., Mazet, N., Mauran, S. et al. (2012). Thermochemical process for seasonal storage of solar energy: characterization and modeling of a high density reactive bed. *Energy* 47: 553–563. https://doi.org/10.1016/J.ENERGY.2012.09.029.

46 Scapino, L., Zondag, H.A., Van Bael, J. et al. (2017). Energy density and storage capacity cost comparison of conceptual solid and liquid sorption seasonal heat storage systems for low-temperature space heating. *Renewable and Sustainable Energy Reviews* 76: 1314–1331. https://doi.org/10.1016/J.RSER.2017.03.101.

47 Stritih, U., Mlakar, U., Stritih, U., and Mlakar, U. (2018). Technologies for seasonal solar energy storage in buildings. In: *Advancements in Energy Storage Technologies*. https://doi.org/10.5772/INTECHOPEN.74404.

48 Abedin, A.H. and Rosen, M.A. (2012). Closed and open thermochemical energy storage: energy- and exergy-based comparisons. *Energy* 41: 83–92. https://doi.org/10.1016/J.ENERGY.2011.06.034.

49 Ghobadpour, A., Kelouwani, S., Amamou, A. et al. (2019). A brief review of plug-in hybrid electric vehicles operation in cold climates. *2019 IEEE Vehicle*

Power and Propulsion Conference, VPPC 2019 – Proceedings. https://doi.org/10.1109/VPPC46532.2019.8952348.

50 Lajunen, A., Yang, Y., and Emadi, A. (2020). Review of cabin thermal management for electrified passenger vehicles. *IEEE Transactions on Vehicular Technology* 69: 6025–6040. https://doi.org/10.1109/TVT.2020.2988468.

51 Zhang, Z., Wang, D., Zhang, C., and Chen, J. (2018). Electric vehicle range extension strategies based on improved AC system in cold climate – a review. *International Journal of Refrigeration* 88: 141–150. https://doi.org/10.1016/j.ijrefrig.2017.12.018.

52 Li, K., Luo, S., Fan, L. et al. (2021). Investigation on the performance and characteristics of a heat pump system for electric vehicles under extreme temperature conditions. *Case Studies in Thermal Engineering* 27: 101308. https://doi.org/10.1016/j.csite.2021.101308.

53 Fopah-Lele, A., Rohde, C., Neumann, K. et al. (2016). Lab-scale experiment of a closed thermochemical heat storage system including honeycomb heat exchanger. *Energy* 114: 225–238. https://doi.org/10.1016/J.ENERGY.2016.08.009.

54 Clark, R.J., Mehrabadi, A., and Farid, M. (2020). State of the art on salt hydrate thermochemical energy storage systems for use in building applications. *Journal of Energy Storage* 27: 101145. https://doi.org/10.1016/J.EST.2019.101145.

55 Fopah Lele, A. (2016). *A Thermochemical Heat Storage System for Households.* Cham: Springer International Publishing. https://doi.org/10.1007/978-3-319-41228-3.

56 Scapino, L., Zondag, H.A., Van Bael, J. et al. (2017). Sorption heat storage for long-term low-temperature applications: a review on the advancements at material and prototype scale. *Applied Energy* 190: 920–948. https://doi.org/10.1016/J.APENERGY.2016.12.148.

57 Zondag, H.A., Van Essen, V.M., Schuitema, R. et al. (2009). Engineering assessment of reactor designs for thermochemical storage of solar heat.

58 Malley-Ernewein, A. and Lorente, S. (2019). Constructal design of thermochemical energy storage. *International Journal of Heat and Mass Transfer* 130: 1299–1306. https://doi.org/10.1016/J.IJHEATMASSTRANSFER.2018.10.097.

59 Li, W., Klemeš, J.J., Wang, Q., and Zeng, M. (2022). Salt hydrate-based gas–solid thermochemical energy storage: current progress, challenges, and perspectives. *Renewable and Sustainable Energy Reviews* 154: 111846. https://doi.org/10.1016/J.RSER.2021.111846.

60 Fopah-Lele, A., Kuznik, F., Osterland, T., and Ruck, W.K.L. (2016). Thermal synthesis of a thermochemical heat storage with heat exchanger optimization. *Applied Thermal Engineering* 101: 669–677. https://doi.org/10.1016/J.APPLTHERMALENG.2015.12.103.

61 Richter, M., Bouché, M., and Linder, M. (2016). Heat transformation based on $CaCl_2/H_2O$–part A: closed operation principle. *Applied Thermal Engineering* 102: 615–621. https://doi.org/10.1016/J.APPLTHERMALENG.2016.03.076.
62 Zondag, H., Kikkert, B., Smeding, S. et al. (2013). Prototype thermochemical heat storage with open reactor system. *Applied Energy* 109: 360–365. https://doi.org/10.1016/J.APENERGY.2013.01.082.
63 Michel, B., Mazet, N., and Neveu, P. (2014). Experimental investigation of an innovative thermochemical process operating with a hydrate salt and moist air for thermal storage of solar energy: global performance. *Applied Energy* 129: 177–186. https://doi.org/10.1016/J.APENERGY.2014.04.073.
64 Michel, B., Mazet, N., and Neveu, P. (2016). Experimental investigation of an open thermochemical process operating with a hydrate salt for thermal storage of solar energy: local reactive bed evolution. *Applied Energy* 180: 234–244. https://doi.org/10.1016/J.APENERGY.2016.07.108.
65 Farcot, L., Le Pierrès, N., and Fourmigué, J.F. (2019). Experimental investigation of a moving-bed heat storage thermochemical reactor with $SrBr_2/H_2O$ couple. *Journal of Energy Storage* 26: 101009. https://doi.org/10.1016/J.EST.2019.101009.
66 Malleswararao, K., Aswin, N., Murthy S, S., and Dutta, P. (2022). Studies on long-term and buffer modes of operations of a thermal energy storage system using coupled metal hydrides. *Energy* 258: 124868. https://doi.org/10.1016/j.energy.2022.124868.
67 Malleswararao, K., Aswin, N., Murthy S, S., and Dutta, P. (2021). Studies on a dynamically coupled multifunctional metal hydride thermal battery. *Journal of Alloys and Compounds* 866: 158979. https://doi.org/10.1016/j.jallcom.2021.158979.

7

State-of-Charge Estimation of Thermal Energy Storage Units

Carlos E. Ugalde-Loo[1], Ivan De la Cruz-Loredo[1], Hector Bastida[1], Arslan Saleem[1], Daniel Morales[1], and Pranaynil Saikia[1]

[1] *School of Engineering, Cardiff University, Cardiff, Wales, UK*

7.1 Introduction

Mitigating climate change through a reduction of greenhouse gas (GHG) emissions is paramount not only to meet legally binding environment targets by 2050 but also to ensure catastrophic effects for nature and the population are not triggered. To prevent global average temperature rise beyond 2 °C compared to preindustrial levels, GHG emissions must peak by 2025 and then decline by around 43% by 2030 [1]. With estimates indicating that the energy sector is responsible for over two-thirds of the total global emissions, decarbonizing energy is thus crucial to meet these ambitious targets.

In June 2019, the UK became the first major economy in the world to pass laws to bring all GHG emissions to net zero by 2050 [2, 3], while the European Union ratified its commitment to fulfill this target through the European Green Deal [4]. Alongside these directives, substantial efforts have been made in developing and deploying innovative low-carbon technologies to adjust how energy is produced and consumed. At the same time, strategies to reduce energy consumption while ensuring energy demand is met are being actively investigated. These are facilitated by a holistic approach in which energy vectors are interfaced through coupling technologies to exploit the interdependencies between energy networks.

Although significant progress has been accomplished in decarbonizing the electricity sector [5] where the grid integration of renewable energy sources has catalyzed change, progress has been more restricted in other areas such as heating. Heating, together with space cooling, accounts for over one-third of the emissions

Advancements in Non-Conventional Cooling and Thermal Storage Strategies, First Edition.
Edited by Bidyut Baran Saha and Dibakar Rakshit.

in the UK [6]. Heating and cooling are essential to our daily lives for comfort, commonplace activities, to facilitate productive workplaces, and to perform several industrial processes. Decarbonizing heating and cooling is thus fundamental to reach net-zero targets.

Thermal energy storage (TES) units, also called thermal stores for brevity, can perform a crucial role in decarbonizing buildings and the industrial sector [7]. Making an analogy with the electric batteries employed in electricity networks, a thermal store acts essentially as a thermal battery in thermal networks [8]. Depending on the particular storage technology adopted, thermal energy may be stored for hours, days, weeks, or even months. TES units are thus ideal to facilitate the effective management and operation of thermal systems as they may act as a buffer between demand and supply. They can also provide flexibility, enabling peak heating and peak cooling demand to be shaved, electrical loads to be shifted, and electricity costs to be decreased [9, 10]. For instance, a thermal store for cooling systems may serve as an energy sink during cooler night hours, when the cost of electricity is reduced, and act as a cooling source during peak demand when the cost of electricity is high.

TES units can use a wide range of technologies to enable thermal energy to be stored and deployed when appropriate. This is done by heating or cooling the storage medium, which modifies its enthalpy [11]. Thermal stores relying on sensible heat, such as hot water tanks, are the simplest and are widely deployed in domestic buildings and large heating systems. The temperature of the storage medium is raised to store heat [12]. In contrast, latent heat TES units use a phase change material (PCM) as the medium to store thermal energy. Compared to sensible heat units, latent heat units store more thermal energy per unit of mass and volume of the storage medium. Thus, they have a higher energy density, making them attractive for applications where space is reduced [13].

Good performance of a TES unit in a thermal network can only be achieved if the charging and discharging operations of the thermal store are regulated effectively. This may require designing control schemes which consider the thermo-hydraulic dynamics of the thermal store. Managing the state of the store though is only possible if the state of charge (SoC) of the unit is known [14]. In practice, a direct assessment of SoC can be made by deploying internal sensors to track the temperature behavior of the storage medium. However, incorporating several sensors throughout the TES unit may lead to high instrumentation costs and demanding technical specifications. In contrast, a reduced number of sensors may lead to results with limited accuracy and large SoC estimation errors, leading, in turn, to a reduced performance of the thermal system supported by the thermal store [15].

To overcome the expensive implementation of internal temperature sensors, state observers (or state estimators) can be adopted. A state observer is a control structure borrowed from modern control theory. It essentially provides an estimation of the internal state of the system under study from measurements of its input and output [16]. In this way, an observer can be employed to estimate the temperature gradient of the storage medium in a thermal store using a limited amount of sensors. Crucially, only input and output temperatures and flow rates are required for this [17].

Following a brief overview of the different TES technologies, this chapter presents a methodology for the quantification of the SoC of thermal stores. Owing to their popularity in heating and cooling systems, the types of TES units under consideration are based on sensible heat and latent heat. The methodology consists of three main stages. Examples for three different storage configurations are discussed to demonstrate the use of the methodology. The chapter is closed with a discussion on the use of artificial intelligence (AI) to support the SoC estimation process alongside concluding remarks.

7.2 Overview of TES Technologies

TES technologies play a crucial role in enhancing energy utilization, managing intermittency, and promoting the integration of renewable energy sources [18]. Thermal stores help bridge the gap between the supply and demand of thermal energy. A typical TES system involves charging, storing, and discharging processes. Each process may occur multiple times during each storage cycle [19].

7.2.1 Classification of TES

This section provides an overview of various TES technologies, including sensible heat storage, latent heat storage, and thermochemical storage [20]. The key principles, operational mechanisms, advantages, and limitations of each technology are discussed. A general classification of the different technologies is shown in Figure 7.1.

7.2.2 Sensible Heat Storage

Sensible heat storage is a widely used, mature, and low-cost TES method that relies on the storage of thermal energy by heating or cooling a material without undergoing a phase change. Solid or liquid materials such as water, ceramics, natural stones, and polymers are commonly adopted as the storage medium [21]. The principle behind sensible heat storage is based on the specific heat capacity of these materials, which determines the amount of energy required to raise or lower their temperature. This concept is shown in Figure 7.2a.

When excess thermal energy is available, the storage medium is heated to store the energy. Conversely, when the stored energy is needed, the medium is cooled,

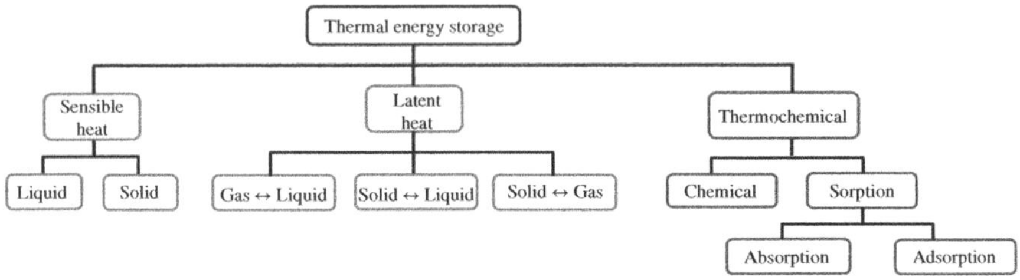

Figure 7.1 Classification of TES methods.

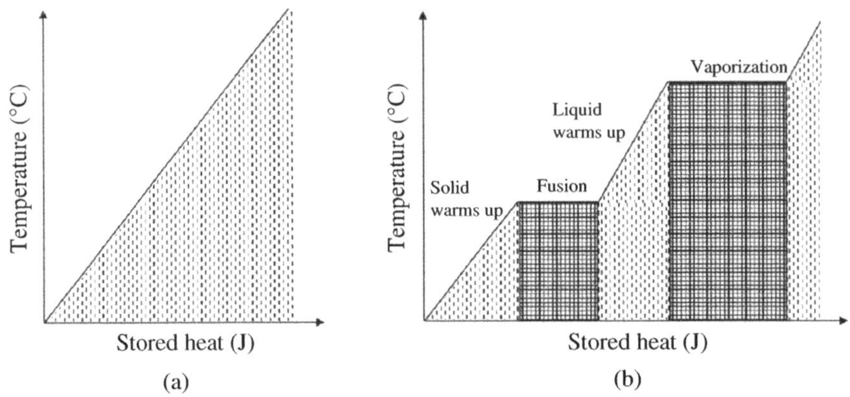

Figure 7.2 Energy stored in (a) sensible heat and (b) latent heat storage methods.

and the stored heat is released. The storage capacity ΔE_{st} (J) of a thermal store is given as

$$\Delta E_{st} = m_{st}\Delta h = m_{st}\int_{T_{min}}^{T_{max}} c dT \quad (7.1)$$

where m_{st} (kg) is the mass of the storage medium. The maximum change in the specific enthalpy of the system Δh (J/kg) is given by the definite integral with respect to a temperature differential dT (°C) of the specific heat capacity of the storage medium c (J/(kg °C)) evaluated in the range of the minimum and maximum allowed temperatures T_{min} (°C) and T_{max} (°C).

7.2.3 Latent Heat Storage

Latent heat storage is an important TES method involving a process requiring a change of phase. PCMs are used as the storage media due to their ability to store and release large amounts of energy during the phase transition between solid and liquid or liquid and gas. This energy is known as latent heat [22]. The principle behind latent heat storage is based on the fact that PCMs have a relatively constant temperature during phase change, enabling isothermal energy storage and release. It is thus possible to store large amounts of energy with only small temperature changes and achieve a high energy density.

When the PCM is exposed to a heat source and heated above its melting point, it absorbs thermal energy and changes its phase from solid to liquid, storing the energy (see Figure 7.2b). For a heating application, this is effectively the charging process. In contrast, in the discharging process, when the PCM is cooled below its freezing point using a heat sink, it releases the stored energy by solidifying and changing its phase from liquid to solid [23, 24]. Heat is transferred into and out of the thermal store through a heat transfer fluid (HTF).

Latent heat storage units primarily store energy as latent heat. However, some portion of energy is also stored as sensible heat during the temperature increase leading up to the phase change. Therefore, the energy stored by a material transitioning between two phases, namely, phases 1 and 2, is given by $\Delta E_{st} = m_{st}\Delta h$ in Eq. (7.1), with Δh expressed as [14]

$$\Delta h = \int_{T_{\min}}^{T_{\max}} c dT = \varepsilon \int_{T_{\min}}^{T_l} c_{p,p1} dT + \varepsilon \Delta h_l + (1-\varepsilon) \int_{T_l}^{T_{\max}} c_{p,p2} dT \tag{7.2}$$

where $c_{p,p1}$ (J/(kg °C)) and $c_{p,p2}$ (J/(kg °C)) are the sensible specific heat capacities of the material in each of the two phases, ε is the mass fraction of the phases (dimensionless), T_l (°C) is the temperature at which the phase change occurs, and Δh_l (J/kg) is the latent specific enthalpy change produced by the phase transition.

7.2.4 Thermochemical Heat Storage

Thermochemical storage is a unique TES method that utilizes chemical reactions to store and release thermal energy. It involves the reversible reaction between reactive materials, which can absorb and release large amounts of heat during the reaction process [25].

Thermochemical storage systems typically consist of two main components: the storage medium and the reactor. The storage medium undergoes a chemical transformation when exposed to heat, storing energy in the form of chemical bonds. The reactor facilitates the reaction and allows for the extraction of the stored energy when needed.

A graphical description of the processes involved in a thermochemical energy storage cycle is shown in Figure 7.3. Common examples of thermochemical reactions include hydration and dehydration of salts, carbonation and decarbonation reactions, and redox reactions.

Thermochemical storage technologies offer advantages compared to other methods. First, they provide high energy density, enabling the storage of large

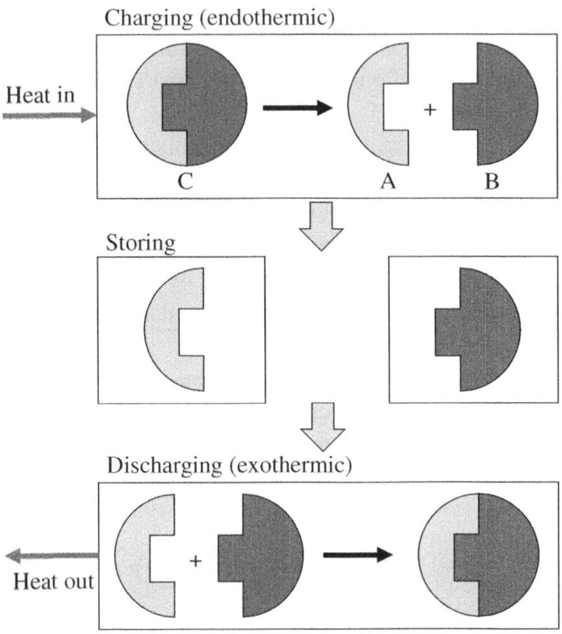

Figure 7.3 Processes involved in a thermochemical energy storage system. Source: Adapted from Refs. [26, 27].

amounts of thermal energy within a relatively small volume. This makes them suitable for applications where space is a constraint. Second, they have the potential for long-term storage over extended durations without significant degradation. Additionally, thermochemical storage systems can operate at high temperatures, making them suitable for applications requiring high-temperature heat sources. However, thermochemical storage also faces limitations. The design and operation of these systems can be complex due to the need for precise control of the reaction conditions, separation processes, and material regeneration. Moreover, the availability and cost of suitable reactive materials may pose challenges for widespread implementation [28].

7.2.5 Sensible-Latent Heat Storage

In addition to the individual technologies of sensible heat storage and latent heat storage, a combined approach known as sensible-latent heat TES has emerged as a promising solution. It employs a combination of materials exhibiting both sensible heat and latent heat storage characteristics. These materials, known as composite PCMs or hybrid materials, consist of a matrix providing sensible heat storage and embedded PCMs offering latent heat storage capabilities [29]. The matrix can be a solid or liquid material with a high specific heat capacity.

The principle behind sensible-latent heat TES systems lies in the sequential use of sensible heat storage followed by latent heat storage. During the charging phase, thermal energy is transferred to the system, initially increasing the temperature of the matrix material until it reaches its upper temperature limit. This is the sensible heat stage. Beyond this limit, the embedded PCMs start to undergo a phase change, absorbing latent heat and storing additional thermal energy. In the discharging phase, the process is reversed: The PCMs release their latent heat, followed by the matrix material releasing its sensible heat [30].

Sensible-latent heat TES units offer advantages. They allow for a higher overall energy storage capacity compared to using either sensible or latent heat storage on their own. In addition, they can enhance the thermal performance and efficiency of energy storage systems [31].

7.3 Methodology for State-of-Charge Estimation

The methodology for SoC estimation of a thermal store is divided into three main steps: The definition of a mathematical model of the TES unit, the design of a model-based state observer to estimate the internal temperatures of the storage medium, and the calculation of SoC using the estimated temperatures [32, 33]. An overview of these steps is briefly provided next.

Figure 7.4 illustrates the requirements for the first step of the methodology: Modeling the thermal store. Since the model must be suitable for designing state observers, a one-dimensional (1-D) representation of the unit is required [32, 33]. This is developed based on a detailed description of the internal structure of the thermal store and the application of the first law of thermodynamics.

A TES unit may be designed either as a closed system, where energy is transferred via an HTF flowing internally across the storage media, or an open system, allowing heat to be transferred by a direct mass transfer across the system boundaries (i.e. energy advection) [34]. To construct the model, the

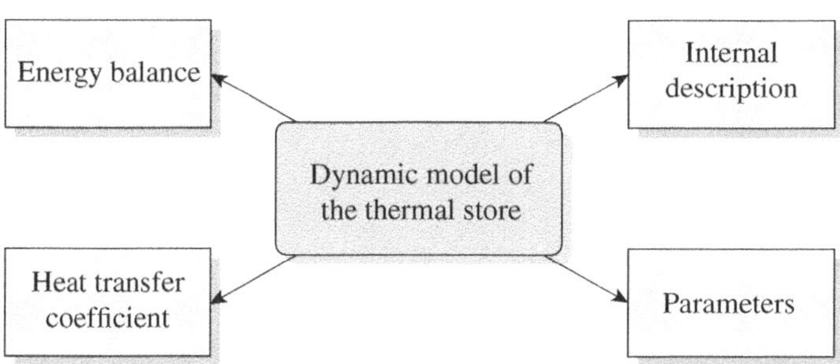

Figure 7.4 Block diagram showing the modeling requirements of the thermal store.

thermophysical properties of the HTF and storage medium are considered alongside other attributes such as the unit size, heat transfer area, and heat transfer coefficient.

Spatial discretization is an essential part of developing 1-D models of TES units. This process involves dividing the thermal store into discrete control volumes along its length to capture the spatial variations of temperature (i.e. thermal gradient) [32, 33]. With discretization, the continuous spatial domain is converted into a finite number of nodes or volumes.

Figure 7.5 illustrates the spatial discretization method for a latent heat storage system. The number of nodes, also defined as the spatial resolution, is directly related to the accuracy and computational efficiency of the model. Once the TES unit is discretized, the governing equations are applied to each control volume considering the interactions between neighboring segments. This allows for calculating heat transfer and energy storage within each segment and modeling the overall behavior of the TES unit over time.

The second step in the methodology is designing a state observer to provide an estimate of the temperature gradient within the thermal store. The design is based on the derived 1-D model and available real-time measurements of the input and output variables of the system. A block diagram illustrating the role of the observer is shown in Figure 7.6, where vector $\mathbf{u}$ contains the system inputs, vector $\mathbf{y}$ contains the system outputs, and vector $\hat{\mathbf{x}}$ contains the estimated internal temperature gradient of the TES unit. Measurements of the HTF's input and output temperatures and its mass flow rate are commonly available for both sensible and latent

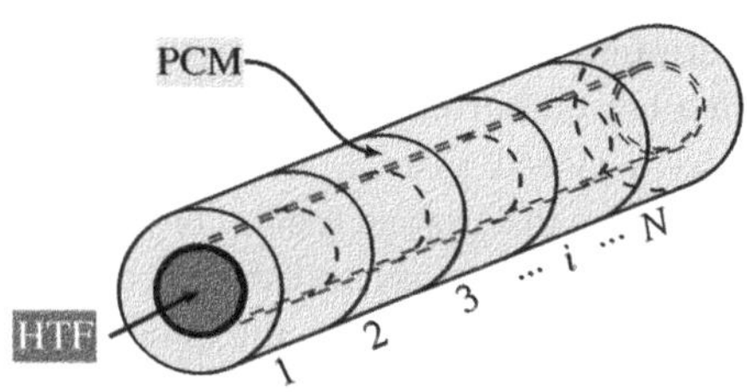

Figure 7.5 Discretization of a latent heat storage system consisting of a PCM surrounding an internal tube carrying an HTF.

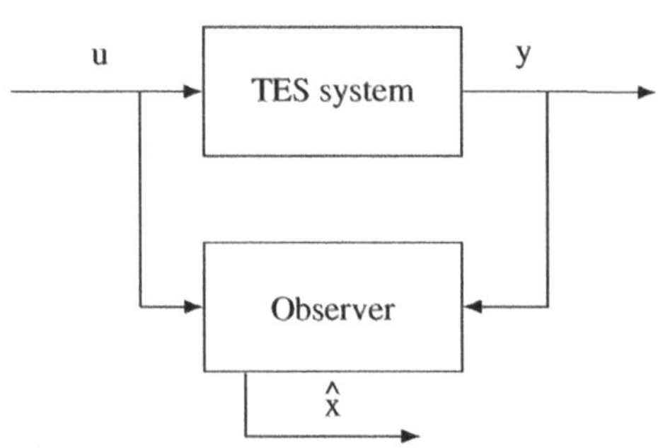

Figure 7.6 Generic block diagram of the state observer for a TES unit.

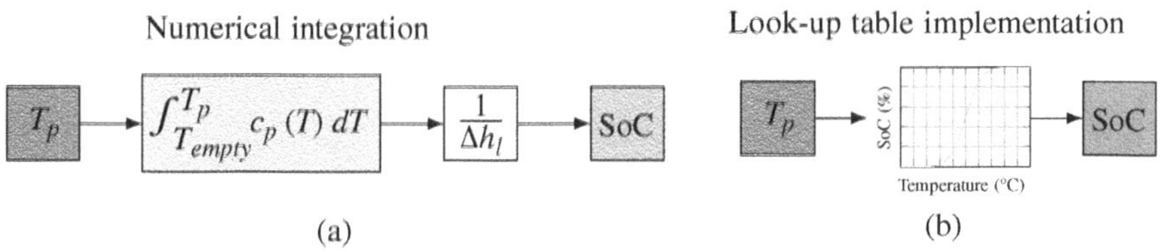

Figure 7.7 SoC calculation with (a) numerical integration and (b) a look-up table.

heat units [32, 33]. As no internal sensors are required, instrumentation costs may be significantly reduced.

The final step of the methodology, the SoC calculation, is illustrated in Figure 7.7a. This requires obtaining a specific heat curve of the storage medium as a function of temperature. This curve is normally provided by the manufacturer of a latent heat storage unit. By integrating this curve within a defined range of temperatures, the specific enthalpy change of the storage medium is obtained. This is then normalized using the maximum specific enthalpy change allowed to calculate the SoC. Alternatively, a look-up table may be adopted to simplify the calculation and reduce the computational cost of the implementation. This process is shown in Figure 7.7b.

Further details for each step in the methodology are provided in the following sections.

7.3.1 Mathematical Modeling of Thermal Stores

Figure 7.8 shows a TES unit modeled as a control volume. By applying the energy conservation law, the rate at which the internal energy of the control volume changes $\dot{E}_{st}$ (W) is defined by

$$\dot{E}_{st} = m_{st} c \dot{T}_{st} = \dot{Q}_{in} - \dot{Q}_{out} \tag{7.3}$$

where m_{st} (kg) is the total mass of the thermal store and c (J/(kg °C)) is the specific heat capacity of the storage medium at a temperature T_{st} (°C). This change in energy is given by the summation of heat flow rates going into and out of the control volume $\dot{Q}_{in}$ (W) and $\dot{Q}_{out}$ (W). In turn, $\dot{Q}_{in}$ and $\dot{Q}_{out}$ result from combining

Figure 7.8 TES system modeled as a control volume with heat transfer effects through open and closed boundaries.

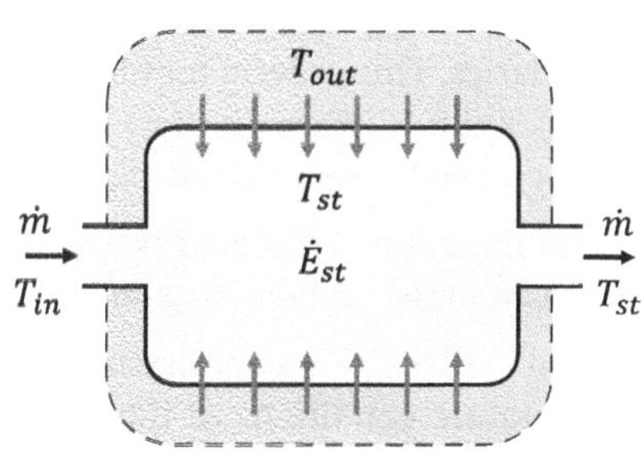

the heat transfer effects taking place in the system boundaries. These are modeled in a general form as

$$m_{st}c\dot{T}_{st} = \dot{m}c_p\left(T_{in} - T_{st}\right) + UA_t\left(T_{out} - T_{st}\right) \tag{7.4}$$

where $\dot{m}c_p\left(T_{in} - T_{st}\right)$ is the heat transferred into and out of the control volume through an open boundary and $UA_t(T_{out} - T_{st})$, considered for sensible heat units only, is the heat transferred through a closed boundary. These are functions of the mass flow rate $\dot{m}$ (kg/s) going in and out of the control volume, the sensible specific heat c_p (J/(kg °C)) of the storage medium, the temperature of the mass T_{in} (°C) entering the control volume through the open boundary, the temperature T_{out} (°C) on the outer side of a closed boundary, and the heat transfer area A_t (m^2) and heat transfer coefficient U (W/(m^2 °C)) through the closed boundary.

For a sensible heat storage unit such as a hot water tank, (7.4) is rewritten as

$$\dot{E}_{HWT} = \dot{m}_{HWT}c_{p,HWT}\left(T_{in} - T_{HWT}\right) + U_aA_w\left(T_a - T_{HWT}\right) \tag{7.5}$$

where $\dot{E}_{HWT}$ (W) is the rate of change in the tank's internal energy, $\dot{m}_{HWT}$ (kg/s) is the mass flow rate through the tank, $c_{p,HWT}$ (J/(kg °C)) is the sensible specific heat capacity of liquid water at a temperature T_{HWT} (°C), T_a (°C) is the ambient temperature outside the tank, U_a (J/(m^2 °C)) is the overall heat transfer coefficient between the ambient and the stored water in the tank, and A_w (m^2) is the heat transfer area of the tank's wall.

For a latent heat storage unit such as the one shown in Figure 7.5, (7.4) must be split into two equations to effectively represent the energy balance in the PCM and HTF,

$$\dot{E}_{PCM} = UA_t\left(T_{HTF} - T_{PCM}\right) \tag{7.6}$$

$$\dot{E}_{HTF} = \dot{m}_{HTF}c_{p,HTF}\left(T_{in} - T_{HTF}\right) + UA_t\left(T_{PCM} - T_{HTF}\right) \tag{7.7}$$

where $\dot{E}_{PCM}$ (W) is the rate of change in the PCM's internal energy, $\dot{E}_{HTF}$ (W) is the rate of change in the HTF's internal energy, T_{PCM} (°C) is the PCM's temperature, T_{HTF} (°C) is the HTF's temperature, U (J/(m^2 °C)) is the overall heat transfer coefficient between the PCM and HTF, and A_t (m^2) is the heat transfer area between the PCM and HTF. Given that the HTF circulates through the latent heat storage system carrying energy as sensible heat, the defined control volume has open boundaries. Thus, a term for the energy advection is included in (7.7), where $\dot{m}_{HTF}$ (kg/s) is the HTF's mass flow rate passing through and $c_{p,HTF}$ (J/(kg °C)) is the HTF's sensible specific heat capacity.

Note: Equation (7.6) could be extended to account for the heat transferred from the PCM to the ambient as done in (7.5) for the hot water tank.

U in (7.6)–(7.7) is a nonlinear parameter, which is calculated from the flow and temperature conditions of the HTF and PCM and the conductivity of the solid conduit or tube where the HTF flows. U is given as

$$U = \frac{1}{U_{cv}^{-1} + U_{cd,PCM}^{-1} + U_{cd,tube}^{-1}} \tag{7.8}$$

where U_{cv} (J/(m^2 °C)) is the convection heat transfer coefficient of the HTF, $U_{cd,PCM}$ (J/(m^2 °C)) is the conduction heat transfer coefficient of the PCM, and $U_{cd,tube}$ (J/(m^2 °C)) is the conduction heat transfer coefficient of the tube. The procedure to calculate U_{cv}, $U_{cd,PCM}$, and $U_{cd,tube}$ is described in detail in [35].

All the expressions previously discussed in this section consider a single control volume of the storage medium and HTF where an energy balance is applied. These expressions can be used for spatially discretized models with N number of nodes or volumes. Thus, the energy stored in a TES unit is given as a summation of the energy stored in each discrete volume as

$$E_{st} = \sum_{i=1}^{N} E_i = \sum_{i=1}^{N} m_i c_i T_i \tag{7.9}$$

where E_i (J) is the internal energy of the discrete volume i, m_i (kg) is its mass, c_i (J/(kg °C)) is its specific heat coefficient, and T_i (°C) is its temperature.

Using a state-space notation, the spatially discretized 1-D model of a TES unit is given as

$$\begin{aligned} \dot{\mathbf{x}} &= f(\mathbf{x}, \mathbf{u}) \\ \mathbf{y} &= j(\mathbf{x}) \end{aligned} \tag{7.10}$$

where $\dot{\mathbf{x}} = f(\mathbf{x}, \mathbf{u})$ is a set of nonlinear ordinary differential equations (ODEs), $\mathbf{x}$ is the state vector defined by the temperatures of each discrete volume, $\mathbf{u}$ is the input vector, and $\mathbf{y}$ is the output vector defined as a function of the model states $j(\mathbf{x})$. For instance, the state vector for a hot water tank model is given as

$$\mathbf{x} = \begin{bmatrix} T_1 & T_2 & \cdots & T_N \end{bmatrix}^{\mathrm{T}} \tag{7.11}$$

On the other hand, the state vector for the latent heat storage unit in Figure 7.5 is given as

$$\mathbf{x} = \begin{bmatrix} T_{HTF,1} & T_{PCM,1} & T_{HTF,2} & T_{PCM,2} & \cdots & T_{HTF,N} & T_{PCM,N} \end{bmatrix}^{\mathrm{T}} \tag{7.12}$$

7.3.2 State Estimation

Given that mathematical models of TES units are highly nonlinear, a nonlinear state observer is necessary for estimating the temperature gradient of the storage medium. The structure of a nonlinear observer is described by [36]

$$\dot{\hat{\mathbf{x}}} = f(\hat{\mathbf{x}}, \mathbf{u}) + \mathbf{L}\left(j(\mathbf{x}) - j(\hat{\mathbf{x}})\right) \tag{7.13}$$

where $\hat{\mathbf{x}}$ is the estimated state vector and $\mathbf{L}$ is a matrix with scalar entries. The estimation error, $\tilde{\mathbf{e}} = \mathbf{y} - \hat{\mathbf{y}}$, satisfies the equation of the error dynamics given as

$$\dot{\tilde{\mathbf{e}}} = f(\mathbf{x}, \mathbf{u}) - f(\hat{\mathbf{x}}, \mathbf{u}) - \mathbf{L}\left(j(\mathbf{x}) - j(\hat{\mathbf{x}})\right) \tag{7.14}$$

Linearizing (7.14) at $\tilde{\mathbf{e}} = 0$ results in

$$\dot{\tilde{\mathbf{e}}} = (\mathbf{A} - \mathbf{LC})\,\tilde{\mathbf{e}} \tag{7.15}$$

where $\mathbf{A}$ and $\mathbf{C}$ are defined as

$$\mathbf{A} = \frac{\partial f}{\partial x}(\mathbf{x}, \mathbf{u}) \tag{7.16}$$

$$\mathbf{C} = \frac{\partial j}{\partial x}(\mathbf{x}) \tag{7.17}$$

From (7.15), $\mathbf{L}$ must be designed such that $(\mathbf{A} - \mathbf{LC})$ is asymptotically stable [36]. Although this task is nontrivial as $\mathbf{L}$ should account for all the combinations of initial conditions and inputs, using positive entries in $\mathbf{L}$ is sufficient to guarantee stability [32, 33].

7.3.3 State-of-Charge Calculation

The expressions for the calculation of the storage capacity of a TES unit given by (7.1)–(7.2) consider an ideal transition between phases taking place at a temperature T_l. However, for many storage materials, this occurs within a specific temperature range. Therefore, the maximum change in the specific enthalpy Δh of the system is given as

$$\Delta h = \int_{T_{\min}}^{T_{\max}} c dT = \varepsilon \int_{T_{\min}}^{T_{l1}} c_{p,p1} dT + \int_{T_{l1}}^{T_{l2}} c_l(T)\, dT + (1 - \varepsilon) \int_{T_{l2}}^{T_{\max}} c_{p,p2} dT \tag{7.18}$$

where T_{l1} (°C) and T_{l2} (°C) define the temperature range where the storage medium changes phase and $c_l(T)$ (J/(kg °C)) is a function describing the behavior of the latent specific heat capacity during phase transition. A probability density function (PDF) may be adopted to represent $c_l(T)$ [14, 33].

The SoC of a TES unit storing heat is defined as

$$\mathrm{SoC}\left(T_{TES}\right) = \begin{cases} 100 & T_{TES} \geq T_{max} \\ \frac{\Delta h_{TES}}{\Delta h} \times 100 & T_{min} < T_{TES} < T_{max} \\ 0 & T_{TES} \leq T_{min} \end{cases} \tag{7.19}$$

where T_{TES} (°C) is the temperature of the storage medium. In (7.19), Δh is obtained using (7.18) directly and Δh_{TES} is calculated evaluating the integral in (7.18) from

T_{min} to T_{TES} (i.e. $\Delta h_{TES} = \int_{T_{min}}^{T_{TES}} c dT$). In contrast, the SoC for a TES unit for cooling is defined as

$$\text{SoC}\left(T_{TES}\right) = \begin{cases} 0 & T_{TES} \geq T_{max} \\ \left(1 - \frac{\Delta h_{TES}}{\Delta h}\right) \times 100 & T_{min} < T_{TES} < T_{max} \\ 100 & T_{TES} \leq T_{min} \end{cases} \tag{7.20}$$

where Δh_{TES} is calculated evaluating the integral from T_{TES} to T_{max} (i.e. $\Delta h_{TES} = \int_{T_{TES}}^{T_{max}} c dT$).

For a spatially discretized model of a thermal store, the previous expressions are employed to quantify the SoC of each discrete volume depending on its temperature. The SoC of the whole unit is then obtained as

$$\text{SoC} = \frac{\sum_{i=1}^{N} \text{SoC}_i}{N} \tag{7.21}$$

7.4 Examples

7.4.1 Sensible Heat Unit

Hot water tanks are extensively used in residential and commercial settings as sensible heat storage units to store and supply hot water. They are designed with inlet and outlet connections to facilitate the flow of water. Thus, a hot water tank has open system boundaries. To maintain a thermally efficient operation, hot water is supplied to the tank by the top inlet, while cold water is removed from the bottom during the charging process. Conversely, during discharging, cold water is supplied to the tank by the bottom inlet and hot water is removed from the top.

To exemplify the SoC estimation methodology for a hot water tank, this section presents the mathematical formulation of a model-based observer for a perfectly insulated vertical hot water cylinder [32]. For simplicity, this assumes there are no heat losses to the ambient.

7.4.1.1 Modeling

The dynamics of a hot water tank discretized into n discrete volumes are given as

$$\rho_1 c_{p,1} \frac{V_t}{n} \dot{T}_1 = \begin{cases} \dot{m} c_{p,1}\left(T_h - T_1\right), \forall\ \dot{m} \geq 0 \\ \dot{m} c_{p,1}\left(T_1 - T_2\right), \forall\ \dot{m} < 0 \end{cases} \tag{7.22}$$

$$\rho_i c_{p,i} \frac{V_t}{n} \dot{T}_i = \begin{cases} \dot{m} c_{p,i}\left(T_{i-1} - T_i\right), \forall\ \dot{m} \geq 0 \\ \dot{m} c_{p,i}\left(T_i - T_{i+1}\right), \forall\ \dot{m} < 0 \end{cases} \tag{7.23}$$

$$\rho_n c_{p,n} \frac{V_t}{n} \dot{T}_n = \begin{cases} \dot{m} c_{p,n}\left(T_{n-1} - T_n\right), \forall\ \dot{m} \geq 0 \\ \dot{m} c_{p,n}\left(T_n - T_c\right), \forall\ \dot{m} < 0 \end{cases} \tag{7.24}$$

where V_t (m^3) is the volume of the tank, ρ (kg/m^3) is the density of liquid water, c_p (J/(kg °C)) is the specific heat coefficient of water, T (°C) is the water temperature, subscripts "1" and "n" stand for the top and bottom discrete nodes of the model, and subscript "i" is used for all the nodes in between (i.e. $i = 2, 3, \ldots, n-1$). At the right-hand side of the equations, $\dot{m}$ (kg/s) is the mass flow rate throughout the tank, T_h (°C) is the hot water temperature supplied at the top of the tank, and T_c (°C) is the cold water temperature supplied at the bottom of the tank. These three variables represent the model inputs. The model is given as a set of piecewise ODEs that distinguish between charging ($\dot{m} \geq 0$) and discharging ($\dot{m} < 0$) processes.

Equations (7.22)–(7.24) are employed to model a 5-node ($n = 5$) hot water tank of a storage capacity of $V_t = 100$ m^3. The state derivatives of the system model are given as

$$\dot{\mathbf{x}} = \begin{bmatrix} \dot{T}_1 \\ \dot{T}_2 \\ \dot{T}_3 \\ \dot{T}_4 \\ \dot{T}_5 \end{bmatrix} = \begin{bmatrix} \dot{x}_1 \\ \dot{x}_2 \\ \dot{x}_3 \\ \dot{x}_4 \\ \dot{x}_5 \end{bmatrix} = f(\mathbf{x}, \mathbf{u}) = \begin{cases} \begin{bmatrix} \dot{m}\left(T_h - x_1\right) / \left(\rho_1 {}^{V_t}\!/_n\right) \\ \dot{m}\left(x_1 - x_2\right) / \left(\rho_2 {}^{V_t}\!/_n\right) \\ \vdots \\ \dot{m}\left(x_4 - x_5\right) / \left(\rho_5 {}^{V_t}\!/_n\right) \end{bmatrix} & \dot{m} \geq 0 \\ \begin{bmatrix} \dot{m}\left(x_1 - x_2\right) / \left(\rho_1 {}^{V_t}\!/_n\right) \\ \vdots \\ \dot{m}\left(x_4 - x_5\right) / \left(\rho_4 {}^{V_t}\!/_n\right) \\ \dot{m}\left(x_5 - T_c\right) / \left(\rho_5 {}^{V_t}\!/_n\right) \end{bmatrix} & \dot{m} < 0 \end{cases} \tag{7.25}$$

and the system outputs are defined as $\mathbf{y} = \begin{bmatrix} x_1 & x_2 & x_3 & x_4 & x_5 \end{bmatrix}^{\mathrm{T}}$.

7.4.1.2 Observer Design

The nonlinear observer is designed considering the measurements for the inputs $\dot{m}$, T_h, and T_c and two available outputs x_1 and x_5. It is defined according to (7.13) as

$$\dot{\hat{\mathbf{x}}} = \begin{bmatrix} \dot{\hat{T}}_1 \\ \dot{\hat{T}}_2 \\ \dot{\hat{T}}_3 \\ \dot{\hat{T}}_4 \\ \dot{\hat{T}}_5 \end{bmatrix} = \begin{bmatrix} \dot{\hat{x}}_1 \\ \dot{\hat{x}}_2 \\ \dot{\hat{x}}_3 \\ \dot{\hat{x}}_4 \\ \dot{\hat{x}}_5 \end{bmatrix} = f\left(\hat{\mathbf{x}}, \mathbf{u}\right) + \begin{bmatrix} l_{11} & l_{12} \\ l_{21} & l_{22} \\ l_{31} & l_{32} \\ l_{41} & l_{42} \\ l_{51} & l_{52} \end{bmatrix} \begin{bmatrix} x_1 - \hat{x}_1 \\ x_5 - \hat{x}_5 \end{bmatrix} \tag{7.26}$$

From the analysis carried out in [32], the observability of the intermediate states (x_2, x_3, and x_4) depends on $\dot{m}$. For example, for $\dot{m} = 0$, the states may only be known through direct measurement. Since the calculation of the intermediate states is limited by the estimated $\hat{x}_1$ and $\hat{x}_5$, these can be estimated by the system model only if l_{21} to l_{42} are set as 0. As $\hat{x}_1$ converges to x_1 and $\hat{x}_5$ to x_5, the same effect occurs with the intermediate states.

States x_1 and x_5 are directly measured, and their estimation relies on their own estimation error only. Thus, l_{12} and l_{51} were set as 0. On the other hand, the values of l_{11} and l_{52} were chosen as 0.1 to ensure a convergence time of around 20 seconds. This was deemed sufficient since the dynamics of a TES tank are slow.

7.4.1.3 SoC Calculation and Control

The SoC calculation employs the estimated temperatures from the observer. From (7.1), the storage capacity is defined using a minimum temperature of 60 °C of the stored water and a maximum temperature of 90 °C. The corresponding specific enthalpies for these temperatures are 251.18 and 377.4 kJ/kg. These enable defining a range for fully discharged (SoC = 0%) and fully charged (SoC = 100%) conditions.

A feedback control scheme was implemented, as shown in Figure 7.9. In the block diagram, $r_{\%}$ is a reference signal corresponding to the SoC previously defined with a value ranging between 0% and 100%. Look-up tables were employed to express $r_{\%}$ as a reference for specific enthalpy r_h and, in turn, as a temperature reference r_T suitable for the model of the hot water tank. The average value of the temperatures estimated by the nonlinear observer was adopted as the feedback signal of the control system, with a proportional controller $k_p = 3.5$.

7.4.1.4 Implementation, Simulations, and Results

The observer was verified through a model-in-the-loop (MiL) co-simulation using two software engines. A high-level schematic is shown in Figure 7.10a. The control scheme shown in Figure 7.9 and the nonlinear observer were implemented in MATLAB/Simulink, while the hot water tank was simulated in Apros using an available high-fidelity model. A screenshot of the hot water tank implementation is shown in Figure 7.10b. Several charging and discharging cycles were simulated, with simulation results provided in Figures 7.11 and 7.12.

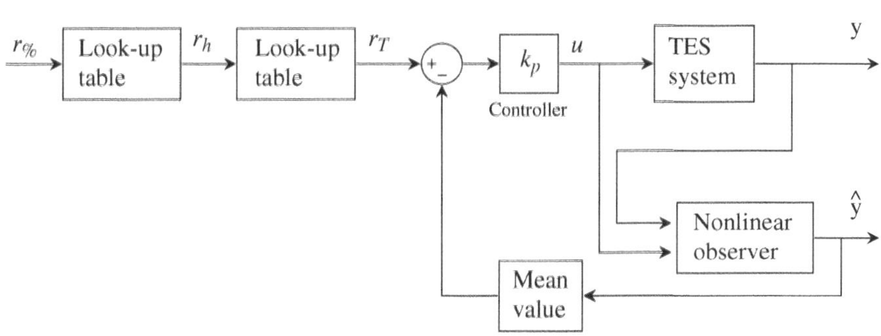

Figure 7.9 SoC control scheme of a hot water tank using a nonlinear state observer.

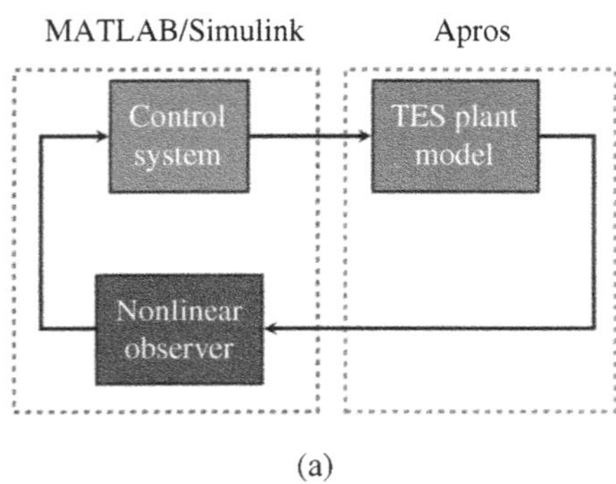

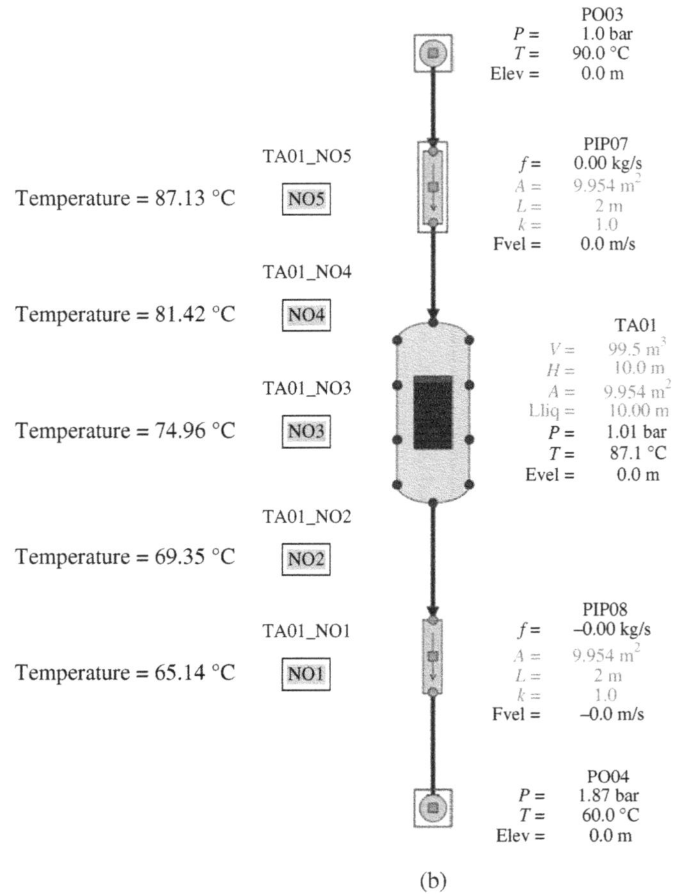

Figure 7.10 (a) MiL configuration for co-simulation. (b) Hot water tank model implemented in Apros.

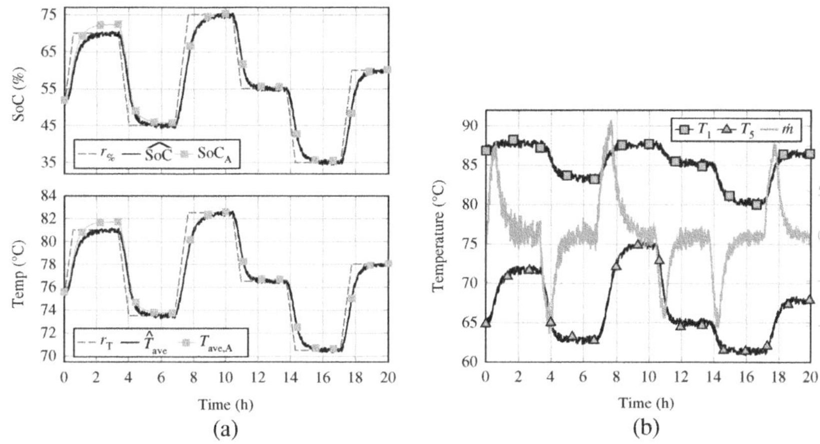

Figure 7.11 (a) SoC regulation for a hot water tank: SoC (top graph) and average temperature (bottom graph). Reference values are shown with a dashed gray trace, data from the hot water tank in Apros with gray trace with square markers, and values from the observer in MATLAB with solid black trace. (b) Measurements from sensors in the hot water tank: temperature of the top and bottom layers T_1 and T_5 (black trace with square markers and black trace with triangle markers) and mass flow rate $\dot{m}$ (gray trace).

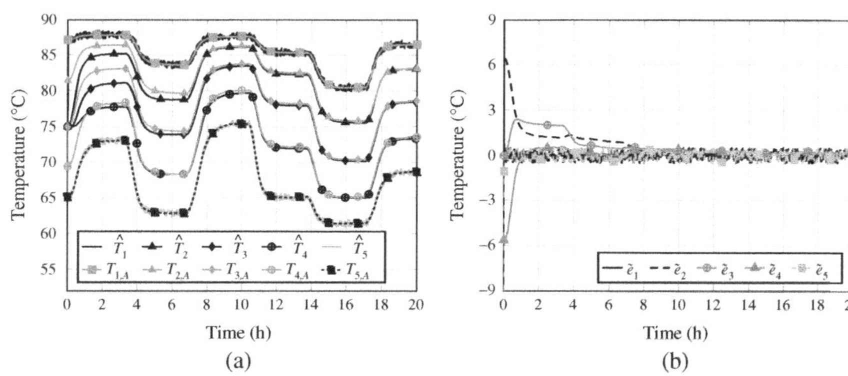

Figure 7.12 (a) Estimated temperature of the layers (MATLAB) compared with the actual temperatures (Apros) for the hot water tank. (b) Temperature estimation error in each discrete volume.

Figure 7.11a shows the performance of the SoC control scheme. Changes in SoC of the thermal store are observed within values of 35–75%. Although the estimated SoC in MATLAB ($\widehat{SoC}$, solid black trace) does not converge initially to the actual SoC obtained in Apros (SoC_A, gray trace with square markers), the tracking performance improves after around seven hours into the simulation. This, in turn, enables an adequate regulation of the system's SoC. Figure 7.11b shows the measured mass flow rate ($\dot{m}$, gray trace) and the top (T_1, black trace with square markers) and bottom (T_5, black trace with triangle markers) temperature measurements of the hot water tank. The model-based observer accurately estimates the temperature distribution throughout the tank's height (see Figure 7.12a) for SoC calculation, as shown in Figure 7.11a.

Figure 7.12a compares graphically the estimated temperatures afforded by the observer in MATLAB with the actual temperatures of the hot water tank in Apros, while Figure 7.12b shows the estimation error. Subscript "A" is used to define the temperatures obtained from the Apros model. The estimated temperatures $\hat{T}_1$ and $\hat{T}_5$ quickly converge to the actual values as a sensor is available for the top and bottom layers. However, temperature estimation for the intermediate states $\hat{T}_2$ to $\hat{T}_4$ is slower. The initial estimation error at the intermediate layers restricts the tracking performance at the beginning of the simulation, but the error becomes negligible aftér seven hours when convergence is achieved for all the estimated temperatures. This behavior is consistent with the results in Figure 7.11a.

7.4.2 Latent Heat Unit for Heating

Figure 7.13 shows the adopted latent heat thermal store for a heating application. This is in essence a shell and tube configuration with submerged tubes carrying a HTF. The shell is filled with the PCM Rigidex HD6070EA – a nonencapsulated high-density polyethylene [38]. This PCM features an enthalpy hysteresis, so the melting temperature for charging (120 °C) is different from that for discharging (135 °C). Marlotherm SH, a synthetic organic transfer medium, is used as the HTF flowing through 72 parallel tubes within the TES unit [33].

7.4.2.1 Modeling

The model was developed according to (7.6)–(7.7). However, an equal temperature distribution for all tubes throughout the thermal store was assumed to simplify the mathematical representation. Thus, the dynamic behavior for all tubes was represented by the energy balance formulation for one tube. The model derivatives

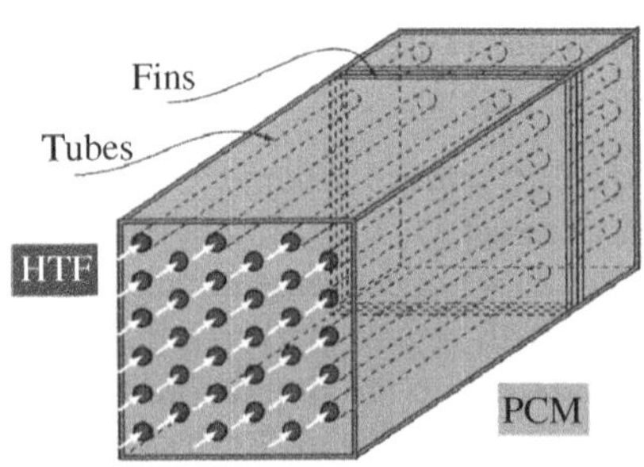

Figure 7.13 Schematic of the shell and tube thermal store [37].

are given as

$$\dot{\mathbf{x}} = \begin{bmatrix} \dot{T}_{f1} \\ \dot{T}_{p1} \\ \vdots \\ \dot{T}_{fi} \\ \dot{T}_{pi} \\ \vdots \\ \dot{T}_{fN} \\ \dot{T}_{pN} \end{bmatrix} = \begin{bmatrix} \dot{x}_1 \\ \dot{x}_2 \\ \vdots \\ \dot{x}_i \\ \dot{x}_{i+1} \\ \vdots \\ \dot{x}_{2N-1} \\ \dot{x}_{2N} \end{bmatrix} = f(\mathbf{x}, \mathbf{u}) = \begin{bmatrix} \frac{\dot{m}_f c_{p,f,1}(T_{f,in} - T_{f,1}) + U(A_{tr}/N)(T_{p,1} - T_{f,1})}{\rho_{f,1} c_{p,f,1}(V_f/N)} \\ \frac{U(A_{tr}/N)(T_{f,1} - T_{p,1})}{\rho_{p,1} c_{p,p,1}(V_p/N)} \\ \vdots \\ \frac{\dot{m}_f c_{p,f,i}(T_{f,i-1} - T_{f,i}) + U(A_{tr}/N)(T_{p,i} - T_{f,i})}{\rho_{f,i} c_{p,f,i}(V_f/N)} \\ \frac{U(A_{tr}/N)(T_{f,i} - T_{p,i})}{\rho_{p,i} c_{p,p,i}(V_p/N)} \\ \vdots \\ \frac{\dot{m}_f c_{p,f,N}(T_{f,N-1} - T_{f,N}) + U(A_{tr}/N)(T_{p,N} - T_{f,N})}{\rho_{f,N} c_{p,f,N}(V_f/N)} \\ \frac{U(A_{tr}/N)(T_{f,N} - T_{p,N})}{\rho_{p,N} c_{p,p,N}(V_p/N)} \end{bmatrix} \tag{7.27}$$

where subscripts "f" and "p" stand for HTF and PCM, while the outputs are given by $\mathbf{y} = \begin{bmatrix} T_{f1} & T_{p1} & T_{f2} & T_{p2} & \cdots & T_{fN} & T_{pN} \end{bmatrix}^{\mathrm{T}}$. The model was parametrized according to Table 7.1.

Table 7.1 Parameters of the latent heat storage unit for heating applications.

Parameters	Symbol	Value	Unit
Tube external radius	r_o	8.25	mm
Tube internal radius	r_i	6.75	mm
Tube hydraulic diameter	D_t	13.5	mm
Tube length	L	2.5	m
Heat transfer area	A_{tr}	0.1060	m^2
Cross-sectional area	A_c	1.4314×10^{-4}	m^2
HTF volume	V_f	3.5785×10^{-4}	m^3
Thermal conductivity of the tube (carbon alloy steel P235GH)	k_t	50	W/(m °C)
Radius of PCM	r_p	19.9774	mm
Cross-sectional area of PCM	A_p	0.00104	m^2
PCM volume	V_p	0.0026	m^3
Number of discrete volumes	N	20	—

Source: Adapted from Refs. [14, 37].

7.4.2.2 Observer Design

This was done considering the measurements of inputs $\dot{m}_f$ and $T_{f,in}$ and the output $T_{f,o} = T_{fN}$. The observer is mathematically expressed according to (7.13) as

$$\dot{\hat{\mathbf{x}}} = \begin{bmatrix} \dot{\hat{T}}_{f1} \\ \dot{\hat{T}}_{p1} \\ \vdots \\ \dot{\hat{T}}_{fi} \\ \dot{\hat{T}}_{pi} \\ \vdots \\ \dot{\hat{T}}_{fN} \\ \dot{\hat{T}}_{pN} \end{bmatrix} = \begin{bmatrix} \dot{\hat{x}}_1 \\ \dot{\hat{x}}_2 \\ \vdots \\ \dot{\hat{x}}_i \\ \dot{\hat{x}}_{i+1} \\ \vdots \\ \dot{\hat{x}}_{2N-1} \\ \dot{\hat{x}}_{2N} \end{bmatrix} = f\left(\hat{\mathbf{x}}, \mathbf{u}\right) + \begin{bmatrix} l_1 \\ l_2 \\ \vdots \\ l_i \\ l_{i+1} \\ \vdots \\ l_{2N-1} \\ l_{2N} \end{bmatrix} \left[T_{fN} - \hat{T}_{fN}\right] \tag{7.28}$$

All the values in $\mathbf{L} = [l_1, l_2, \cdots l_{2N}]^{\mathrm{T}}$ were heuristically defined as 0.1.

7.4.2.3 SoC Calculation

The SoC calculation approach is based on Eqs. (7.18)–(7.19). However, for simplicity, this example considers the latent heat storage capacity only and the sensible heat storage capacity is disregarded (i.e. $\Delta h = \Delta h_l = \int_{T_{l1}}^{T_{l2}} c_l(T)\,dT$). A Gumbel minimum distribution PDF was employed to model $c_l(T)$ as

$$c_l(T) = 1000\left[a_0 + a_1 T_p + b_1 \varphi^G\right] \tag{7.29}$$

$$\varphi^G(T, \eta, \theta) = (1/\theta)\exp\left(\frac{T-\eta}{\theta}\right)\exp\left[-\exp\left(\frac{T-\eta}{\theta}\right)\right] \tag{7.30}$$

where $a_0 = 0.8405$, $a_1 = 6.5655 \times 10^3$, $b_1 = 261.55$, $\eta = 126.5$, and $\theta = 9.3897$.

The PCM's specific heat curve as a function of temperature and SoC calculation via a look-up table are shown in Figure 7.14a. The specific enthalpy calculation is shown in Figure 7.14b.

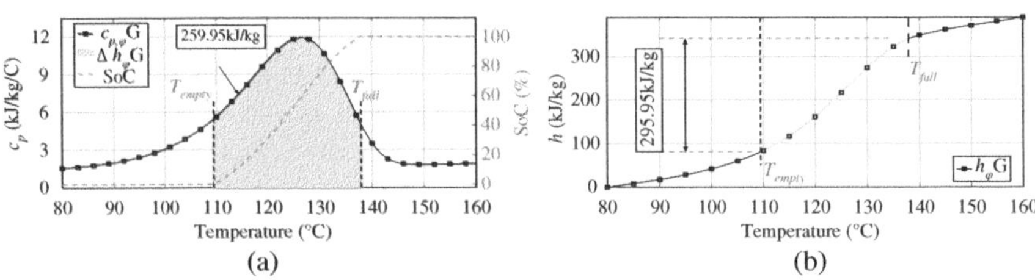

Figure 7.14 (a) PCM's specific heat-temperature curve for the latent heat thermal store for heating applications and SoC calculation via a look-up table. (b) Calculation of the specific enthalpy.

7.4.2.4 Implementation, Simulations, and Results

A series of charging and discharging cycles of the latent heat thermal store at a constant $\dot{m}_f = 1.09$ kg/s was simulated. A comparison of the simulated and experimental data from [14] was carried out for the HTF's output temperature, the total supplied energy by the thermal store, and its SoC. The initial temperature conditions for the model of the TES unit and the observer states were set to 80 and 50 °C.

An error analysis was carried out to numerically verify the simulation results with the experimental data. To this end, the root mean square error (RMSE) and mean absolute error (MAE) for each variable were calculated with respect to the experimental data.

Figure 7.15 shows the HTF's input and output temperature profiles $T_{f,in,exp}$ (°C) (dashed gray trace) and $T_{f,o,exp}$ (°C) (solid light gray trace with square markers) employed for the experiments, where "*exp*" refers to experimental data. $T_{f,in,exp}$ was fed to the model to obtain an HTF output temperature $T_{f,o,sim}$ through simulation (solid black trace with diamond markers, where "*sim*" refers to simulation results). The graph also includes the temperature of the last discrete volume estimated by the observer $T_{f,o,obs} = \hat{T}_{fN}$ (solid gray trace with diamond markers). The RMSE and MAE of $T_{f,o,sim}$ with respect to $T_{f,o,exp}$ are provided in Table 7.2. The results demonstrate the accuracy of the thermal store's model, with an MAE of 1.81 °C and an RMSE of 2.13 °C, which correspondingly represent 1.13% and 1.33% of the maximum allowable temperature (i.e. 160 °C). The observer can estimate $T_{f,o,sim}$ with an error $T_{f,o,sim} - T_{f,o,obs}$.

Figure 7.16 presents the estimation error for all 40 states, with Figure 7.16a depicting the errors in HTF temperatures ($\tilde{e}_i = T_{f,i} - \hat{T}_{f,i}$) and Figure 7.16b the errors in PCM temperatures ($\tilde{e}_i = T_{p,i} - \hat{T}_{p,i}$). The observer exhibits an effective performance, decreasing all the estimation errors to zero in less than 300 s – despite the 30 °C difference in the initial conditions of the thermal store

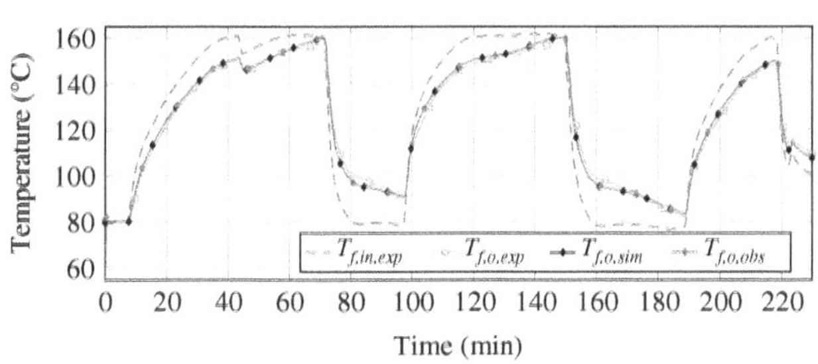

Figure 7.15 HTF's experimental input temperature ($T_{f,in,exp}$), and HTF's experimental ($T_{f,o,exp}$), simulated ($T_{f,o,sim}$), and estimated ($T_{f,o,obs}$) output temperature for the shell and tube thermal store.

Table 7.2 RMSE and MAE with respect to experimental data of the HTF's simulated and observed output temperature ($T_{f,o,sim}$, $T_{f,o,obs}$), SoC, and total energy (E_T).

	$T_{f,o,sim}$ (°C)	$T_{f,o,obs}$ (°C)	SoC (%)	E_T (kWh)
RMSE	2.13	2.21	4.58	0.57
MAE	1.81	1.84	3.61	0.49

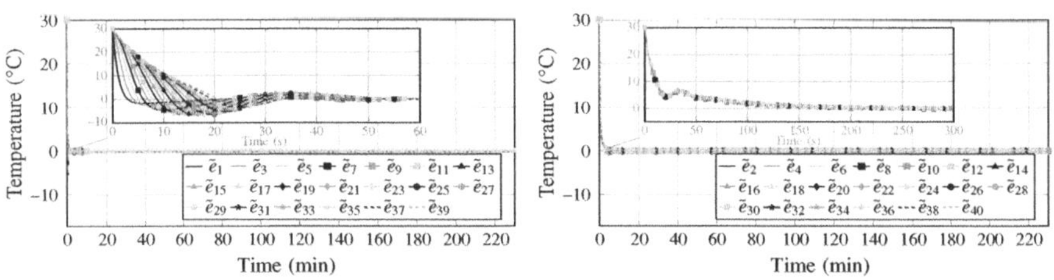

(a) (b)

Figure 7.16 Estimation errors for the shell and tube thermal store during a sequence of charging and discharging operations. Temperature plots for (a) HTF and (b) PCM.

model and its observer. The agreement of $T_{f,o,obs}$ (solid gray trace with diamond markers) with $T_{f,o,exp}$ (solid light gray trace with square markers) in Figure 7.15 is further evaluated by calculating the MAE and RMSE, yielding 1.84 and 2.21°C, which correspondingly represent 1.15% and 1.38% of the HTF's maximum allowable temperature.

The total stored energy E_T (kWh) and the SoC calculated during simulation were also compared against the experimental data available in [14]. The experimental and simulated total energies $E_{T,exp}$ (kWh) and $E_{T,sim}$ (kWh) are shown in Figure 7.17a. The MAE and RSME for the stored energy are 0.49 and 0.57 kWh, which represent 1.6% and 3% of the maximum value for this variable (18.5 kWh). These reduced errors demonstrate the high accuracy of the thermal store model. The contributions of sensible heat $E_{s,sim}$ (kWh) and latent heat $E_{l,sim}$ (kWh) to the simulated total energy $E_{T,sim}$ are also presented in Figure 7.17a.

As shown in Figure 7.17b, the experimental profile for SoC reported in [14] (SoC_{exp}, black trace with square markers) and the SoC obtained with the method presented here (SoC_{obs}, gray trace with triangle markers) are significantly close, exhibiting an MAE of 3.61% and an RMSE of 4.58%. The energy contribution in the form of latent heat ($E_{l,sim}$, dashed black trace) is also provided in Figure 7.17b for a graphical comparison. The good agreement between experimental data

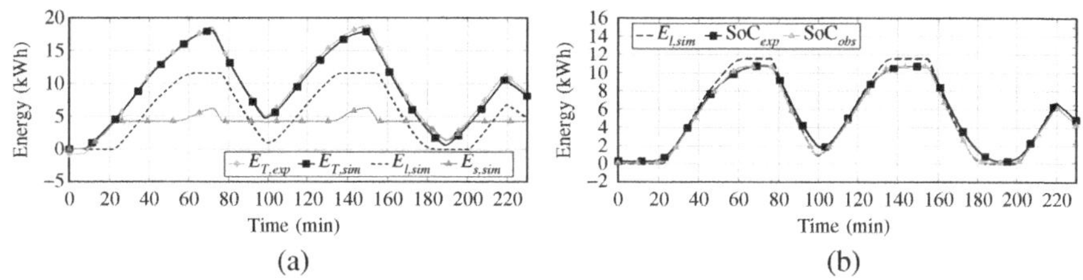

Figure 7.17 (a) Experimental and simulated total stored energy by the shell and tube thermal store, and energy contributed by latent heat and sensible heat. (b) Comparison of SoC and energy contributed by latent heat reported by [14] with the SoC obtained by the presented method. Source: Adapted from Barz et al. [14].

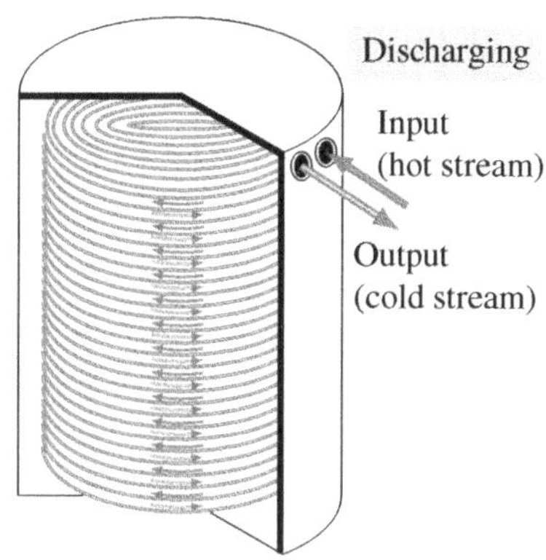

Figure 7.18 Schematic of the ice thermal store showing its internal geometry. Source: Bastida et al. [35]/John Wiley & Sons/CC BY 4.0.

and simulation results and the small errors instill confidence in the presented methodology for SoC calculation.

7.4.3 Latent Heat Unit for Cooling

The latent heat storage unit for cooling applications adopted here is the ICEBANK 1098 model manufactured by CALMAC (see Figure 7.18). The PCM is water, which is solidified into ice for storing cooling energy. The unit has an internal configuration with 34 pairs of spiraled polyethylene tubes. The HTF circulating through the tubes is a mix of glycol and water (34%) [33].

7.4.3.1 Modeling

As in the latent heat TES unit for heating, the model was simplified by considering an equal internal flow and an equal temperature distribution for every pair of tubes within the thermal store. This enabled representing the dynamic behavior of all the tubes with a single pair of tubes. The model derivatives are given by

$$
\dot{\mathbf{x}} = \begin{bmatrix} \dot{T}_{f,1,a} \\ \dot{T}_{w,1,a} \\ \dot{T}_{f,1,b} \\ \dot{T}_{w,1,b} \\ \vdots \\ \dot{T}_{f,i,a} \\ \dot{T}_{w,i,a} \\ \dot{T}_{f,i,b} \\ \dot{T}_{w,i,b} \\ \vdots \\ \dot{T}_{f,N,a} \\ \dot{T}_{w,N,a} \\ \dot{T}_{f,N,b} \\ \dot{T}_{w,N,b} \end{bmatrix} = \begin{bmatrix} \dot{x}_1 \\ \dot{x}_2 \\ \dot{x}_3 \\ \dot{x}_4 \\ \vdots \\ \dot{x}_i \\ \dot{x}_{i+1} \\ \dot{x}_{i+2} \\ \dot{x}_{i+3} \\ \vdots \\ \dot{x}_{4N-3} \\ \dot{x}_{4N-2} \\ \dot{x}_{4N-1} \\ \dot{x}_{4N} \end{bmatrix} = \begin{bmatrix}
\frac{\dot{m}_f c_{p,f,1,a}(T_{f,in}-T_{f,1,a})+U\left(\frac{A_{tr}}{N}\right)(T_{w,1,a}-T_{f,1,a})}{\rho_{f,1,a}c_{p,f,1,a}\left(\frac{V_f}{N}\right)} \\
\frac{U(A_{tr}/N)(T_{f,1,a}-T_{w,1,a})+U_w(A_e/N)(T_{w,1,b}-T_{w,1,a})+U_l(A_e/N)(0-T_{w,1,a})}{\rho_{w,1,a}c_{p,w,1,a}\left(\frac{V_w}{N}\right)} \\
\frac{\dot{m}_f c_{p,f,1,b}(T_{f,2,b}-T_{f,1,b})+U\left(\frac{A_{tr}}{N}\right)(T_{w,1,b}-T_{f,1,b})}{\rho_{f,1,b}c_{p,f,1,b}\left(\frac{V_f}{N}\right)} \\
\frac{U(A_{tr}/N)(T_{f,1,b}-T_{w,1,b})+U_w(A_e/N)(T_{w,1,a}-T_{w,1,b})+U_l(A_e/N)(0-T_{w,1,b})}{\rho_{w,1,b}c_{p,w,1,b}\left(\frac{V_w}{N}\right)} \\
\vdots \\
\frac{\dot{m}_f c_{p,f,i,a}(T_{f,i-1}-T_{f,i,a})+U\left(\frac{A_{tr}}{N}\right)(T_{w,i,a}-T_{f,i,a})}{\rho_{f,i,a}c_{p,f,i,a}\left(\frac{V_f}{N}\right)} \\
\frac{U(A_{tr}/N)(T_{f,i,a}-T_{w,i,a})+U_w(A_e/N)(T_{w,i,b}-T_{w,i,a})+U_l(A_e/N)(0-T_{w,i,a})}{\rho_{w,i,a}c_{p,w,i,a}\left(\frac{V_w}{N}\right)} \\
\frac{\dot{m}_f c_{p,f,i,b}(T_{f,i+1,b}-T_{f,i,b})+U\left(\frac{A_{tr}}{N}\right)(T_{w,i,b}-T_{f,i,b})}{\rho_{f,i,b}c_{p,f,i,b}\left(\frac{V_f}{N}\right)} \\
\frac{U(A_{tr}/N)(T_{f,i,b}-T_{w,i,b})+U_w(A_e/N)(T_{w,i,a}-T_{w,i,b})+U_l(A_e/N)(0-T_{w,i,b})}{\rho_{w,i,b}c_{p,w,i,b}\left(\frac{V_w}{N}\right)} \\
\vdots \\
\frac{\dot{m}_f c_{p,f,i,a}(T_{f,N-1}-T_{f,N,a})+U\left(\frac{A_{tr}}{N}\right)(T_{w,N,a}-T_{f,N,a})}{\rho_{f,N,a}c_{p,f,N,a}\left(\frac{V_f}{N}\right)} \\
\frac{U(A_{tr}/N)(T_{f,N,a}-T_{w,N,a})+U_w(A_e/N)(T_{w,N,b}-T_{w,N,a})+U_l(A_e/N)(0-T_{w,N,a})}{\rho_{w,N,a}c_{p,w,N,a}\left(\frac{V_w}{N}\right)} \\
\frac{\dot{m}_f c_{p,f,N,b}(T_{f,N+1,b}-T_{f,N,b})+U\left(\frac{A_{tr}}{N}\right)(T_{w,N,b}-T_{f,N,b})}{\rho_{f,N,b}c_{p,f,N,b}\left(\frac{V_f}{N}\right)} \\
\frac{U(A_{tr}/N)(T_{f,N,b}-T_{w,N,b})+U_w(A_e/N)(T_{w,N,a}-T_{w,N,b})+U_l(A_e/N)(0-T_{w,i,b})}{\rho_{w,N,b}c_{p,w,N,b}\left(\frac{V_w}{N}\right)}
\end{bmatrix} \tag{7.31}
$$

where subscripts "w" and "f" stand for PCM (water) and HTF and "a" and "b" stand for each individual tube in every tube pair in the tank. The outputs of the model are given as $\mathbf{y} = \left[T_{f,1,a}\ T_{w,1,a}\ T_{f,1,b}\ T_{w,1,b}\ \cdots\ T_{f,N,a}\ T_{w,N,a}\ T_{f,N,b}\ T_{w,N,b}\right]^{\mathrm{T}}$.

The model was parametrized according to Table 7.3.

7.4.3.2 Observer Design

The nonlinear observer was designed considering the measurements for the inputs $\dot{m}_f$ and $T_{f,in}$ and the outputs $T_{f,N,a}$ and $T_{f,N,b}$. It is defined according to (7.13) as

$$\dot{\mathbf{x}} = \begin{bmatrix} \dot{\hat{T}}_{f,1,a} \\ \dot{\hat{T}}_{w,1,a} \\ \dot{\hat{T}}_{f,1,b} \\ \dot{\hat{T}}_{w,1,b} \\ \vdots \\ \dot{\hat{T}}_{f,i,a} \\ \dot{\hat{T}}_{w,i,a} \\ \dot{\hat{T}}_{f,i,b} \\ \dot{\hat{T}}_{w,i,b} \\ \vdots \\ \dot{\hat{T}}_{f,N,a} \\ \dot{\hat{T}}_{w,N,a} \\ \dot{\hat{T}}_{f,N,b} \\ \dot{\hat{T}}_{w,N,b} \end{bmatrix} = \begin{bmatrix} \dot{\hat{x}}_1 \\ \dot{\hat{x}}_2 \\ \dot{\hat{x}}_3 \\ \dot{\hat{x}}_4 \\ \vdots \\ \dot{\hat{x}}_i \\ \dot{\hat{x}}_{i+1} \\ \dot{\hat{x}}_{i+2} \\ \dot{\hat{x}}_{i+3} \\ \vdots \\ \dot{\hat{x}}_{4N-3} \\ \dot{\hat{x}}_{4N-2} \\ \dot{\hat{x}}_{4N-1} \\ \dot{\hat{x}}_{4N} \end{bmatrix} = f\left(\hat{\mathbf{x}}, \mathbf{u}\right) + \begin{bmatrix} l_{1,1} & l_{1,2} \\ l_{2,1} & l_{2,2} \\ l_{3,1} & l_{3,2} \\ l_{4,1} & l_{4,2} \\ \vdots & \vdots \\ l_{i,1} & l_{i,2} \\ l_{i+1,1} & l_{i+1,2} \\ l_{i+2,1} & l_{i+2,2} \\ l_{i+3,1} & l_{i+3,2} \\ \vdots & \vdots \\ l_{4N-3,1} & l_{4N-3,2} \\ l_{4N-2,1} & l_{4N-2,2} \\ l_{4N-1,1} & l_{4N-1,2} \\ l_{4N,1} & l_{4N,2} \end{bmatrix} \begin{bmatrix} T_{f,N,a} - \hat{T}_{f,N,a} \\ T_{f,N,b} - \hat{T}_{f,N,b} \end{bmatrix} \tag{7.32}$$

All the values in **L** were heuristically defined as 0.5.

Table 7.3 Parameters of the latent heat storage unit for cooling applications.

Parameters	Symbol	Value	Unit
Tube external radius	r_o	7.9375	mm
Tube internal radius	r_i	6.35	mm
Tube hydraulic diameter	D_t	12.7	mm
Tube length	L	32.5581	m
Heat transfer area	A_{tr}	1.6238	m^2
Cross-sectional area	A_c	1.266×10^{-4}	m^2
HTF volume	V_f	0.0041	m^3
Thermal conductivity of the tube (polyethylene)	k_t	0.33	W/(m°C)
Water/ice thermal conductivity	k_w	1.35	W/(m°C)
Radius of PCM	r_w	23.9395	mm
Cross-sectional area of PCM	A_w	0.0016	m^2
Surface area	A_{ex}	4.9	m^2
PCM volume	V_w	0.0026	m^3
Number of discrete volumes	N	20	—

Source: Adapted from Bastida et al. [35].

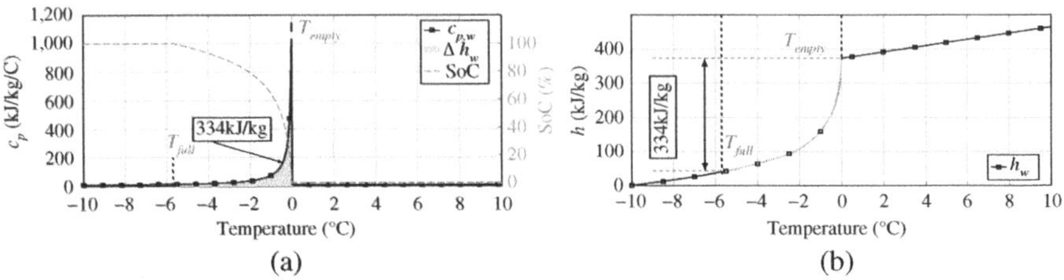

Figure 7.19 (a) PCM's specific heat-temperature curve (water) of the latent heat thermal store for cooling applications and SoC calculation. (b) Specific enthalpy.

7.4.3.3 SoC Calculation

Similar to the previous example, the SoC for the ice tank considers the latent heat storage capacity only and disregards the sensible heat storage capacity. A lognormal distribution PDF was employed to model $c_l(T)$ as

$$c_l(T) = 1000\left[a_0 + b_1\left(\varphi^L - a_1\right)\right] \tag{7.33}$$

$$\varphi^L(T,\gamma,\eta,\theta) = \begin{cases} \dfrac{\exp\left[-\left(\left(\ln[-T-\eta/\theta]\right)^2/(2\gamma^2)\right)\right]}{-(-T-\eta)\gamma\sqrt{2\pi}} & T < \eta \\ 0 & T \geq \eta \end{cases} \tag{7.34}$$

where $a_0 = 2.903$, $a_1 = 6.5655\times10^3$, $b_1 = 220.0744$, $\eta = 126$, $\theta = 4.7951$, and $\gamma = 2.118$.

The SoC calculation included a look-up table, as shown in Figure 7.7b. The PCM's specific heat-temperature curve (i.e. liquid water/ice) and SoC are shown in Figure 7.19a, while the specific enthalpy is shown in Figure 7.19b.

7.5 Implementation, Simulations, and Results

The effectiveness of the observer and SoC calculation were evaluated for two charging and two discharging cycles for varying mass flow rates ($\dot{m}_f$). Simulation results are provided in Figure 7.20. Large values of $\dot{m}_f = 22$ kg/s and then $\dot{m}_f = 19$ kg/s were adopted to achieve charging within a practical time duration (see Figure 7.20a). Figure 7.20b shows how the error between the HTF's output temperature ($T_{f,o}$, black trace with square markers) and the estimated value ($T_{f,o,obs}$, gray trace with triangle markers) converges to zero early in the first charging cycle. The reduced $\dot{m}_f$ for the second charging operation ($\dot{m}_f = 19$ kg/s), starting from ~24 hours into the simulation, moderately increases the time for fully charging the thermal store. Despite the sudden change in the HTF's input temperature when discharging (at ~17 and ~34 hours), the observer produces an accurate estimation of the HTF and PCM node temperatures during the whole simulation.

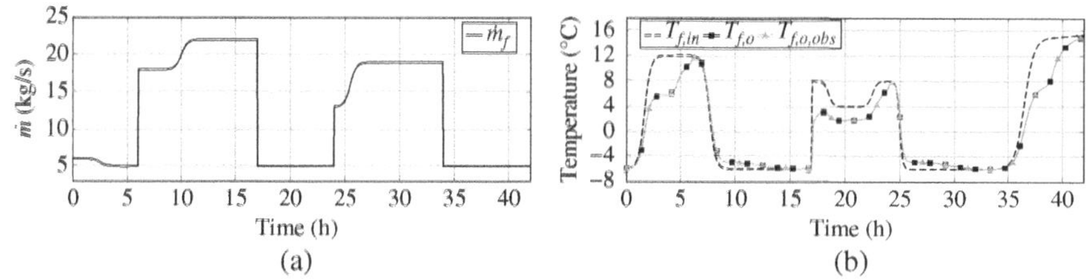

Figure 7.20 Results for the ice tank. (a) Mass flow rate. (b) HTF's input and output temperature given by the thermal store model, and HTF's estimated output temperature.

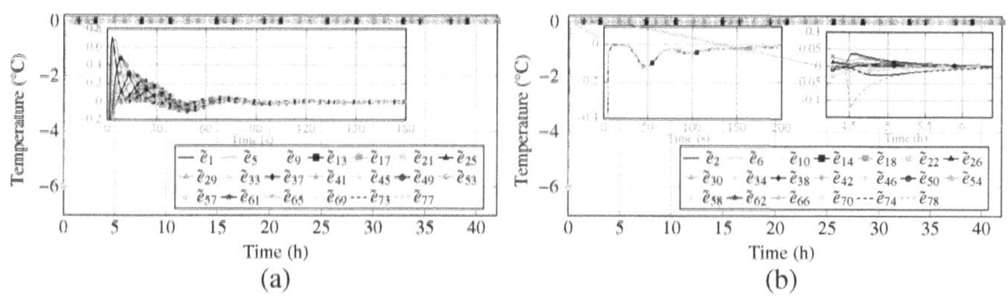

Figure 7.21 Estimation errors for the ice tank during a series of charging and discharging operations. Temperatures of the (a) HTF and (b) PCM.

Figure 7.21 depicts the estimation errors afforded by the observer. Its overall performance is satisfactory, as evidenced by the errors reducing to zero within 200 seconds. However, the errors for PCM temperature slightly increase at ~4.5 hours (see the zoomed-in enclosed graph within Figure 7.21b). These error increments take place when the PCM's phase transition has concluded and sensible heat is present only, producing a sharp temperature change. However, the observer reduces all errors to zero after ~1.5 hours. The maximum value of these errors is ~0.1 C – without causing major issues with SoC calculation.

The total energy $E_{T,sim}$ (solid gray trace with triangle markers), latent heat contribution $E_{l,sim}$ (dashed black trace), and sensible heat contribution $E_{s,sim}$ (solid black trace with square markers) are shown in Figure 7.22a. By the completion of each discharging operation, the total latent heat (delivered as cooling) is 310 kWh. This implies that the stored energy increases to release cooling. In contrast, the energy stored in the unit reduces during charging as the temperature of the PCM is decreased to produce ice. Thus, for comparison with the SoC curve, the energy contributed by latent heat is multiplied by −1. Figure 7.22b shows profiles of $E_{l,sim}$ (gray trace with triangle markers) and SoC (dashed black trace). Since the initial conditions of the observer were set as 0 °C, the SoC is initially 0% as well. However, once the estimation error converges to zero, the SoC and the latent heat profiles agree on well. This provides confidence in the presented method for SoC calculation.

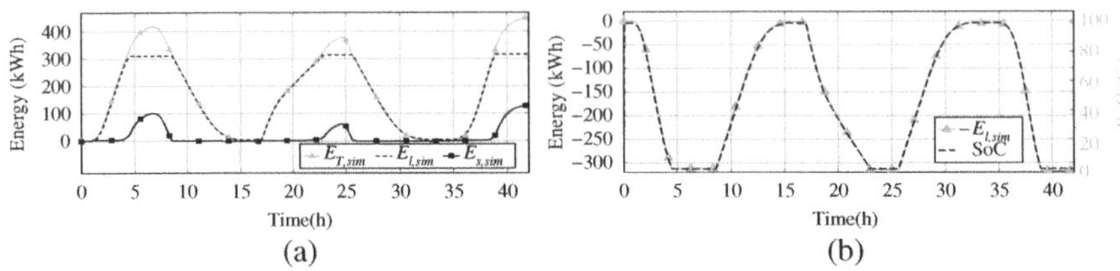

Figure 7.22 (a) Total stored energy by the thermal store and energy contributed by latent heat and sensible heat. (b) SoC calculation compared against latent heat energy.

7.6 Conclusions and Further Discussion

Thermal stores may play a crucial role in decarbonizing heat, cooling, buildings, and the industrial sector. They are ideal to facilitate the effective management and operation of thermal systems. They can also provide operational flexibility. For all this, the effective control of the charging and discharging operations of the TES unit is essential, which can be facilitated by monitoring the SoC of the unit.

This chapter presented an innovative methodology for calculating the SoC of thermal stores based on sensible heat or latent heat. The approach consists of three main steps and works well as long as a good model of the TES unit is available and the thermophysical properties of the storage medium (and the HTF for latent heat units) are known. The methodology was verified with three examples, where simulation results were compared with experimental data. A good SoC estimation was achieved in all cases.

The presented methodology has a number of advantages. It is based on simplified mathematical models of the thermal store. There is no need for costly sensors to measure internal temperatures as these are estimated. Only three measurements are required, which are the HTF's mass flow rate and input and output temperatures. Since a look-up table may be adopted for SoC calculation, computational burden is reduced.

The methods discussed in this chapter can underpin further advancements with the integration of AI. The SoC estimation is essentially a time-series mapping task between inputs (HTF's mass flow rate and temperature) and outputs (SoC of the thermal store). In recent years, recurrent neural networks with long short-term memory cells and nonlinear autoregressive networks with exogenous inputs have achieved great success in mapping time-series inputs and outputs that are interlinked with complex relationships. Interested readers are referred to [39, 40] for relevant examples. Additionally, these AI models have demonstrated, in diverse fields of applications, the capability of retaining high prediction accuracy while improving on the execution speed compared to physics-based iterative solvers such as the ones presented in this chapter.

Such capabilities could be translated to the task of SoC estimation if sufficient data on the variation of SoC with respect to the variations in the HTF's mass flow rate and temperature are available to train AI models. These data can in turn be sourced from high-fidelity physics-based models and experiments. The trained AI models may then offer reductions in the requirement of computational resources as well as hardware instrumentation.

Nomenclature

1–D	one–dimensional
AI	artificial intelligence
CFD	computational fluid dynamics
GHG	greenhouse gas
HTF	heat transfer fluid
MAE	mean absolute error
MiL	model–in–the–loop
ODE	ordinary differential equation
PCM	phase change material
PDF	probability density function
RMSE	root mean square error
SoC	state of charge
TES	thermal energy storage

References

1 United Nations Framework Convention on Climate Change (2016). Report of the Conference of the Parties on its 21st session, Addendum Part 2, p. 01195, January 2016.

2 Department for Business Energy & Industrial Strategy (2019). *UK Becomes First Major Economy to Pass Net Zero Emissions Law*, 2–3. Department for Business, Energy & Industrial Strategy.

3 Office of Gas and Electricity Markets (2020). *Ofgem Decarbonisation Action Plan*, 13. Ofgem.

4 Fetting, C. (2020). The European Green Deal. ESDN Report 53 (9).

5 Elliot, L., Walker, P., and Harvey, F. (2020). *Chancellor Set to Announce £3bn Green Investment Package*. The Guardian.

6 Department for Business Energy & Industrial Strategy (2018). Clean Growth – Transforming Heating: Overview of Current Evidence, December 2018.

7 BEIS (2016). *Evidence Gathering: Thermal Energy Storage (TES) Technologies*. Department for Business, Energy & Industrial Strategy.

8 Jiang, Z.Y., Qu, Z.G., Zhang, J.F., and Rao, Z.H. (2020). Rapid prediction method for thermal runaway propagation in battery pack based on lumped thermal resistance network and electric circuit analogy. *Applied Energy* 268: 115007.

9 Catapult (2020). *Storage & Flexibility Net Zero Series Thermal Energy Storage for Heat Networks*. Catapult.

10 Lepiksaar, K., Mašatin, V., Latõšov, E. et al. (2021). Improving CHP flexibility by integrating thermal energy storage and power-to-heat technologies into the energy system. *Smart Energy* 2: 100022.

11 Cabeza, L.F., Martorell, I., Miró, L. et al. (2015). *Introduction to Thermal Energy Storage (TES) Systems*. Woodhead Publishing Limited.

12 Kalaiselvam, S. and Parameshwaran, R. (2014). Sensible thermal energy storage. In: *Thermal Energy Storage Technologies for Sustainability* (ed. L.F. Cabeza), 65–81. Elsevier B.V.

13 Noël, J.A., Kahwaji, S., Desgrosseilliers, L. et al. (2022). *Phase Change Materials*. Elsevier Inc.

14 Barz, T., Seliger, D., Marx, K. et al. (2018). State and state of charge estimation for a latent heat storage. *Control Engineering Practice* 72: 151–166.

15 Beyne, W., Couvreur, K., T'Jollyn, I. et al. (2022). Estimating the state of charge in a latent thermal energy storage heat exchanger based on inlet/outlet and surface measurements. *Applied Thermal Engineering* 201 (PB): 117806.

16 Yuan, H., Dai, H., Wei, X., and Ming, P. (2020). Model-based observers for internal states estimation and control of proton exchange membrane fuel cell system: a review. *Journal of Power Sources* 468: 228376.

17 Dong, X., Griffo, A., Hewitt, D., and Wang, J. (2020). Reduced-order thermal observer for power modules temperature estimation. *IEEE Transactions on Industrial Electronics* 67 (12): 10085–10094.

18 Aneke, M. and Wang, M. (2016). Energy storage technologies and real life applications – a state of the art review. *Applied Energy* 179: 350–377.

19 Cabeza, L.F. and Palomba, V. (2022). *Introduction to Thermal Energy Storage and Technologies Definition*. Elsevier Inc.

20 Kousksou, T., Bruel, P., Jamil, A. et al. (2014). Energy storage: applications and challenges. *Solar Energy Materials and Solar Cells* 120 (Part A): 59–80.

21 Fernandez, A.I., Martnez, M., Segarra, M. et al. (2010). Selection of materials with potential in sensible thermal energy storage. *Solar Energy Materials and Solar Cells* 94 (10): 1723–1729.

22 Cabeza, L.F., Castell, A., Barreneche, C. et al. (2011). Materials used as PCM in thermal energy storage in buildings: a review. *Renewable and Sustainable Energy Reviews* 15 (3): 1675–1695.

23 Sharshir, S.W., Joseph, A., Elsharkawy, M. et al. (2023). Thermal energy storage using phase change materials in building applications: a review of the recent development. *Energy and Buildings* 285: 112908.
24 De Gracia, A. and Cabeza, L.F. (2015). Phase change materials and thermal energy storage for buildings. *Energy and Buildings* 103: 414–419.
25 Pardo, P., Deydier, A., Anxionnaz-Minvielle, Z. et al. (2014). A review on high temperature thermochemical heat energy storage. *Renewable and Sustainable Energy Reviews* 32: 591–610.
26 Abedin, A.H. and Rosen, M.A. (2011). A critical review of thermochemical energy storage systems. *The Open Renewable Energy Journal* 4 (1): 42–46.
27 Abedin, A.H. (2010). Thermochemical energy storage systems: modelling, analysis and design. MSc. Mechanical Engineering.
28 Agrafiotis, C., Roeb, M., Schmücker, M., and Sattler, C. (2014). Exploitation of thermochemical cycles based on solid oxide redox systems for thermochemical storage of solar heat. Part 1: testing of cobalt oxide-based powders. *Solar Energy* 102: 189–211.
29 Zhang, P., Xiao, X., and Ma, Z.W. (2016). A review of the composite phase change materials: fabrication, characterization, mathematical modeling and application to performance enhancement. *Applied Energy* 165: 472–510.
30 Wang, C., Lin, T., Li, N., and Zheng, H. (2016). Heat transfer enhancement of phase change composite material: copper foam/paraffin. *Renewable Energy* 96: 960–965.
31 Zauner, C., Hengstberger, F., Mörzinger, B. et al. (2017). Experimental characterization and simulation of a hybrid sensible-latent heat storage. *Applied Energy* 189: 506–519.
32 Morales Sandoval, D.A., De la Cruz-Loredo, I., Bastida, H. et al. (2021). Design and verification of an effective state-of-charge estimator for thermal energy storage. *IET Smart Grid* 4 (2): 202–214.
33 Bastida, H., De la Cruz-Loredo, I., and Ugalde-Loo, C.E. (2023). Effective estimation of the state-of-charge of latent heat thermal energy storage for heating and cooling systems using non-linear state observers. *Applied Energy* 331: 120448.
34 Incropera, F.P., DeWitt, D.P., Bergman, T.L., and Lavine, A.S. (1996). *Fundamentals of Heat and Mass Transfer*. New York: Wiley.
35 Bastida, H., Ugalde-Loo, C.E., Abeysekera, M., and Jenkins, N. (2022). Dynamic modelling of ice-based thermal energy storage for cooling applications. *IET Energy Systems Integration* 4 (3): 317–334.
36 Khalil, H.K. (2015). *Nonlinear Control*. New York: Pearson.
37 Bastida, H., De la Cruz-Loredo, I., and Ugalde-Loo, C.E. (2023). Effective one-dimensional dynamic modelling of latent heat thermal energy storage

units for heating applications. IET Energy Systems Integration, Early View. https://ietresearch.onlinelibrary.wiley.com/doi/full/10.1049/esi2.12128

38 Ineos (2022). Rigidex HD6070EA data sheet. https://www.ineos.com/showdocument/?grade=HD6070EA&bu=INEOS+O+%2526+P+Europe&documentType=TechnicalData+Sheet&docLanguage=EN&version=2566d5566193b465bb3681d8abe64fea (accessed 6 November 2022).

39 Bastida, H., De la Cruz-Loredo, I., Saikia, P., and Ugalde-Loo, C.E. (2024). Discrete-time state-of-charge estimator for latent heat thermal energy storage units based on a recurrent neural network. *Applied Energy* 371: 123526.

40 Saikia, P., Bastida, H., and Ugalde-Loo, C.E. (2024). An effective predictor of the dynamic operation of latent heat thermal energy storage units based on a non-linear autoregressive network with exogenous inputs. *Applied Energy* 360: 122697.

8

Beyond Conventional Cooling: Unveiling the Potential of Adsorption Cooling

A brief investigation of basic concepts, modeling, performance enhancements, applications, and future frontiers of adsorption cooling

Jorge Duarte Benther, Xiaolin Wang, and Evan Franklin

School of Engineering, University of Tasmania, Hobart, Tasmania 7005, Australia

8.1 Introduction

Cooling can be traced back centuries, as civilizations sought relief from the unpleasant heat [1–7]. From ancient methods of utilizing evaporative cooling to the advent of mechanical refrigeration systems, our ability to control temperature has advanced significantly. However, conventional cooling technologies, such as vapor compression systems, have inherent limitations, including reliance on ozone-depleting refrigerants and high energy consumption. Moreover, conventional cooling technologies depending on electricity consumption have contributed to escalating energy consumption and greenhouse gas emissions, exacerbating environmental challenges [8]. The search for sustainable and energy-efficient cooling technologies has intensified by rapid industrialization, urbanization, and climate change, where energy consumption and environmental impact are of critical concern, and exploring sustainable cooling alternatives becomes imperative. Therefore, the demand for efficient cooling and environmentally friendly systems has significantly increased [9, 10].

The increasing global demand for cooling and heating, driven by temperature changes, population growth, urbanization, and increasing living standards, has resulted in a substantial energy burden [11]. The residential sector represents a significant portion of global cooling energy demand, particularly in regions with hot climates. With the growing middle class and increasing access to cooling technologies, the demand for residential air-conditioning has been rising steadily across the globe. The commercial and industrial sectors also contribute notably to the global energy demand for cooling. Commercial buildings, shopping centers, hospitals, laboratories, data centers, and manufacturing facilities rely heavily on cooling systems to maintain comfortable conditions and preserve product quality.

Advancements in Non-Conventional Cooling and Thermal Storage Strategies, First Edition.
Edited by Bidyut Baran Saha and Dibakar Rakshit.

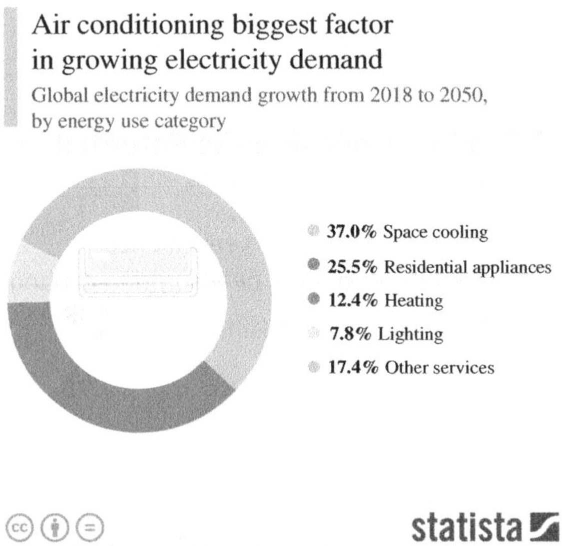

Figure 8.1 Worldwide electricity demand growth trend by energy use category (2018–2050) [13].

Statistical data from the International Energy Agency indicates a rapid global cooling energy demand increase [12] (see Figure 8.1). This data underscores the urgent need for energy-efficient cooling solutions. The cooling sector accounts for a substantial portion of global greenhouse gas emissions [14]. Adopting sustainable cooling technologies like adsorption cooling can contribute to reducing these emissions [15]. Equally important, integrating renewable energy sources, such as solar thermal, waste thermal, and geothermal energy, with cooling systems presents an opportunity to reduce reliance on fossil fuels and mitigate greenhouse gas emissions [14, 16, 17].

Cooling is vital in numerous applications. Research is in progress in low-grade thermal sources for cooling to address environmental concerns and energy costs [18]. The United States, China, Russia, Brazil, Indonesia, India, and European countries contribute to significant carbon dioxide (CO_2) emissions and energy consumption, with a substantial portion attributed to air-conditioning systems [19]. For instance, buildings account for a significant portion of energy consumption for cooling worldwide [20]. Moreover, renewable energy-driven air-conditioning systems have the potential to save energy in warm weather conditions. Hence, implementing low-grade thermal energy-driven cooling technologies can save fuel and reduce CO_2 emissions.

Opposed to vapor compression, cooling systems based on absorption, adsorption, and desiccant are available, with absorption being the most widely used [21]. However, adsorption refrigeration technology, driven by low-temperature heat sources, shows promise in areas with limited electricity access and aligns with solar power availability, promoting energy self-sufficiency and resilience [22]. It can effectively cool while reducing greenhouse gas emissions. The adsorption cooling versatility extends its applicability to various sectors, including residential and commercial air-conditioning, industrial process cooling, refrigerated transportation, data centers, and remote off-grid cooling applications. The ability to harness renewable energy sources further strengthens the potential of adsorption cooling in remote and decentralized cooling scenarios.

Adsorption cooling systems have gained significant attention as a sustainable and environmentally friendly alternative to conventional cooling technologies with industrial applications. These systems utilize the adsorption process, where a solid material, known as an adsorbent, is used to adsorb a refrigerant from a gaseous mixture. The heat required for desorption is then supplied, generating cooling effects [23]. This fundamental principle, combined with advancements in materials science and thermodynamics, led the way for modern adsorption cooling technology [24]. While adsorption cooling systems offer several advantages, including low energy consumption and the use of nontoxic refrigerants, their performance can be further enhanced through various techniques.

Adsorption cooling offers several advantages over conventional cooling technologies. It eliminates ozone-depleting refrigerants and reduces energy consumption from conventional sources, making it an environmentally friendly and energy-efficient option [16, 25]. Additionally, adsorption cooling systems can operate with low-grade waste heat or solar energy, making them suitable for remote or off-grid applications. This flexibility improves energy efficiency and contributes to reducing greenhouse gas emissions and dependence on conventional energy sources. Furthermore, adsorption cooling systems are characterized by simplicity, enhancing system reliability and reduced maintenance requirements.

Adsorption cooling represents a ground-breaking thermally driven cooling technology that harnesses the power of adsorption to achieve efficient and sustainable cooling. With its ability to utilize waste heat sources, low environmental impact, and reliable operation, adsorption cooling holds immense potential in various applications, including industrial processes, air-conditioning, and refrigeration, contributing to a more sustainable and energy-efficient future. Improving adsorption cooling technology involves reducing costs, enhancing performance, developing standardized systems, and raising awareness of its benefits. Enhancements in specific cooling power (SCP) and coefficient of performance (COP) are crucial for maximizing the potential of adsorption refrigeration systems [14, 26]. Continuous innovation can address the global demand for cooling while promoting sustainability and energy efficiency. Moreover, adsorbent performance improvements include cycle modifications, adsorbent materials, heat integration, and hybridization.

This chapter briefly examines adsorption cooling, encompassing its fundamentals, mathematical modeling, performance enhancements and optimization techniques, and applications. Discussions, including challenges and opportunities, are provided, closing with conclusions and potential future studies. Many publications on topics covered in this chapter were set together, allowing a quick and easy way to find specific literature on adsorption cooling.

8.2 Basic Concept

Adsorption cooling relies on the adsorption phenomenon. This process occurs due to attractive forces, such as *van der Waals* interactions or chemical bonding between the gas molecules and the adsorbent surface (the working pair adsorbent–refrigerant [27]). The adsorbent material is carefully selected for high affinity for a specific refrigerant. The adsorbate, or the gas being adsorbed, can be a refrigerant selected based on desirable adsorption characteristics. The refrigerant is adsorbed onto the surface of the adsorbent, effectively storing thermal energy. Adsorption cooling uses adsorption and desorption processes to achieve cooling effects. It relies on the reversible adsorption of a refrigerant onto a solid adsorbent material and subsequent desorption to release the stored heat. The cyclic process involves alternating adsorption and desorption stages to create a continuous cooling effect.

During the adsorption stage, a refrigerant vapor is introduced into an adsorption bed containing a porous adsorbent material. The vapor molecules adhere to the surface of the adsorbent, forming a thin film or monolayer, as shown in Figure 8.2a. This adsorption process is typically exothermic, resulting in heat release and a decrease in the temperature of the adsorption bed. The system enters

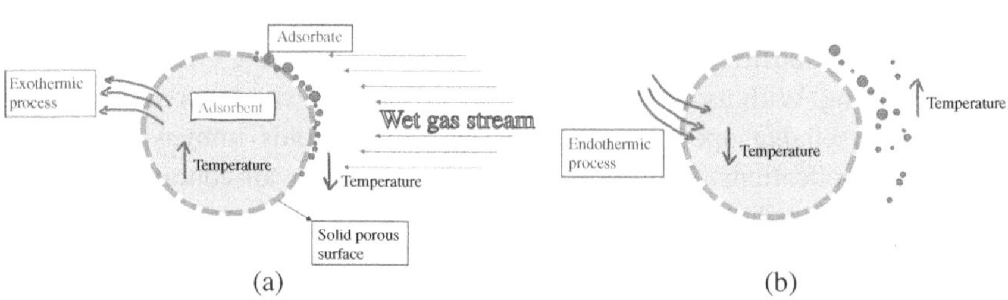

Figure 8.2 Working pair simplified processes: (a) adsorption and (b) desorption (or regeneration).

the desorption stage once the adsorption bed is saturated or reaches a desired capacity. Heat is applied to the adsorption bed, causing the adsorbed refrigerant molecules to desorb and return to the gas phase, as seen in Figure 8.2b. The desorbed refrigerant vapor is then condensed and collected for further cooling cycles.

The continuous cycling between adsorption and desorption stages allows for a sustained cooling effect, making adsorption cooling an attractive alternative to conventional cooling methods. Heat is supplied to the adsorbent material, often through a heat source like solar energy or waste heat from industrial processes. The heat causes the refrigerant (water, ethanol, ammonia, and others) molecules to evaporate from the liquid phase and adhere or adsorb onto the surface of the adsorbent material (e.g. silica gel, activated carbon, and zeolite, among others). This adsorption process is highly endothermic, absorbing heat from its surroundings. Once the adsorption process reaches its maximum capacity or is driven by a specific temperature limit, the system transitions to the desorption process. The heat supply is discontinued, and a heat sink, such as ambient air or water, is introduced to cool down the adsorbent material. As the adsorbent material cools, the refrigerant molecules desorb or release from the surface. This desorption process is exothermic, releasing heat into its surroundings. Consequently, the temperature of the refrigerant drops, and it condenses back into a liquid phase. The condensed refrigerant is then throttled (generally via an expansion valve) and passed through an evaporator, which absorbs heat from the surrounding environment, producing the desired cooling effect. The adsorption cycle steps are illustrated in Figure 8.3.

The heat exchange occurs during the transition between the adsorption and desorption processes in an adsorption cooling system. As the refrigerant molecules adsorb onto the adsorbent material, heat is transferred from the surrounding environment to the adsorbent material, resulting in cooling. Conversely, heat is transferred from the adsorbent material to the heat sink during desorption, leading to further system cooling. In this way, the adsorption cooling system acts as a heat exchanger. It receives heat from one medium (surrounding environment) during the adsorption process and transfers it to another medium (heat sink) during the desorption process. This transfer of heat allows for cooling effects to be achieved (see Figure 8.4).

During the adsorption stage, the adsorbate gas (or vapor) is introduced into the adsorption chamber, where it adheres to the surface of the adsorbent material, releasing heat and causing a reduction in the temperature of the surrounding environment. Subsequently, the desorption stage involves applying heat to the adsorbent, which causes the adsorbate to detach from the surface and become a vapor. The adsorbate vapor can then be condensed and recycled for further cooling cycles. This desorption process releases the stored thermal energy, causing the refrigerant to transition from a liquid or solid phase to a vapor phase. Once the refrigerant has

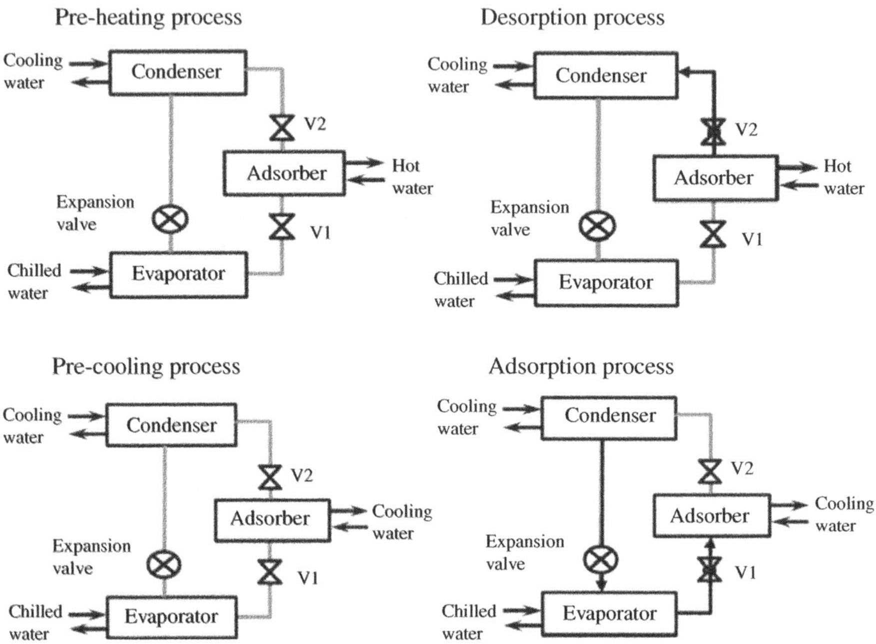

Figure 8.3 Adsorption cooling system's flow diagram for one cycle. Source: Ref. [28] / IntechOpen / CC BY 3.0.

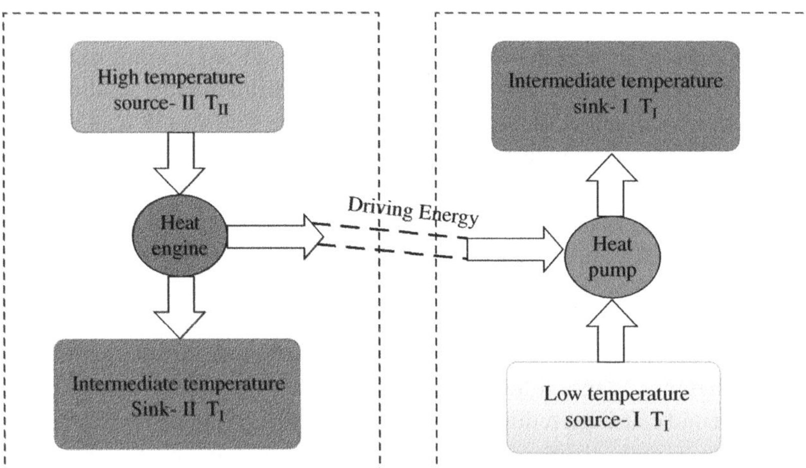

Figure 8.4 Thermodynamic representation of an ideal adsorption cooling cycle. Source: Ref. [28] / IntechOpen / CC BY 3.0.

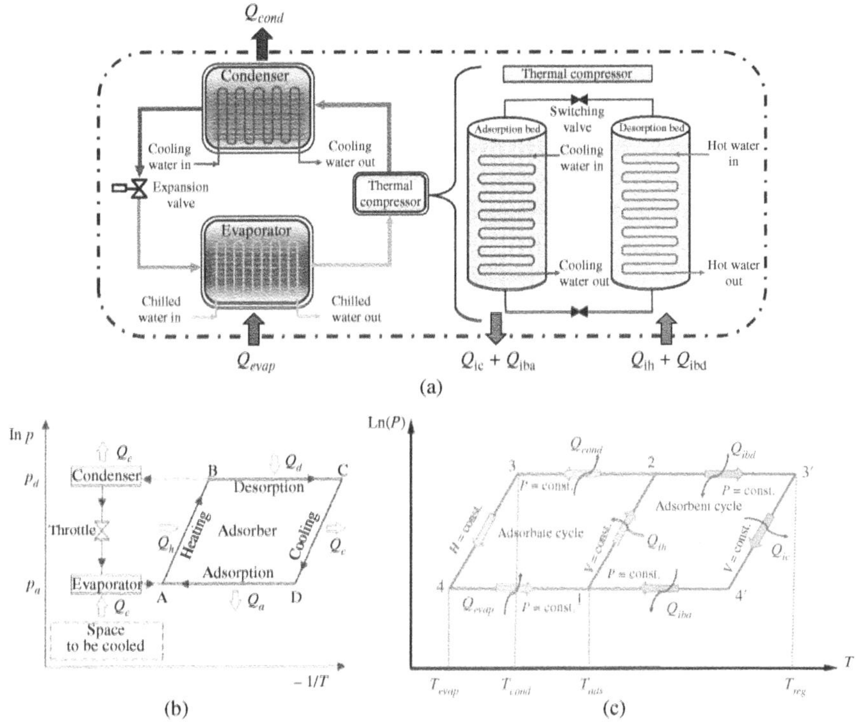

Figure 8.5 (a) Schematic of an adsorption cooling system. Source: Ref. [29] / with permission of Elsevier, and working principle of physisorption refrigeration cycle (*Clapeyron* diagrams): (b) simplified cycle. Source: Ref. [30] / IAEME Publication and (c) adsorbent–adsorbate cycles. Source: Adapted from Refs. [17, 31].

been desorbed, it is then transported to a condenser, where it undergoes a phase change, releasing the absorbed heat to the surroundings.

From a thermodynamic cycle perspective, the *Clapeyron* diagram (ln P versus $-T^{-1}$ or T, logarithm of pressure versus temperature) of the working pair represents the ideal thermodynamics occurring in the adsorption/desorption and evaporator/condenser processes, as seen in Figure 8.5a and b. In more detail [31, 32] for Figures 8.3 and 8.5a, c, the adsorbent takes in heat Q_{ih} from an external heat source through an isosteric process known as isosteric heating in step 1–2.

During this step, both the adsorbent and adsorbate are heated, leading to an increase in the pressure and temperature of the adsorbent bed. This heating process continues until the pressure in the adsorber bed matches the pressure in the condenser. At this point, the entrance valve of the condenser is opened. Moving on to steps 2–3′, the external heat source continues to provide heat to the adsorbent (Q_{lbd}) during an isobaric desorption process, causing the adsorbate to be released from the adsorbent. Simultaneously, the condenser cools the adsorbate through an isobaric cooling process (steps 2–3). In this stage, the gaseous adsorbate loses heat (Q_{cond}) and condenses into a liquid state. The adsorbent is heated until it reaches point 3′, which is the maximum temperature of the cycle. At this point, the valve connecting the adsorber bed and the condenser is closed. Then, during an isosteric cooling process in steps 3′–4′, the adsorbent releases its heat (Q_{ic}) as it encounters a heat sink. During steps 3–4, the adsorbate passes through the expansion valve and, in a constant enthalpy process, expands into a saturated mixture of vapor and liquid. In steps 4–1, the adsorbate absorbs heat Q_{evap} from the surroundings, converting it into vapor. Simultaneously, the entrance valve to the adsorber bed is opened, and the adsorbent adsorbs the vaporized adsorbate during isobaric adsorption (step 4′–1), releasing its own heat (Q_{iba}).

The COP is an important metric for the adsorption cooling system's efficiency, representing the cooling capacity (CC) ratio to the input heat or energy. The definition of COP can vary but is commonly described as in Eq. (8.1) for an adsorption refrigeration chiller [28]. The adsorption capacity of a chosen working pair can be expressed in terms of the specific cooling effect (SCE) and the SCP as in Eq. (8.2) and Eq. (8.3) [21, 28, 33] with Q_{evap} (CC or evaporator cooling output), Q_{heat} or Q_{regen} (added/supplied heat or regeneration heat), M_s, and t_{cyc} being the heat from evaporator, the heat from the heat source (sensible and latent), the total amount of adsorbent packed in the adsorbers, and the shorter cycle time, respectively. Also, the system's heat and mass transfer rates, including adsorption and desorption rates, are essential factors to consider. Furthermore, thermodynamic analyses involving pressure–temperature diagrams and adsorption isotherms provide valuable insights into the system's operational characteristics and efficiency (see Section 8.3),

$$COP = \frac{Q_{evap}}{Q_{heat}} = \frac{Q_{evap}}{Q_{regen}} \tag{8.1}$$

$$SCE = \frac{Q_{evap}}{M_s} \tag{8.2}$$

$$SCP = \frac{SCE}{t_{cyc}} = \frac{Q_{evap}}{M_s\, t_{cyc}} \tag{8.3}$$

Adsorption cooling as a cycle offers a unique and efficient way to achieve cooling effects by leveraging an adsorbent material's adsorption and desorption processes. However, selecting the optimal configuration for adsorption cooling depends on factors such as cooling requirements, available space, energy source, and system complexity [14]. The one-bed, two-bed, multi-bed, and multistage configurations are used for adsorption cooling applications.

8.2.1 One-bed Configuration

The one-bed system is the most straightforward configuration in adsorption cooling, where a single adsorption bed is utilized, including patented systems [25, 34–42]. In this scenario, the adsorption and desorption processes occur sequentially in the same bed. During the adsorption stage, the refrigerant vapor is introduced into the adsorption bed, and cooling is achieved through heat release during adsorption. Once the adsorption bed is saturated, the desorption stage begins by applying heat, causing the refrigerant to desorb and release as vapor.

A one-bed adsorption refrigeration configuration offers advantages in its simple and compact structure and low initial cost. It finds practical applications where continuous cooling or higher CC is not required, such as cooling beverages or storing vaccines, as in the early 1990s by Critoph et al. [37], where he developed a solar-powered refrigerator endorsed by the United Nations for vaccine storage in impoverished regions.

A one-bed layout commonly consists of a refrigeration chamber, an evaporator, and an adsorbent bed, as depicted in Figures 8.6 and 8.7. A thermally conductive element encompasses the refrigeration chamber, while the evaporator is situated on the outer side surrounding the periphery of the refrigeration chamber.

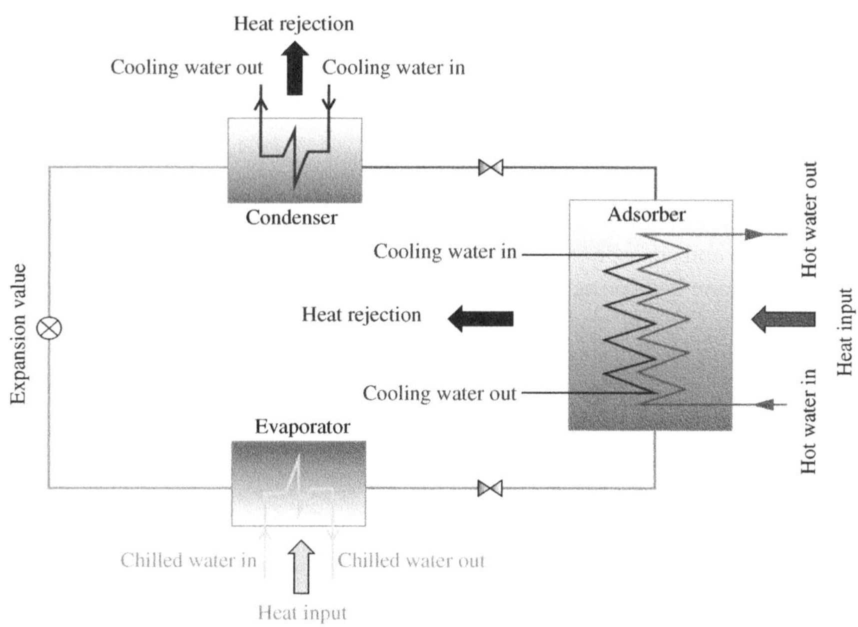

Figure 8.6 Example of a one-bed schematic diagram system during adsorption. Source: Ref. [43] / with permission of Elsevier.

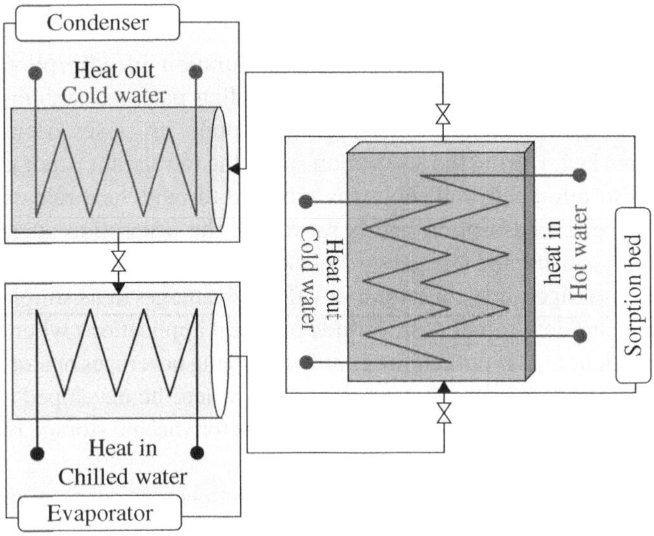

Figure 8.7 Example of a one-bed schematic diagram sorption system [44]. Source: Adapted from Ref. [45].

By utilizing the thermally conductive member to transfer cold generated in the evaporator to the refrigeration chamber, the evaporator cooling range is significantly expanded, enabling rapid initial cooling of the refrigeration chamber, or, as depicted in Figure 8.7, the containers are interconnected via a line containing a blocking valve. The adsorber and evaporator possess at least one orifice and can be selectively connected to an overpressure generator and a condenser. Individual components or component groups are selectively separated and connected for regeneration and adsorption cycles in a pressure-tight or vacuum-tight manner.

While the one-bed configuration is easy to implement, it needs to be improved regarding the CC and cycle duration (interrupted by switching from adsorption/desorption and vice versa). The cooling output is limited by the size and capacity of the single adsorption bed, and the desorption process may require a substantial amount of heat, which affects system efficiency.

8.2.2 Two-bed Configuration

The two-bed configuration was introduced to overcome the limitations of the one-bed system, and it employs two adsorption beds that operate in parallel, allowing for continuous cooling without interruption [46–60]. It is the most used configuration. While one bed undergoes the adsorption process, the other bed undergoes the desorption process, and the roles are switched periodically. The

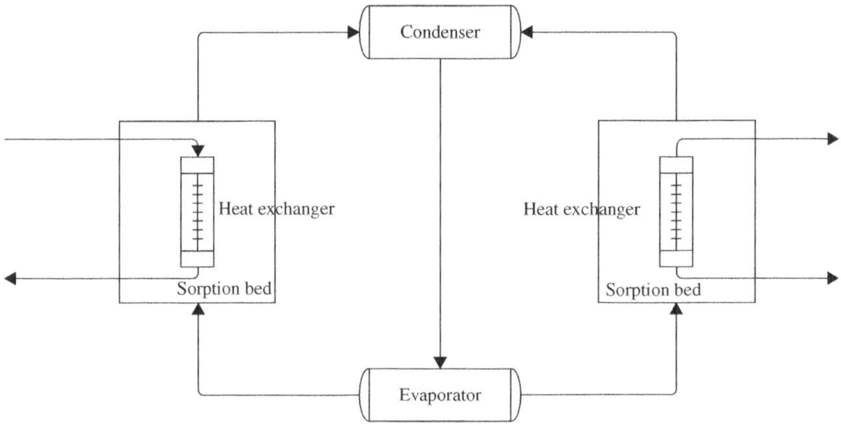

Figure 8.8 Simple two-bed adsorption cooling illustration. Source: Ref. [61] / with permission of Elsevier.

two-bed system enables a higher CC and longer cycle duration than the one-bed configuration. By distributing the adsorption and desorption processes between two beds, the cooling output can be increased, and the system efficiency can be improved.

The two-bed configuration of adsorption cooling has two separate adsorbent beds, as seen in the simplified version of Figure 8.8. Valves regulate the connections between the adsorption beds, the condenser, and the evaporator. During the adsorption phase of one bed, its connection to the evaporator is opened. However, the connection with the condenser is closed, enabling the adsorbate from the evaporator to be adsorbed by the bed. Concurrently, another bed is connected to the condenser and disconnected from the evaporator, facilitating the condensation of the desorbed adsorbate from the desorption process. The cycles are continued when two beds reverse these processes using piping and instrumentation, as represented by Figures 8.9 and 8.10. By employing additional beds following the same principle, the switching time, during which no cooling can be obtained, can be minimized.

8.2.3 Multi-bed and Multistage Configuration

More complex configurations involve multiple beds (more than two beds), as seen in Figures 8.11 and 8.12. Multi-bed or multistage configuration can further enhance the performance of adsorption cooling systems [21], and they are sometimes considered as performance enhancements [65]. Two adsorbent beds are dedicated to adsorption in a four-bed system, while the others are employed for desorption. This arrangement allows for increased CC and better heat utilization.

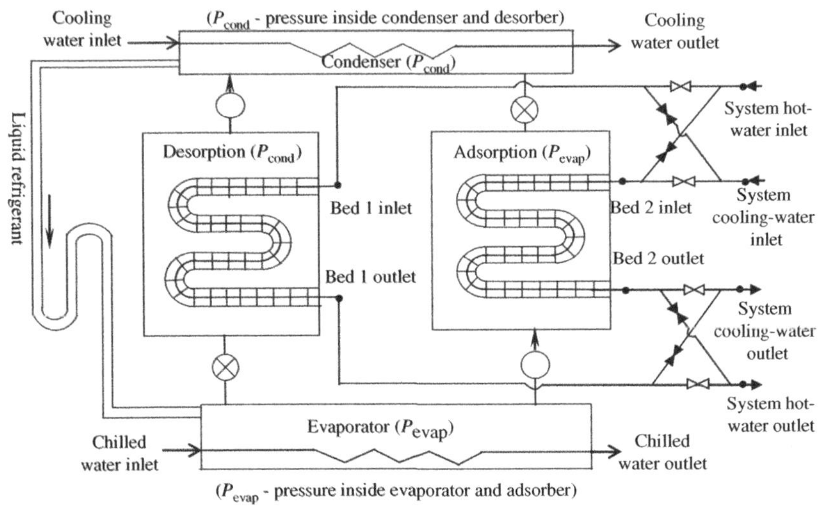

Figure 8.9 Schematics of two-bed adsorption chiller (⋈: opened water valve; ⧓: closed water valve; ⊗: closed gas valve; ○: opened gas valve). Source: Ref. [48] / with permission of Elsevier.

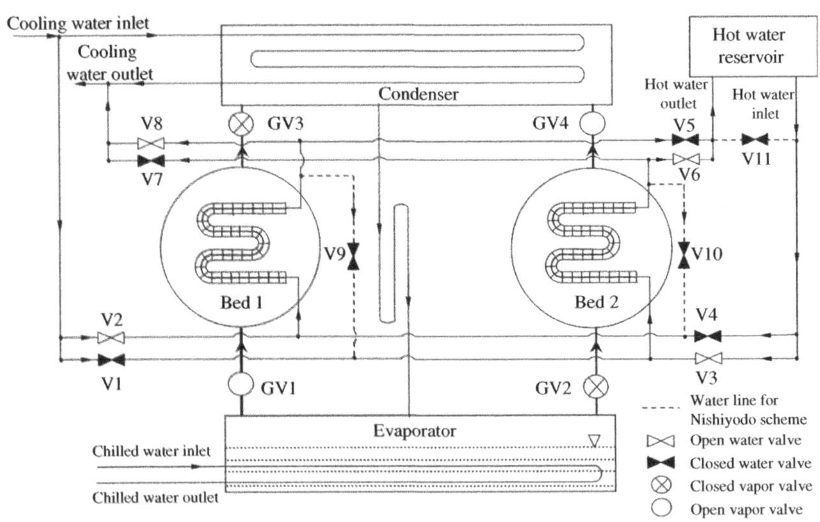

Figure 8.10 Schematic of a two-bed silica gel adsorption chiller. Source: Ref. [62] / with permission of Elsevier.

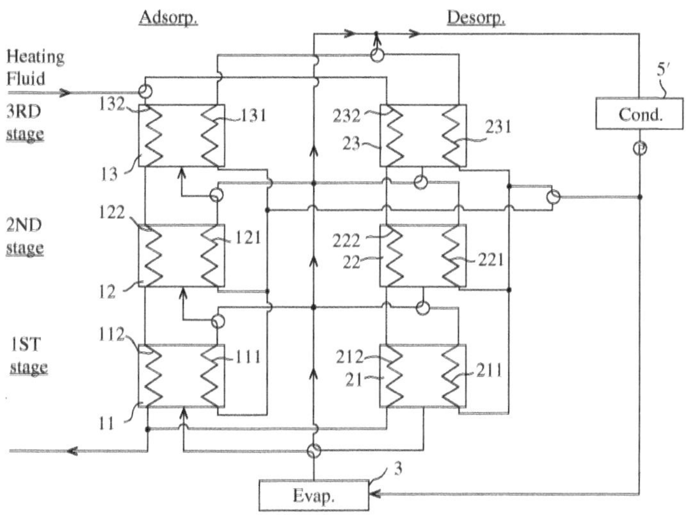

Figure 8.11 Diagram for multistage and multi-bed adsorptive type refrigeration apparatus. Source: Ref. [63] / with permission of Elsevier.

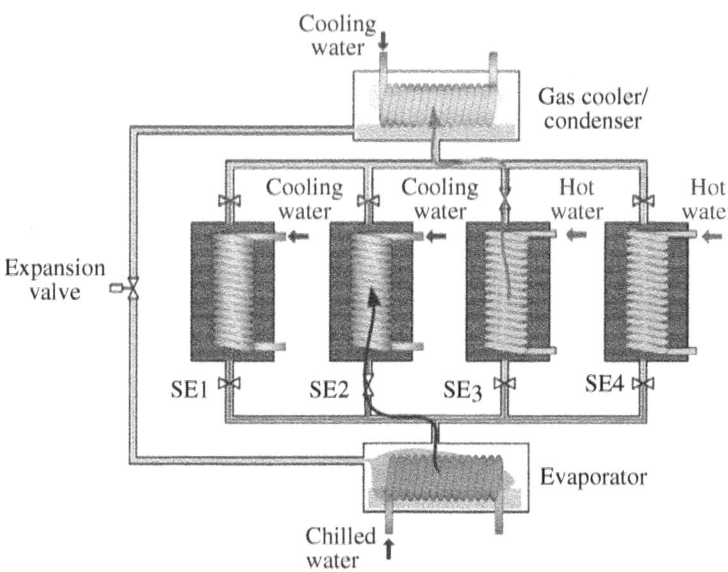

Figure 8.12 Schematic diagram of the 4-bed-adsorption cooling system. Source: Ref. [64] / with permission of Elsevier.

The multi-bed configuration involves more than a pair of adsorption beds, offering further scalability and enhanced CC [66, 67]. In multi-bed/multistage systems, multiple adsorption beds are arranged in a series or parallel configuration. In the series layout, the beds are connected in series, where the refrigerant flows sequentially through each bed, allowing for a more efficient adsorbent capacity usage and better utilization of performance enhancement techniques. In the parallel arrangement, each bed operates independently, and the refrigerant is distributed evenly across all beds simultaneously, providing redundancy and allowing for greater flexibility in system operation. Hybrid arrangements are also used to balance system efficiency, capacity, and flexibility, depending on specific application requirements. This hybrid configuration can be advantageous for large-scale applications requiring higher cooling demands, with more flexible control of the adsorption and desorption stages, leading to improved cooling performance, higher energy efficiency, and better temperature control.

Multi-bed configurations are more complex than one-bed configurations, requiring additional components, such as valves, switches, and control mechanisms, to manage the operation of each bed. This complexity results in a relatively more expensive upfront cost (parts, installation, adsorbent material, etc.) and can impact system reliability and maintenance requirements. Multi-bed adsorption cooling systems provide numerous advantages, such as maximizing CC, minimizing fluctuations in the chilled water outlet temperature and enhancing the quality of rapid cooling compared to two-bed chillers with equivalent CC [14].

Solar thermal-driven adsorption chillers face a significant challenge in operating throughout the year due to low solar radiation during specific periods [14, 68]. Multiple-stage adsorption cooling systems employ a sequence of adsorption beds to improve cooling performance and efficiency. Each stage comprises an adsorbent bed and operates on the same cyclic adsorption and desorption principles as single-stage systems. The operating concept is to pass the refrigerant through numerous stages successively, allowing for progressive cooling and increased overall performance. The refrigerant vapor reaches the first step, which is absorbed into the adsorbent material and then cooled. Following that, the partially adsorbed refrigerant moves on to the next stage of cooling. This process is repeated in phases, with each stage extracting more heat from the refrigerant until the required cooling effect is obtained. Multistage configurations can improve the COP by extracting more heat from the refrigerant, resulting in a more efficient cooling process.

8.3 Mathematical Modeling

Accurate experimental performance evaluation and simulation of adsorption cooling systems are crucial for their design, optimization, and performance enhancement. Numerical models and simulation tools have been developed to predict the system behavior, including the adsorption/desorption process, heat and mass transfer, and system performance. These models can evaluate the effects of different enhancement techniques and guide the design process. Experimental validation of the simulation results is essential to ensure the accuracy and reliability of the developed models.

Simulations assist in understanding and optimizing the performance of adsorption cooling systems. By employing computational models and numerical methods, simulations enable the analysis of complex phenomena, prediction of system behavior, and exploration of various design and operating parameters [69]. Process modeling involves developing, or simplifying and improving, mathematical models that represent the physical and thermodynamic processes occurring within an adsorption cooling system. These models describe the heat and mass transfer phenomena and the adsorption and desorption kinetics. They are typically based on fundamental principles of thermodynamics, fluid dynamics, and chemical engineering.

Process models capture the interactions between the adsorbent, refrigerant, and heat transfer fluid, providing insights into the performance and behavior of the system. They enable the estimation of critical parameters such as CC, COP, adsorbent utilization, and system efficiency. Various modeling approaches, such as equilibrium-based models, kinetic models, and empirical correlations, can be employed depending on the complexity of the system and the available data.

Heat and mass transfer simulations are essential for understanding the transport phenomena within an adsorption cooling system. These simulations analyze the system's temperature, pressure, and concentration distribution, providing insights into the thermal and adsorption dynamics.

Computational fluid dynamics (CFD) techniques are commonly used for heat and mass transfer simulations in adsorption cooling systems. These techniques solve the governing equations for fluid flow, heat transfer, and mass transfer within the system. CFD simulations provide detailed information about flow patterns, temperature distributions, and adsorbent loading by discretizing the domain into a mesh and solving the equations numerically.

Additionally, finite element or finite volume methods can be utilized for simulations involving heat conduction within the adsorbent bed. These methods solve the transient heat conduction equation, considering the variation in bed properties, such as thermal conductivity and heat capacity, during the adsorption and desorption processes.

Simulations are valuable tools for evaluating the performance of adsorption cooling systems under different operating conditions and design parameters. Performance evaluations analyze key metrics such as CC, COP, SCP, and thermal efficiency.

By incorporating the validated process models and heat and mass transfer simulations, system performance evaluations can provide insights into the effects of variables such as adsorbent properties, heat transfer fluid characteristics, cycle time, and operating conditions. These evaluations aid in optimizing system performance, determining the optimal operating parameters, and guiding system design decisions.

Validation against experimental data and case studies is essential to ensure the accuracy and reliability of simulations. Adsorption relations and mathematical models are frequently used in simulations. Comparing simulation results with experimental measurements helps verify the models' accuracy and applicability to real-world systems. Case studies involving different adsorbent materials, system configurations, and operating conditions provide valuable insights into system behavior and performance. Validation of simulations against experimental data and case studies strengthens confidence in the simulation results and allows for extrapolating the findings to different scenarios and applications.

8.3.1 Adsorption Equilibrium Relations, Diffusion, Adsorption Kinetics, and Heat of Sorption Models

In practice, using the capacity of an adsorbent is unattainable because of mass transfer effects in the process of fluid–solid contact. The adsorption equilibrium is a fundamental property that forms the basis for more practical analysis. Adsorption equilibrium is a vital fundamental adsorbent property for a practical standpoint analysis of adsorption cooling, which can be influenced by concentration, temperature, and pressure.

Numerous studies have been conducted to find the relationship between the adsorbed substance on an adsorbent (working pair) under specific conditions, including temperature, pressure, and concentration [70]. The correlation between the concentration of the substance in the fluid or gas phase at a particular temperature and the amount adsorbed is known as an isotherm [61, 69, 71]. The general expression for an isotherm is usually presented as

$$q = q(C) \text{ at constant } T \tag{8.4}$$

where q is the amount of species adsorbed, C is the concentration of the species in the fluid or gas phase, and T is the temperature during adsorption. Also, the relation between the concentration and temperature to yield a specific amount of adsorbed species can be expressed as

$$C = C(T) \text{ at constant } q \tag{8.5}$$

Since the number of molecules adsorbed onto the adsorbent is essential, isotherms are commonly used to describe the adsorption behavior, as summarized in Table 8.1. Many known isotherms are utilized for adsorption models. It is assumed that an energetically uniform surface is caused by adsorption on an adsorbent surface without interaction with adsorbed molecules. However, novel working pairs need further research [85].

Several diffusion models are known for porous particle-made adsorbents. Many adsorption kinetic models are also available in the literature. Table 8.2 summarizes the most common diffusion and adsorption kinetic models used in the adsorption cooling literature. The appropriate model selection is based on the dominant phenomena or application, with the central parameter based on the adsorbent–refrigerant pair behavior. The isotherm, diffusion, and adsorption kinetic models identified herein can describe the adsorption mechanism of different materials in distinct levels of complexity and accuracy [83] with many others available in the literature [95–99]. However, one should be open to those isotherms, primarily due to advancements in materials science.

Different models and empirical and experimental values for the heat of adsorption are also available in the literature. The heat of sorption can be derived from sorption isotherms at many temperatures using the basic C–C model: $\Delta H_{is} = -\mathfrak{R}\left(\frac{\partial \ln a_w}{\partial T^{-1}}\right)_X$ [100], where $\mathfrak{R}$ is the gas constant, a_w is the water activity of material, ΔH_{is} is the isosteric enthalpy of sorption, and X is the moisture content. An example for estimating the heat of adsorption using the Clausius–Clapeyron (C–C) equation and isotherms is written in Eq. (8.6) [101, 102], where Q_{st} is the heat of adsorption, T is the gas constant, T is the temperature, P is the pressure, E is the adsorption characteristic parameter, n is the heterogeneity parameter, and W_0 and W are the maximum and transient adsorbate mass per unit of adsorbent mass, respectively. Other models are available in the literature for isosteric heat/heat of adsorption models, such as Chakraborty et al., Rahman et al., Askalany et al., Tian et al., and Chakraborty–Saha–Koyama (C–S–K) models [103–112],

$$Q_{st} = R\,T^2 \left(\frac{\partial ln(P_s)}{\partial T} + \frac{E}{R\,T^2} \left(ln \left(\frac{W_0}{W} \right) \right)^{\frac{1}{n}} \right) \tag{8.6}$$

Isotherm, diffusion, and adsorption kinetic models are frequently integrated with mathematical modeling to solve adsorption cooling simulations. Mathematical modeling plays a crucial role in understanding and optimizing the performance of adsorption cooling systems. One can analyze system behavior, predict performance, and guide system design and control strategies by developing mathematical equations describing the physical and thermodynamic processes. Various mathematical modeling approaches for adsorption cooling, including the simple method, lumped capacitance, and distributed parameter models, are available and identified next.

Table 8.1 Examples of isotherms used in adsorption cooling.

References	Isotherm	Formulation		Characteristics
[23, 61, 72, 73]	Langmuir	$\theta = \frac{q^*}{q_m}, \theta = \frac{KP}{1+KP}$	(8.7)	Ratio of adsorption capacity in equilibrium to adsorption monolayer capacity.
	Henry-type	$\theta = KP$	(8.8)	Adsorbed amount ≪ adsorbent adsorption capacity. Short-period adsorption.
	Modified Langmuir	$P = \frac{1}{K}\left(\frac{\theta}{1-\theta}\right) e^{\left[\left(\frac{\theta}{1-\theta}\right)+\frac{2u_c}{k_B T}\right]}$	(8.9)	Interaction between the adsorbing molecules. Moving adsorbed molecules across the surface of the adsorbent.
[74]	Freundlich	$\frac{q}{M_s} = K_F P^{\frac{1}{n_F}}$	(8.10)	Adsorbent's surface roughness.
[75]	Tóth	$\theta = \frac{K_T P}{\left[1+(K_T P)^t\right]^{\frac{1}{t}}}$	(8.11)	Adsorption behavior for both low and high concentrations of adsorbate.
[76]	Dubinin and Astakhov	$\theta_{d-a} = e^{\left(-\frac{A}{E_c}\right)^n}$	(8.12)	Adsorption occurs at micropores.
[77, 78]	Dubinin–Raduskevich	$\frac{q}{q_0} = e^{-D\left(T \ln \frac{P_s}{P}\right)^n}$	(8.13)	Similar to the Dubinin and Astakhov model, adsorption occurs at micropores.

[79]	Sun and Chakraborty	$$\theta = \frac{K\left(\frac{P}{P_s}\right)^m}{1+(K-1)\left(\frac{P}{P_s}\right)^m} \quad (8.14)$$ $$K = \propto e^{\left[\frac{m\left(Q_{st}^{*}-h_{fg}\right)}{RT}\right]} \quad (8.15)$$	Thermodynamically consistent boundary conditions throughout the entire adsorption range. Adsorbent porous structure.
[80–82]	Do–Do	$$q = q_s \frac{K_s \sum_{n=1}^{N} n\left(P/P_0\right)^n}{1+K_s \sum_{n=1}^{N} n\left(P/P_0\right)^n} + q_m \frac{K_m \left(P/P_0\right)^M}{1+K_m\left(P/P_0\right)^M} \quad (8.16)$$ $$q_s = q_1\, e^{\left(\frac{-\alpha}{RT}\right)} \quad (8.17)$$ $$K_s = K_{0s}\, e^{\left(\frac{-\Delta H_s}{RT}\right)} \quad (8.18)$$ $$q_m = q_2\, e^{\left(\frac{-\beta}{RT}\right)} \quad (8.19)$$ $$K_m = K_{0m}\, e^{\left(\frac{-\Delta H_m}{RT}\right)} \quad (8.20)$$	Extreme microporous adsorbent, strong bond to the primary adsorption sites.
[82]	Hybrid Langmuir–Sips isotherm models	$$q = q_{LS}\left(\frac{K_L\left(P/P_0\right)}{1+K_L\left(P/P_0\right)} + \frac{K_S\left(P/P_0\right)^\theta}{1+K_S\left(P/P_0\right)^\theta}\right) \quad (8.21)$$ $$q_{LS} = q_{max}\, e^{\left(\frac{-\alpha}{RT}\right)} \quad (8.22)$$	Combination of Langmuir and Freundlich models.

(Continued)

Table 8.1 (Continued)

References	Isotherm	Formulation	Characteristics
		$K_L = K_{0L}\, e^{\left(\frac{-\Delta H_{LS}}{R\,T}\right)}$ (8.23)	
		$\theta = \theta_0\, e^{\left(\frac{-\beta}{R\,T}\right)}$ (8.24)	
		$K_S = K_{0S}\, e^{\left(\frac{-\Delta H_{LS}}{R\,T}\right)}$ (8.25)	
[83, 84]	The Universal isotherm model	$\theta_t = \frac{q}{q_0} = \sum_{i=1}^{n} \alpha_i \left[\frac{\left\{ \frac{P}{P_s} e^{\left(\frac{\varepsilon_{0i}}{R\,T}\right)} \right\}^{\frac{R\,T}{mi}}}{1 + \left\{ \frac{P}{P_s} e^{\left(\frac{\varepsilon_{0i}}{R\,T}\right)} \right\}^{\frac{R\,T}{mi}}} \right]_i$ (8.26)	The adsorbent surface is an heterogenous agglomeration of many homogenous patching having equal energy level for each site.

Nomenclature: q^* is the adsorbed amount in equilibrium, q_m is the adsorption monolayer capacity of the adsorbent, K is the adsorption equilibrium constant, and P is the partial pressure of adsorbate in the gas phase; $2\,u_e$ is the energy of pair interaction and k_B is the Boltzmann constant; M_s is the weight of the adsorbent, while K_F and n_F are empirical constants; K_T is an equation constant, and t is Toth's constant; θ_{d-a} is the degree of filling of adsorbate in adsorption space, E_c is the characteristic energy of adsorption, A is the loss in Gibb's free energy or $A = -\Delta G$, and n is an integer to determine whether the adsorption site scale (surface, micropore, or ultra-micropore); P_s is the saturated pressure, m is the heterogeneity factor, K is the isotherm coefficient, $\propto$ is the pre-exponential coefficient correlated to the adsorption entropy, m is the heterogeneity factor, Q^*_{st} is the isosteric heat of adsorption at zero surface coverage, h_{fg} is the evaporation enthalpy, and R is the gas constant. q_s is the water vapor uptake of surface active groups, R is the universal gas constant, K_s is the equilibrium constant for chemisorption and hydrogen bonding at the primary sites, K_0 is the pre-exponential factor, $N(N \geq M+1)$ is the maximum number of water molecules adsorbed on the surface sites, P is the partial pressure, P_0 is the saturation pressure at the adsorbent temperature, q_m is the water vapor uptake by micropores, K_m is the micropore equilibrium constant, and ΔH_s (C–C equation) and ΔH_m (Henry's equation) are isosteric enthalpies for chemical reaction and micropore filling, respectively, and q_1 and q_2 are the initial water vapor uptake of each bed; q_{max} is the maximum water vapor uptake, K_L and K_S are the equilibrium constants, ΔH_{LS} is the adsorption isosteric enthalpy, θ is the adsorbent structural heterogeneity parameter, and n is the energy sites' number, α is the probability factor for a set of energy sites, ε_{0i} is the energy value at adsorption sites at maximum probability, and m is the surface heterogeneity parameter.

Table 8.2 Examples of diffusion and adsorption kinetic models.

References	Model	Formulation	Dominant phenomena or application
[61, 86]	Diffusion	$D_m = \frac{0.00158\, T^{3/2} \left({}^{1}/_{M_1} + {}^{1}/_{M_2} \right)^{1/2}}{P\, \sigma^2{}_{12}\, \Omega \left({}^{\varepsilon}/_{k_B\, T} \right)}$ (8.27)	Intermolecular collision.
	Diffusion	$D_k = \left(\frac{2}{3} \right) \sqrt{\frac{8\, R\, T}{\pi\, M}}\, R_p$ (8.28)	Knudsen diffusivity.
[61, 87]	Diffusion	$\frac{1}{D_c} = \frac{1}{D_m} + \frac{1}{D_k}$ (8.29)	Intermolecular collision and Knudsen diffusivity combined.
[61]	Diffusion	$D_s = D_{s0}\, e^{-\left({}^{E_a}/_{R\, T} \right)}$ (8.30)	Surface diffusion.
[88]	Diffusion	$\frac{\partial}{\partial t}(\rho_w\, \gamma) + \nabla \left[\rho_w \left(-\frac{k_\varepsilon}{\mu} \nabla P_{ads} \right) \right] = \frac{\partial}{\partial t}(\rho_{ads}\, w)$ (8.31)	Darcy, adsorbent as a block of porous medium.
[61, 89–91]	Adsorption kinetics	$\frac{\partial \overline{q}}{\partial t} = \frac{15\, D_s}{r_p{}^2} (q^* - \overline{q})$ (8.32)	Linear driving force model and adsorption uptake.

(Continued)

Table 8.2 (Continued)

References	Model	Formulation		Dominant phenomena or application
[61, 92]	Adsorption kinetics	$\frac{\partial c}{\partial t} = \frac{\varepsilon\, D_e}{\varepsilon + (1-\varepsilon)\, K_f} \left(\frac{\partial^2 c}{\partial R^2} - \frac{2}{R} \frac{\partial c}{\partial R} \right)$	(8.33)	Fick's law of diffusion.
[61, 93, 94]	Adsorption kinetics	$\overline{q}(t) = q_0 \pm \Delta q [1 - e^{(-t/\tau)}]$	(8.34)	Real operating conditions for adsorption uptake rate.
[61, 74]	Adsorption kinetics	$\frac{\Delta \theta_t}{\Delta \theta} = \frac{\theta_t - \theta_{ini}}{\theta - \theta_{ini}} = 1 - e^{-t/\tau}$	(8.35)	Real operating conditions for adsorption uptake rate.

Nomenclature [61]: D_m is the molecular diffusivity, T is the temperature, M_1 and M_2 are the molecular weights, P is the pressure, σ_{12} is the collision diameter, Ω is a function of ${}^{\varepsilon}/_{k\,T}$ with ε as the Lennard–Jones force constant, D_k is the Knudsen diffusivity, R_p is the mean pore diameter, M is the diffusing gas molar mass, R is the gas constant, D_c is the combined diffusivity, D_s is the surface diffusivity and D_{s0} is the adsorbent pre-exponential efficient diffusivity factor, and E_a is the activation energy; $\overline{q}$ is the mean adsorption capacity, t is the time, r_p is the adsorbent particle radius, c is the adsorbate in gas phase concentration, D_e is the effective pore diffusivity, K_f is the local slope of adsorption uptake versus gas phase concentration, R is the radial coordinate, τ is the temperature-invariant characteristic time dependent on the uptake of adsorbent and $\Delta q = q^* - q_0$, and θ is the ratio of adsorption capacity in equilibrium to adsorption monolayer capacity and $1/\tau$ is the time constant.

8.3.2 Simple Model

The simple model is a straightforward approach to modeling the performance of adsorption cooling systems. This simple model is the most basic approach used in adsorption cooling studies. It assumes simplified geometries and simplifies the governing equations to make them analytically solvable. The model usually assumes a uniform temperature and concentration distribution within the adsorbent bed, while heat and mass transfer resistances are neglected, and equilibrium conditions are assumed to be achieved instantaneously. This model estimates the adsorption equilibrium of the refrigerant within the adsorbent bed based on several simplifying assumptions, such as assuming an ideal adsorbed phase, a constant adsorption equilibrium, and a uniform temperature within the bed. Despite these simplifications, the simple model offers a valuable approximation for preliminary analysis and system design. The equations derived from the model of Saha et al. [113] allow for estimating the adsorption capacity, equilibrium pressure, and temperature within the adsorbent bed.

The simple model utilizes mass and energy balance equations to describe the adsorption and desorption processes. The bed temperature and refrigerant concentration are typically the key variables of interest. Solving these equations can determine the time-dependent temperature and concentration profiles within the adsorption bed.

According to the simple model of Saha et al. [113] (effects of non-condensable gases and heat losses to the environment are neglected), the mass balance can be expressed as

$$\frac{dW_w}{dt} + W_s\left(\frac{dq_{ads}}{dt} + \frac{dq_{des}}{dt}\right) = 0 \tag{8.36}$$

where t is the time, W_w and W_s are the weight of the refrigerant in the liquid phase and the weight of the solid adsorbent (silica gel in Ref. [113]), respectively. The adsorption $\left(\frac{dq_{ads}}{dt}\right)$ and desorption $\left(\frac{dq_{des}}{dt}\right)$ rates, controlled by macroscopic diffusion into the adsorbent bed particle, are described by the adsorbent–refrigerant properties (equilibrium process without hysteresis and an isobaric adsorption–desorption process) and the operating conditions.

The adsorption and desorption heat balances are expressed as

$$T_{water\,out} = T_{ad} + (T_{water\,in} - T_{ad})\,e^{-\left[\frac{U_{ad}\,A_{Hex}}{(m_{water}\,c_{p,water})}\right]} \tag{8.37}$$

$$\begin{aligned}&\frac{d}{dt}[W_s\,(c_{ps} + c_{pw}\,q) + (c_{p\,Cu}W_{k\,Hex} + c_{p\,Al}W_{f\,Hex})]\,T_{ad}\\&= Q_{st}\,W_s\frac{dq}{dt} + m_{water}\,c_{p\,water}\,(T_{water\,in} - T_{water\,out})\end{aligned} \tag{8.38}$$

where $T_{water\,out}$ is the temperature, ad is the adsorber/desorber bed, U is the heat transfer coefficient, A is the area, Hex (or HEX) is the heat exchanger, m is the mass, c_p is the heat capacity at constant pressure, q is the amount adsorbed, Cu stands for copper, Al stands for aluminum, and Q_{st} is the isosteric heat of adsorption. Refer to the work of Saha et al. [113] for more details and units.

The heat balances for the evaporator and condenser are similarly described as

$$T_{water\,out} = T_{eva,cond} + (T_{water\,in} - T_{eva,cond})\, e^{-\left[\frac{U_{eva,cond}\, A_{eva,cond}}{(m_{water}\, c_{p,water})}\right]} \tag{8.39}$$

$$\begin{aligned} &\frac{d}{dt}[(c_{p\,w}\, W_{ew} + c_{p\,Cu}\, W_{eva,cond})\, T_{eva,cond}] \\ &\quad = -L_w\, W_s\, \frac{dq_{ads,des}}{dt} + m_{water}\, c_{p\,water}\, (T_{water\,in} - T_{water\,out}) \\ &\quad + c_{p\,w}\, T_{cond}\, W_s\, \frac{dq_{des}}{dt} \end{aligned} \tag{8.40}$$

where L_w is the latent heat of vaporization, $_{ew}$ stands for the liquid refrigerant in the evaporator, $_{cond}$ is related to the condenser, and $_{eva}$ stands for the evaporator. Numerical methods (finite difference, etc.) can be applied to solve the set of equations with appropriate simulation parameters.

As in the example of Reference [113], the equations derived from this model capture the simplified system's behavior, requiring less computational effort for the solution. However, they need more detailed spatial (including bed and adsorbent porosities) and temporal information and have limitations with relatively less agreement with experimental results.

System thermodynamic models are also frequently used to investigate the system's performance with limited application [34, 114–122]. Thermodynamic models, which consider no heat and mass transfer from/to the bed/heat source, rely on the first and second laws of thermodynamics. Also, they can use experimental or numerical data achieved from the three models described before for simplified adsorption cooling system analysis.

8.3.3 Lumped Parameter Model

The lumped parameter model is another widely used modeling approach for adsorption cooling systems [69]. This model improves upon the simple model by considering thermal capacitance effects [71]. It divides the adsorbent bed into multiple discrete sections, treating each section as a lumped heat transfer element with its temperature. It allows for simulation using one-dimensional models, which are preferred due to their simplicity and reduced computational time. Moreover, this model treats the adsorption bed as a single control volume with well-mixed conditions. It assumes uniform temperature and concentration within the bed. It represents the heat and mass transfer processes through lumped parameters, such as the overall heat and mass transfer coefficients.

In the review by Teng et al. [61], the mathematical modeling evolution for adsorption systems is presented, including the lumped parameter model and heat and mass transfer model (distributed parameter model). For the lumped parameter model, Sakoda and Suzuki [61, 69, 71, 123] proposed the first lumped parameter model for the mass and energy balances of the adsorbate as

$$M_s \frac{d\overline{q}}{dt} + \frac{dM_a}{dt} = 0 \tag{8.41}$$

$$M_s\, c_{p,s} \frac{dT_s}{dt} = M_s\, Q_{st} \frac{d\overline{q}}{dt} - A_s\, h_s\, (T_s - T_f) + A_s\, Q \tag{8.42}$$

where M_s is the weight of the adsorbent (it can incorporate the aspects of the heat exchanger/adsorbent [71]), M_a is the mass of the adsorbate in the liquid phase, $c_{p,s}$ is the specific heat of the adsorbent, and A_s, h_s, and T_s are the heat transfer area, overall heat transfer coefficient, and temperature in the bed, respectively. T_f is the atmospheric temperature or temperature of cooling or heating medium in more general cases. Q is the effective heat influx into the beds by radiation, which is sometimes neglected, resulting in the simplified version of Eq. (8.42) by Chihara and Suzuki [124], and $\frac{d\overline{q}}{dt}$ is the rate of adsorption ($\overline{q}$ is the mean amount of adsorbed molecules). The set of equations can be solved using the finite difference method with proper boundary and initial conditions.

Equations (8.41) and (8.42) evolved, with many authors contributing to simplified or more complex and realistic scenarios [36, 46, 61, 123–134] with some examples summarized in Table 8.3. The lumped parameter model in Eqs. (8.42) and (8.41) assumed that the adsorbate temperature and the beds are uniform because the packed beds are small (negligible variation in temperature distribution). Also, the lumped parameter model accounts for heat transfer resistances and transient effects, providing more accurate results compared to the simple model. However, it still considers uniform temperature and concentration within each section and neglects spatial variations within the adsorbent bed. Also, it frequently overestimates COP.

The lumped parameter model solves the energy and mass balance equations of the adsorption bed. These equations consider heat transfer, mass transfer, and the kinetics of adsorption and desorption processes. The review by Teng et al. [61] presents mathematical modeling evolution for adsorption systems. While this model reasonably estimates system behavior, it neglects spatial variations within the adsorption bed.

Table 8.3 Summary of other lumped parameter models.

References	Formulation
[61, 131]	$M_s c_{p,s} \frac{dT_s}{dt} = -M_s \Delta H \frac{d\overline{q}}{dt} - E \dot{m}_f c_{p,f} (T_b - T_s)$ (8.43)
[61, 126]	$(M_s c_{p,s} + q M_s c_{p,a} + M_{hex} c_{p,hex}) \frac{dT_s}{dt} = -M_s \Delta H \frac{d\overline{q}}{dt} - E \dot{m}_f c_{p,f} (T_s - T_f) + M_s c_{p,v} (T_{eva} - T_s) \frac{d\overline{q}}{dt}$ (8.44)
[46, 61]	$(M_s c_{p,s} + M_{hex} c_{p,hex}) \frac{dT_{bed}}{dt} + q M_s \frac{dh_{ads}}{dt} = M_s \frac{d\overline{q}}{dt} \{\delta[h_g (T_{eva}) - h_g (P_{eva}, T_{bed}) + \Delta H] + (1-\delta)\Delta H\} - U_{cooling} \frac{A_{bed}}{N} \sum_{i=1}^{N} (T_{bed} - T_i)$ (8.45)
[135, 136]	$(M_{ad} c_{p,ad} + q_{bed} M_{ad} c_{p,w} + M_{hex} c_{p,hex}) \frac{dT_{bed}}{dt} = \Phi M_{ad} Q_{st} \frac{dq_{bed}}{dt} - (M_{bed,w} c_{p,w}) (T_{bed,out} - T_{bed,in})$ (8.46)
	$T_{bed,out} = T_{bed} + (T_{bed,in} - T_{bed}) e^{\left(-\frac{U_{bed} A_{bed}}{m_{bed,w} c_{p,w}}\right)}$ (8.47)
	$M_{con} c_{p,con} \frac{dT_{con}}{dt} = -\Phi L_H M_{ad} \frac{dq_{ads}}{dt} - M_{ads,w} c_{p,w} \frac{dq_{ads}}{dt} - M_{cooling,w} c_{p,w} (T_{cooling,out} - T_{cooling,in})$ (8.48)
	$T_{cooling,out} = T_{con} + (T_{cooling,in} - T_{con}) e^{\left(-\frac{U_{con} A_{con}}{m_{cooling,w} c_{p,w}}\right)}$ (8.49)
	$M_{eva} c_{p,eva} \frac{dT_{eva}}{dt} = -\Phi L_H M_{ad} \frac{dq_{ads}}{dt} + M_{cooling,w} c_{p,w} (T_{chill,in} - T_{chill,out})$ (8.50)

Table 8.3 (Continued)

References	Formulation	
	$T_{chill,out} = T_{eva} + (T_{chill,in} - T_{eva})\, e^{\left(-\frac{U_{eva} A_{eva}}{m_{chill,w}\, c_{p,w}}\right)}$	(8.51)
	$\frac{dM_w}{dt} = -M_{ad}\left(\frac{dq_{ads}}{dt} + \frac{dq_{des}}{dt}\right)$	(8.52)

Nomenclature: $\Delta H = -Q_{st}$ (heat of adsorption), E is the effectiveness of heat exchanger, T_b is the boiler temperature in the heat recovery process and fluid temperature for the non-recovery process, $\dot{m}_f$ is the fluid mass flow rate, and $c_{p,f}$ is the fluid specific heat; $c_{p,a}$ is the refrigerant specific heat, M_{hex} is the heat exchanger mass, $c_{p,hex}$ is the specific heat of the heat exchanger, T_{eva} is the evaporator temperature, and $M_s\, c_{p,v}\,(T_{eva} - T_s)\frac{d\bar{q}}{dt}$ is the sensible heat of adsorbate evaporated from the evaporator term; h_{ads} is the adsorbate partial enthalpy, h_g is the water vapor specific enthalpy, δ is the bed state (adsorption $\delta = 1$ or desorption $\delta = 0$), $U_{cooling}$ is the overall heat transfer coefficient of the adsorber, and N is the number of discrete element in the adsorber heat exchanger tubes; and Refs. [135, 136] include formulations for condenser (subscript: con), evaporator (subscript: eva), and heat exchanger (subscript: hex), where other subscripts are chiller water (subscript: chill), adsorbent (subscript: ad), adsorbate (subscript: ad), adsorption (subscript: ads), and desorption (subscript: des); L_H is the latent heat of water vaporisation, and Φ is factor for when the switch between beds occur ($\Phi = 1$ at operating time and $\Phi = 0$ at switching time).

8.3.4 Distributed Parameter Model

The distributed parameter model is the most comprehensive and accurate approach for modeling adsorption cooling systems. It considers the spatial variations in temperature and concentration within the adsorbent bed, allowing for a detailed analysis of heat and mass transfer phenomena. The distributed parameter model offers a more detailed and accurate modeling approach by considering spatial variations within the adsorption bed. It divides the bed into small segments or control volumes and accounts for heat and mass transfer at each location. This model captures the effects of temperature and concentration gradients within the bed, providing a more realistic representation of the system's behavior. It also includes changes in the fluid/solid properties with temperature in each element.

The distributed parameter model requires solving a set of partial differential equations describing the energy and mass conservation for each segment of the adsorption bed. These equations incorporate terms representing heat conduction, convection, and adsorption and desorption kinetics. Numerical methods, such as finite difference or finite element techniques, are commonly employed to solve the governing equations.

The general mass and momentum balances for one-dimensional (axial) used in the distributed parameter model are written respectively as [71]

$$\varepsilon \frac{\partial \rho_v}{\partial t} + \nabla \cdot (\rho_v u) + \rho_z \frac{\partial q}{\partial t} = 0 \tag{8.53}$$

$$\frac{\partial(\rho u_v)}{\partial t} + \frac{\partial(\rho u_v u_v)}{\partial x} = -\frac{\partial p}{\partial x} - \frac{\partial}{\partial x}\left(\mu \frac{\partial u_v}{\partial x}\right) \tag{8.54}$$

where subscript v is for the refrigerant vapor phase and z is for the adsorbent, t is the time, ε is the adsorbent bed porosity, x is the axial cartesian coordinate, u is the axial velocity, q is the mass adsorption capacity, p is the pressure, μ is the fluid viscosity, and the mass balance over the particle is

$$\frac{\partial q}{\partial t} = \frac{\partial}{\partial x}\left(\rho D_s \frac{\partial q}{\partial x}\right) \tag{8.55}$$

with D_s as the surface diffusion coefficient.

The heat transfer balances for vapor and solid adsorbent are [71]

$$\varepsilon \frac{\partial(\rho_v c_{pv} T_v)}{\partial t} + \frac{\partial(\rho_v c_{pv} u_v T_v)}{\partial x} = \frac{\partial}{\partial x}\left(\varepsilon k_v \frac{\partial T_v}{\partial x}\right) + \frac{\partial(\rho_v c_{pv} q)}{\partial t}(T_v - T_z) - h_{vz}(T_v - T_z) \tag{8.56}$$

$$\rho_z(c_{pz} + c_{pa} q)\frac{\partial T_z}{\partial t} = \frac{\partial}{\partial x}\left[k_z(1-\varepsilon)\frac{\partial T_z}{\partial x}\right] + \rho_z \Delta H \frac{\partial q}{\partial t} + h_{vz}(T_v - T_z) \tag{8.57}$$

respectively, with c_{pv} and c_{pa} as the specific heat capacity as constant pressure for the vapor phase and for the adsorbent respectively, T_v as the vapor temperature, k_v and k_z as the vapor and adsorbent thermal conductivities, respectively, T_z as the adsorbent temperature, h_{vz} as the vapor heat transfer coefficient, and ΔH as the heat of adsorption.

While showing a similar mass conservation as in Eq. (8.53) for the adsorbent bed, Pesaran et al. [69] extended the one-dimensional heat model analysis for the heat transfer fluid as

$$\frac{\partial(\rho_f c_{pf} T_f)}{\partial t} + u_f \frac{\partial(\rho_f c_{pf} T_f)}{\partial Z} = \frac{\partial}{\partial Z}\left(K_f \frac{\partial T_f}{\partial Z}\right) \pm 4\frac{h_{ft}}{D_{in}}(T_f - T_t) \tag{8.58}$$

with subscript f for the fluid, t for the tubes, tf for the fluid-tube metal, in for inside, out for outside, ads for the absorbent bed, $adst$ for the adsorbent bed-tube metal, and fin for fins; Z as the axial coordinate; K as the thermal conductivity; D is the tube diameter; h is the heat transfer coefficient; δ for the fin thickness; and q_{ext} is the heat conduction from the pipe to the fins (or to the extended surfaces). For the heat exchanger pipes and extended surface (fin), the energy conservation is written as

$$\rho_t\, c_{pt}\,\frac{\partial T_t}{\partial t} - K_t\,\frac{\partial^2 T_t}{\partial Z^2} + \frac{4\,D_{out}\,h_{adst}}{D^2{}_{out} - D^2{}_{in}}(T_t - T_{ads}) + \frac{4\,D_{in}\,h_{ft}}{D^2{}_{out} - D^2{}_{in}}(T_t - T_f) + q_{ext} = 0 \tag{8.59}$$

with a bidimensional approach in the case of small fin thickness as

$$\delta\rho_{fin}\,c_{pfin}\frac{\partial T_{fin}}{\partial t} = K_{fin}\,\delta\,\nabla^2 T_{fin} - 2\,h_{adst}\,(T_{fin} - T_{ads}) \tag{8.60}$$

The energy conservation for the adsorbent bed is written as

$$\frac{\partial(\rho\,c_p\,T_{ads})}{\partial t} + \nabla(\rho_v\,c_{pv}\,u\,T_{ads}) = K_{bed}\,\nabla^2 T_{ads} + \rho_{ads}\,Q_{st}\,\frac{\partial q}{\partial t} \tag{8.61}$$

with subscript v as the vapor phase and a as the adsorbate, K_{bed} is the bed effective thermal conductivity, Q_{st} is the adsorption heat, q is the adsorbate uptake, and the adsorbent bent heat capacity relation as

$$\rho\,c_p = \varepsilon_t\,\rho_v\,c_{pv} + \rho_{ads}\,(c_{pads} + c_{pa}\,q) \tag{8.62}$$

For a bidimensional scenario considering the cylindrical coordinate, the following set of equations from Leong and Liu [137] for the heat transfer model included in the publication by Teng et al. [61] is written as for the thermal-fluid system with r as the radial coordinate

$$\frac{\partial}{\partial t}(\rho_f\,c_{pf}\,T_f) + u_f\,\frac{\partial}{\partial z}(\rho_f\,c_{pf}\,T_f) = \frac{\partial}{\partial z}\left(k_f\,\frac{\partial T_f}{\partial z}\right) + \frac{1}{r}\frac{\partial}{\partial t}\left(r\,k_f\,\frac{\partial T_f}{\partial r}\right) \tag{8.63}$$

for the metal tube with m subscript for metal tube as

$$\frac{\partial}{\partial t}(\rho_m\,c_{pm}\,T_m) = \frac{\partial}{\partial z}\left(k_m\,\frac{\partial T_m}{\partial z}\right) + \frac{1}{r}\frac{\partial}{\partial t}\left(r\,k_m\,\frac{\partial T_m}{\partial r}\right) \tag{8.64}$$

for the adsorbent as

$$\begin{aligned}(\rho_s\,c_{ps} + \rho_s\,c_{pa}\,q + \varepsilon\,\rho_g\,c_{pg})\frac{\partial T_s}{\partial t} + \frac{\partial(\rho_g\,c_{pg}\,u\,T_s)}{\partial z} + \frac{1}{r}\frac{\partial(r\,\rho_g\,c_{pg}\,v\,T)}{\partial t}\\ = \frac{\partial}{\partial z}\left(k_{eq}\frac{\partial T_s}{\partial z}\right) + \frac{1}{r}\frac{\partial}{\partial t}\left(r\,k_{eq}\,\frac{\partial T_s}{\partial r}\right) + \rho_s\Delta H\frac{\partial\overline{q}}{\partial t}\end{aligned} \tag{8.65}$$

with subscript s for the adsorbent and g for the adsorbate in gas phase, v as the radial fluid velocity, k_{eq} as the equivalent thermal conductivity of the adsorbent, and where the mass conservation model is similarly written as in Eq. (8.53) with the appropriated nomenclature as follows (**u** as the gaseous velocity vector):

$$\frac{\partial}{\partial t}(\varepsilon\,\rho_g) + \nabla\cdot(\rho_g\,\boldsymbol{u}) + \rho_s\frac{\partial\overline{q}}{\partial t} = 0 \tag{8.66}$$

knowing that proper boundary conditions and initial conditions must be included for the specificity of each analyzed case.

Several studies used the distributed parameter model [51, 137–169] to investigate adsorption solving the heat and mass transfer governing equations. The complexity and nonlinearity of coupled heat and mass transfer models exclude the feasibility of an analytical solution. Thus, numerical approaches are the only viable option for meeting the requirements for simulating adsorbent bed dynamics. The finite difference method, finite-volume method, method of lines, and finite-element method are among the numerical approaches employed [71]. The finite difference method is widely used because of its simplicity, efficiency in solving parabolic equations, and ease of altering initial and boundary conditions [61, 69].

The choice of modeling approach depends on the level of detail required, computational resources, and the specific goals of the analysis. The simple model offers a quick and precise system performance estimation, making it suitable for initial system design and analysis. The lumped-capacitance model balances accuracy and computational complexity, making it useful for system optimization and control strategies. The distributed parameter model, while computationally intensive, offers the highest level of accuracy and is suitable for in-depth investigations, advanced system design, and optimization studies [170]. Combining or using more than one mathematical modeling can also be a solution in some analysis [171].

8.4 Performance Enhancement

While adsorption cooling systems offer several advantages, including low energy consumption and the use of nontoxic refrigerants, their performance can be further enhanced through various techniques. Performance enhancement techniques can be applied to many applications to improve energy efficiency, reduce environmental impact, and enhance CC. The development and implementation of advanced adsorption cooling systems have the potential to contribute significantly to sustainable cooling solutions and the reduction of greenhouse gas emissions.

Performance enhancement techniques are crucial in optimizing adsorption cooling systems. Adsorption cooling systems frequently use renewable and waste energy as heat sources, which improves the cooling system's overall performance. However, recovery techniques, cycle modifications, and advanced adsorbent materials can be explored to enhance the adsorption system performance further. Accurate performance evaluation and simulation tools are essential for system design and optimization. The continued research and development in this field hold great promise for the advancement and widespread adoption of adsorption cooling systems, contributing to a sustainable and energy-efficient future.

Figure 8.13 illustrates the categorization of various solid–gas sorption refrigeration cycles. Within these cycles (not all covered here, referred to Ref. [68]),

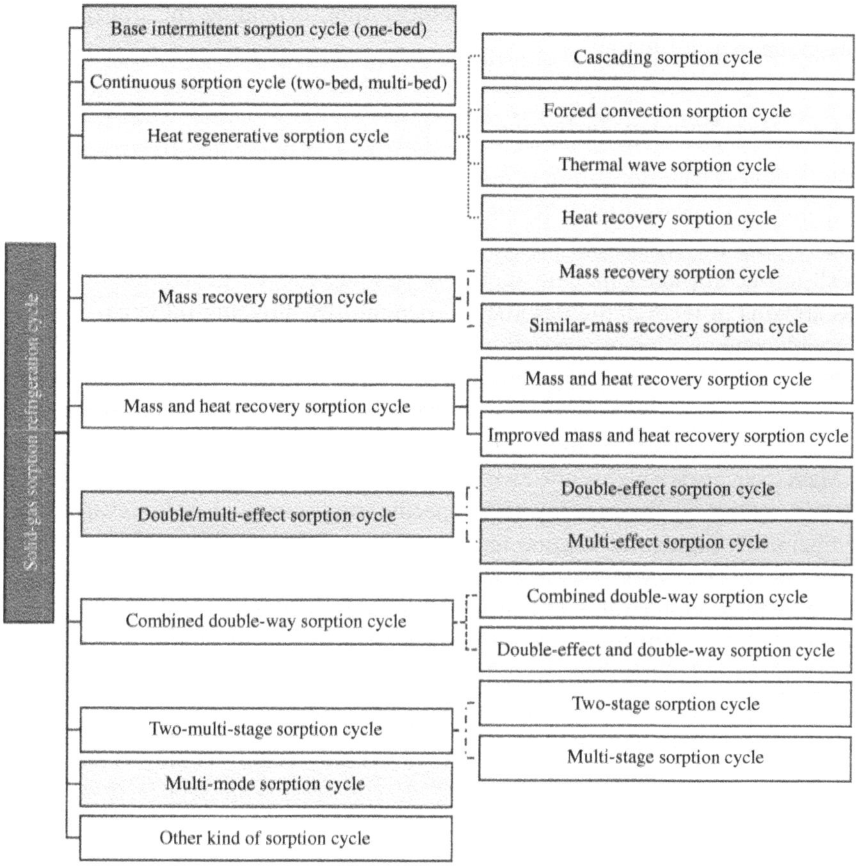

Figure 8.13 Classification solid–gas sorption refrigeration cycles. Source: Adapted from Ref. [68].

two operational approaches have been suggested and employed to enhance the system performance. The first approach focuses on improving working performance by increasing the cycled mass of refrigerant during the adsorption phase, accomplished through a mass recovery process between different sorption beds. The second approach aims to enhance working performance by reducing heat consumption during the desorption phase through heat recovery between different sorption beds. Subsequently, these two strategies are combined to optimize the system performance by simultaneously increasing CC and reducing heat consumption. These advanced sorption refrigeration cycles can significantly enhance thermodynamic efficiency, thus improving the overall working performance. Additionally, specific advanced solid sorption refrigeration cycles have been proposed to lower the driving temperature, extending the utilization of low-grade thermal energy.

8.4.1 Mass Recovery Technique

Mass recovery techniques aim to maximize the utilization of adsorbent materials and improve the system's efficiency [68]. These techniques involve capturing and reusing the refrigerant desorbed during desorption, reducing the need for fresh refrigerant makeup. One approach uses adsorbent regeneration or purges gas streams to recover the desorbed refrigerant. By directing these gas streams through condensation or adsorption processes using piping and instrumentation (see Figure 8.14), the refrigerant can be separated and collected for reuse. This technique reduces refrigerant losses, increases the system efficiency, and minimizes the environmental impact.

Mass recovery is a highly effective technique for enhancing the performance of a system, mainly when operating under specific conditions such as low evaporation temperature, high condensation temperature, and low-temperature heat source [35]. The mass recovery process can be adopted to boost the CC of a traditional two-bed adsorption chiller. Usually, during the pressurization and depressurization steps in the switching process, the adsorbers are heated and cooled without being connected to the evaporator or condenser [39]. However, partial pressurization and depressurization occur without heating and cooling the bed in the mass recovery approach. Through the connection of the adsorption and desorption

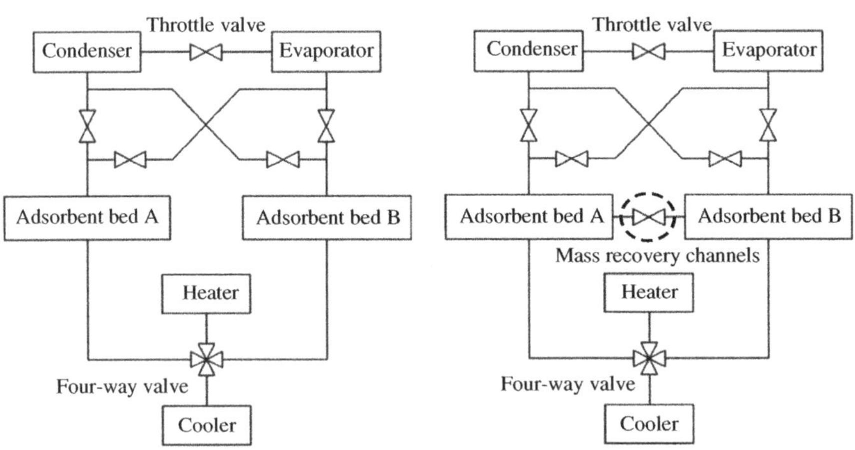

Figure 8.14 Schematic of a two-bed continuous cycle to mass recovery cycle by connecting Adsorbent bed A to Adsorbent bed B via a valve (dashed circle). Source: Adapted from Refs [65, 172].

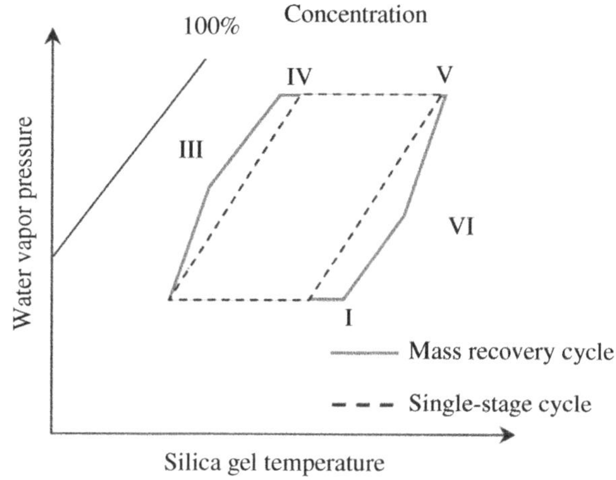

Figure 8.15 Conceptual schematic water–silica gel *Dühring* diagram for a mass recovery cycle. Source: Ref. [173] / with permission of Elsevier.

beds during the switching process, the refrigerant flows from the desorption bed to the adsorption bed due to a pressure drop. This connection results in a higher circulation of refrigerant within the bed, leading to an overall enhancement in CC. The basic principle of the mass recovery effect compared to the conventional cycle is demonstrated in Figure 8.15 using a *P–T–X* (pressure–temperature–uptake, sometimes uptake *X* is referenced as *W* or *q* – concentration) *Dühring* diagram.

The pressure difference in an adsorption cooling system operates the mass recovery technique. Before transitioning to the next phase, the two beds are connected in a mass recovery cycle. It induces the refrigerant vapor flow from the high-pressure bed (after desorption) to the low-pressure bed (after adsorption). This pressure drop promotes the desorption phase in the desorbed bed and the adsorption phase in the beds. This fact makes both processes more efficient, improving COP and SCP in the system [60]. Pressure equalization techniques are used in adsorption cooling systems to improve mass recovery. These strategies try to equalize pressures between multiple adsorption beds or stages, allowing the adsorbed refrigerant to be released more efficiently and boosting total mass recovery efficiency [58, 133, 174–177]. Pressure equalization is accomplished using valves, pressure equalization lines, or fluid flow manipulation. By equalizing pressures, desorption becomes more effective and efficient, resulting in improved mass recovery. Connecting the adsorption beds or stages via equalization lines often achieves pressure equalization. During the desorption phase, the pressure within the beds is equalized, allowing the refrigerant to be released from the adsorbent material more uniformly and quickly. To ensure the proper sequence and timing of pressure equalization, valves and flow control devices can be used to manage the equalization process. Mass recovery can be improved by optimizing pressure equalization processes, resulting in increased system performance and efficiency.

The mass recovery cycle can be divided into adiabatic and isothermal cycles. In the adiabatic mass recovery cycle, the adsorption bed remains thermally isolated. This leads to a temperature decrease in the desorbed bed due to intensified desorption and an increase in the adsorption bed due to intensified adsorption. On the other hand, the isothermal mass recovery cycle exchanges heat effectively with an external source and maintains the temperature. Both adiabatic and isothermal mass recovery cycles ensure equivalent pressures between the two beds at the end of the operation, with the decrease in adsorbate in the high-pressure bed matching the increase in adsorbate in the low-pressure bed [65, 172].

Advanced cycle topologies can be combined with heat recovery techniques to improve mass recovery in adsorption cooling systems. These designs attempt to enhance adsorbent capacity utilization and improve the efficiency of the adsorption and desorption processes.

8.4.2 Heat Recovery Technique

Heat recovery techniques focus on capturing and utilizing waste heat generated during desorption [178]. The heat is recovered and utilized for various purposes, instead of simply dissipating it, such as preheating the incoming adsorbent or providing heat for other processes [68]. The heat recovery efficiency (ε) is given by Eq. (8.67) [68], where Q_{rec} is the recovered heat and deducing it from Q_{heat} in Eq. (8.1) results in Eq. (8.68) [68] for the heat recovery (rec) coefficient of performance (COP_{rec}). Heat recovery can be achieved through heat exchangers that transfer the heat from the desorption stream to a separate fluid or process stream [62, 179]. The recovered heat can be used for preheating the adsorbent, heating water, or providing thermal energy for other industrial applications. The system's energy efficiency is improved by utilizing waste heat effectively,

$$\varepsilon = \frac{Q_{rec}}{Q_{heat}} \tag{8.67}$$

$$COP_{rec} = \frac{Q_{evap}}{Q_{heat} - Q_{rec}} = \frac{COP}{1 - \varepsilon} \tag{8.68}$$

Heat integration techniques have been widely investigated to improve the efficiency of adsorption cooling systems. These techniques involve recovering and utilizing waste heat or low-grade heat sources to regenerate the adsorbent [180]. Heat exchangers, heat pipes, and heat recovery units are commonly employed to transfer heat effectively between different process streams. Heat can be passively or actively transferred in the heat recovery technique, and other means are also in the literature (double effect and thermal wave cycles and circular and serial heat recovery, among others) using equipment, piping, and instrumentation [14]. By utilizing waste heat sources such as industrial processes, solar energy,

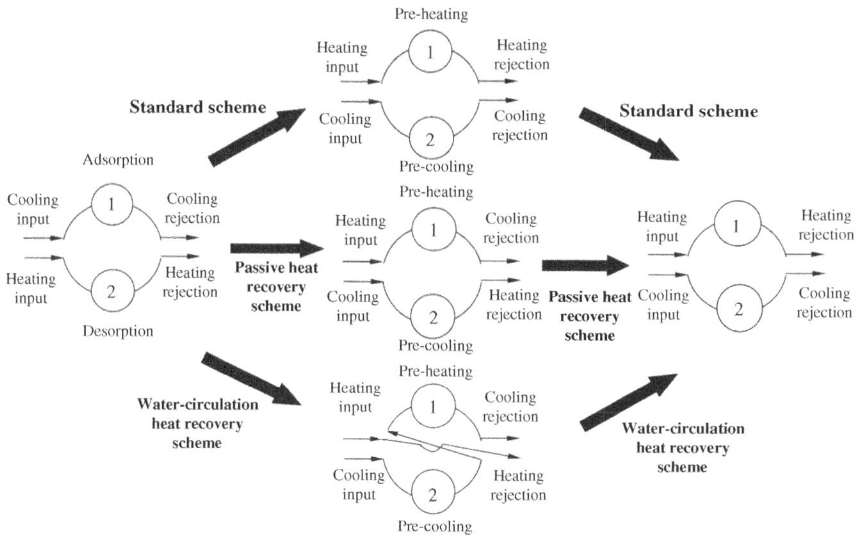

Figure 8.16 Schematic of regenerative schemes with heating/cooling inputs/rejections. Source: Ref. [62] / with permission of Elsevier.

or geothermal energy, the system's energy consumption can be reduced as well, resulting in higher performance.

The heat recovery cycle is proposed by Wang and Chua [62] as the scheme seen in Figure 8.16. The heat recovery cycle is used in a twin-bedded system. One adsorber bed is in the desorption phase, and the other is in the adsorption phase. In this case, it is possible to consider that the desorption bed temperature is relatively high, whereas the opposite is verified for the adsorption bed. Therefore, heat is released from one bed to another due to the temperature difference. The heat generation in the adsorber is used to heat the adsorbent for desorption. In the next phase, each bed is switched to the role of heating and cooling, allowing for continuous operation of the system, and significantly reducing its energy consumption. By transferring heat from one bed to another as in Figure 8.17, the heat recovery cycle can maximize thermal energy utilization. This fact can improve the COP of the adsorption refrigeration cycle.

8.4.3 Other Recovery Techniques

Combined mass and heat recovery techniques integrate both mass recovery and heat recovery strategies to maximize the efficiency of the adsorption cooling system [21, 137, 181]. These techniques aim to recover the refrigerant and the waste heat simultaneously, ensuring a comprehensive utilization of resources.

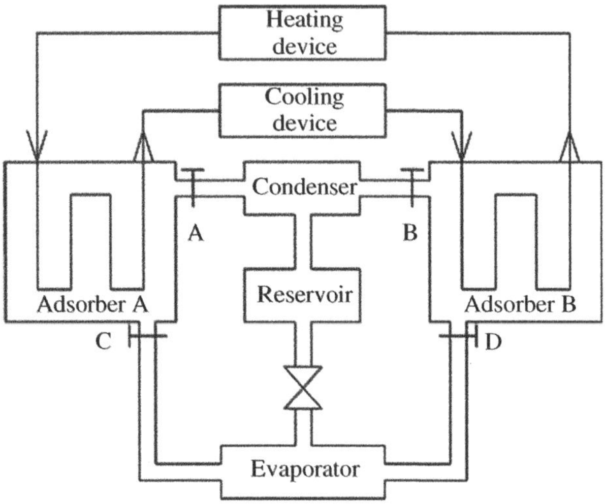

Figure 8.17 Schematic of the continuous adsorption cycle with heat recovery. Source: Ref. [65] / with permission of Elsevier.

The system achieves higher energy efficiency and reduces environmental impact by combining the recovery of desorbed refrigerant and waste heat through dedicated processes or integrated heat exchangers. This integrated approach enhances the performance of the adsorption cooling system. Usually, heat recovery and mass recovery are used together and complement each other to enhance cycling efficiency and improve the system's performance. Combining mass and heat recovery techniques and optimizing the duration of both techniques, the cooling cycle can improve the COP and SCP. The heat recovery process provides the additional heat required for the mass recovery process. This combined approach improved refrigeration performance and reduced cycle time [65, 177].

In a two-bed adsorption cooling example, one of the adsorbent beds is chilled at position a_2, while the other is situated at g_2 with elevated temperatures after each half cycle. This temperature disparity facilitates efficient heat recovery between these two adsorbent beds. By adiabatically circulating thermal fluid between the two adsorbent beds, a substantial enhancement in energy efficiency can be achieved. In an optimal heat recovery process, the two beds eventually attain uniform temperature levels; one corresponds to point e, and the other aligns with e' (refer to Figure 8.18). The schematic representation of this system is presented in Figure 8.19. Consequently, the thermal load required for heating can be lessened by the amount in the process $a_2 - g_1 - e$.

One bed is heated, leading to high pressure (P_c), while the other bed is cooled, resulting in low pressure (P_e) (depicted as a_2 and g_2 in Figure 8.18). Later, the

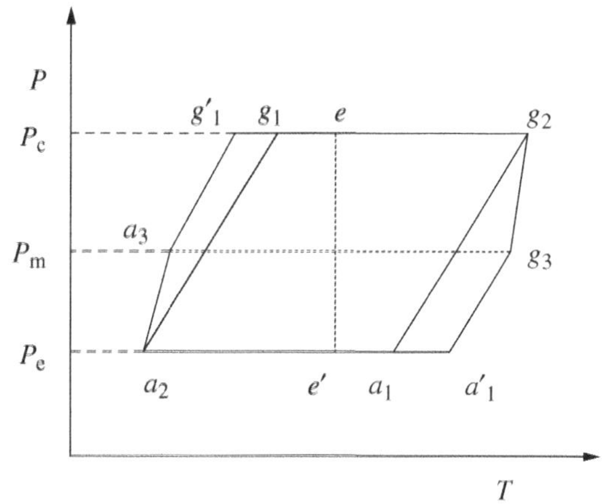

Figure 8.18 Diagram of the mass recovery process and combined cycle with heat and mass recovery. Source: Ref. [178] / with permission of Elsevier.

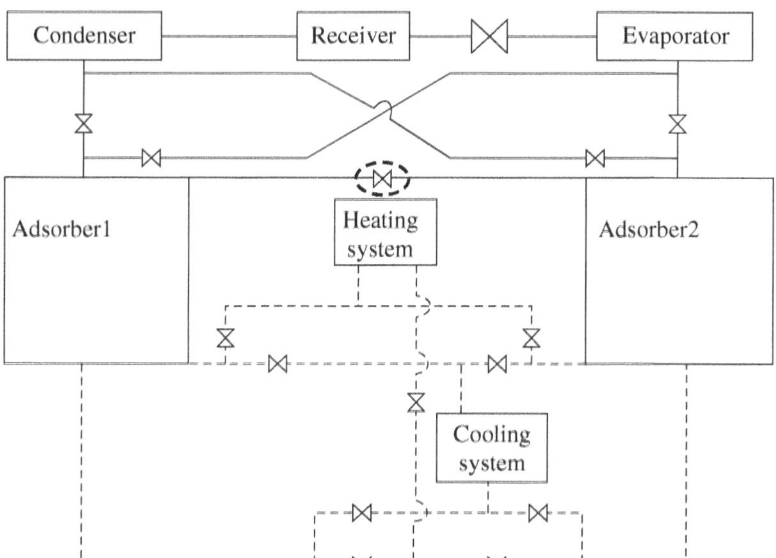

Figure 8.19 Schematic system of the mass recovery/heat and mass recovery cycle. The dashed line represents the thermal fluid flow; the solid line and valve between the beds process mass recovery, and the dashed line between beds processes heat recovery. Source: Ref. [178] / with permission of Elsevier.

high-pressure adsorber must be cooled and depressurized, while the low-pressure adsorber needs to be heated and pressurized. These two adsorbent beds can then be directly linked using a valve, causing the refrigerant vapor to flow from the high-pressure bed to the low-pressure one (mass recovery illustrated in

Figure 8.19, the red circle is the valve connecting both beds). During this process, the pressure of Adsorber 1 decreases due to mass outflow, leading to desorption, while the pressure of Adsorber 2 increases due to mass inflow, resulting in further adsorption. This cycle continues until both beds reach the same pressure (P_m in Figure 8.18). Afterward, the connection is ceased, and each bed continues its heating and cooling process, like the basic cycle. The mass recovery process accelerates circulation and enhances the cooling power of the cycle, as it only involves direct mass flow, which achieves a quicker pressure balance compared to the temperature balance achieved through a heat-transfer medium.

A heat regeneration process can also be implemented after the mass recovery, capitalizing on the temperature difference between the two beds. This can further decrease the thermal load, thus enhancing the overall cycle performance. In Figure 8.18, this process begins with a_3 and g_3 and concludes at e–e', where the temperatures are equal, as depicted in Figure 8.19. The mass recovery process extends the concentration change of adsorbate in a cycle, as indicated by the Dühring diagram in Figure 8.18. Throughout this discourse, the mass recovery process is considered adiabatic and is split into two segments, conducted in the two beds labelled a_2–a_3 and g_2–g_3. The process can be effectively simulated and calculated using models summarized in the section "Mathematical modeling."

One of the critical approaches to enhancing adsorption cooling systems' performance is cycle modification [182]. This cycle modification involves optimizing the operating parameters and configurations of the system to improve its energy efficiency and CC. Techniques such as multistage adsorption, cascade, and intermittent cycles have been investigated for higher performance [116, 183, 184]. By incorporating multiple adsorption beds or stages, the system can benefit from better utilization of the available heat source and increased CC. Additionally, intermittent cycles allow for better temperature lift, reduced heat loss, and improved cooling performance.

The dual-temperature cycle is one example of an advanced cycle configuration. During the desorption phase, two distinct temperature levels are used in this design, allowing for better utilization of the adsorbent capacity. The first temperature level desorbs a large percentage of the refrigerant. The second, higher temperature level, on the other hand, accelerates the desorption process and yields more significant mass recovery.

The multi-bed cascaded cycle is another sophisticated cycle configuration, as shown in Figure 8.20. This configuration comprises cascading numerous adsorption beds, with the partially desorbed refrigerant from one bed utilized to warm the incoming refrigerant in the following bed. This cascade effect boosts heat utilization and mass recovery efficiency in the system. It can be applied particularly when the heat source temperatures exceed 150 °C. In the cascading cycle, various working pair materials are utilized for consecutive cycles operating at different

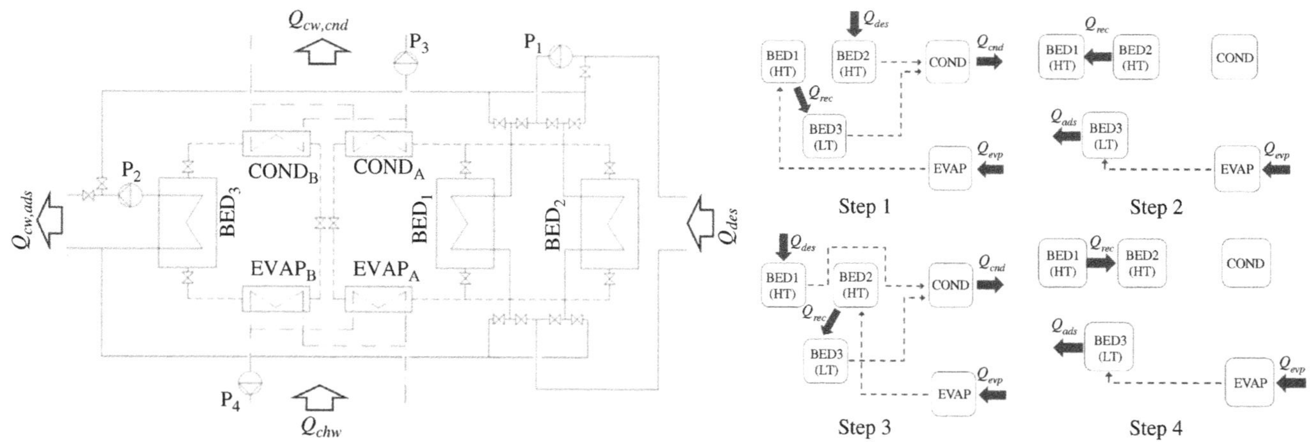

Figure 8.20 Schematics of a three-bed cascade cycle operating in four steps. Source: Ref. [183] / with permission of Elsevier.

temperature levels. Thus, the adsorption configurations become more intricate when compared to continuous cycles.

Multi-effect cycles can also be used to promote mass/heat recovery. These cycles employ numerous adsorption and desorption stages, each operating at a distinct temperature. The heat created during one desorption stage fuels the desorption process in succeeding stages, resulting in higher overall mass/heat recovery and increased COP [185].

Thermal regeneration cycles (thermal and forced convective/wave cycle) are also used to boost adsorption refrigeration output. The thermal regeneration cycles minimize operational inefficiencies in the adsorption cooling cycle. Two adsorbent beds are thermally interconnected in the thermal wave cycle using a heat transfer fluid, and this fluid generates a temperature variation or wave within the beds by passing through each adsorbent bed. In the convective thermal wave cycle, a refrigerant is employed instead of an external heat transfer fluid loop to achieve the same objective but with higher efficiency. The key to enhancing heat transfer rates within the bed is to induce forced convection between the adsorbent and the refrigerant itself. However, the main drawback lies in transferring recovered energy between the fluid loops, necessitating an extensive and costly gas-to-gas heat exchanger [14, 61, 65, 68].

8.4.4 Advanced Adsorbent Materials and Heat Exchanger Design

The choice of adsorbent–refrigerant materials plays a crucial role in the performance of an adsorption cooling system [29, 186]. Researchers have focused on developing and utilizing advanced adsorbent materials [187] with high adsorption capacity, fast adsorption and desorption kinetics, and good thermal stability. Various adsorbents, such as zeolites, activated carbon, metal–organic frameworks (MOFs), and silica gels, have been explored for their potential application in adsorption cooling systems [188]. These materials exhibit unique characteristics that can improve the system performance [189], including higher refrigerant uptake, better heat transfer properties, and enhanced cycle stability.

Optimizing the adsorbent materials is another venue for enhancing the performance of adsorption cooling systems. Selecting adsorbents with desirable properties, such as high surface area, adsorption capacity, selectivity, and thermal conductivity, can significantly impact the system's performance. Research and development efforts focus on designing and synthesizing novel adsorbent materials with improved characteristics. By optimizing the adsorbent properties, the system's capacity, cycle performance, and energy efficiency can be enhanced. Outside the adsorption cooling system approach, the adsorbent material and heat exchangers' design is the most investigated topic with high potential for research and improvement [8, 22, 43, 44, 61, 64, 67, 78, 82, 83, 88, 101, 135, 136, 142, 147, 163, 182, 190–225]. Some references are summarized in Table 8.4.

Table 8.4 Summary of the adsorbent material and heat exchangers' design (NA means not applied).

References	Year	Working pair	Adsorbent composition	Study characteristics	Heat exchanger layout
[101]	2023	Activated carbon-based R32 refrigerant	H25 Graphene nanoplatelets (GNPs) 1-Hexyl-3-methylimidazolium bis(trifluoromethyl-sulfonyl)imide ([HMIM][Tf2N]) ionic liquid Polyvinyl alcohol (PVA)	Material only. No modeling. Isotherm and diffusion/adsorption kinetic models from Eq. (8.12) and Eq. (8.32). Anisotropic thermal conductivity. HEX configuration dependency. Better cooling capacities and compact design than HFCs (hydrofluoro carbons). SCP = 100.6 W/kg (80 : 0 : 10 : 10) and 55.4 W/kg (40 : 40 : 10 : 10). COP = 0.13 (80 : 0 : 10 : 10) and 0.14 (40 : 40 : 10 : 10).	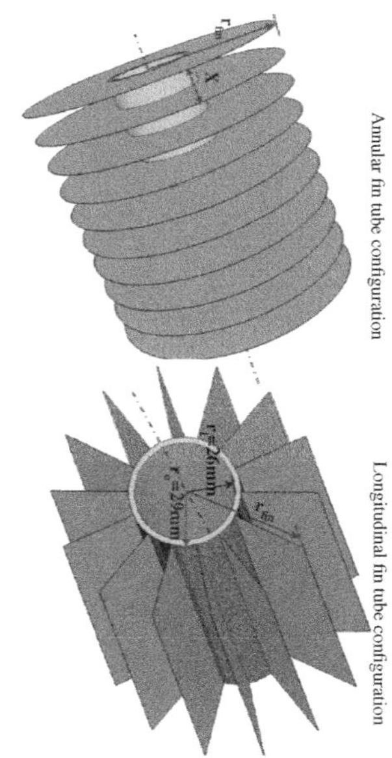
[135]	2023	MOF Water	MIL-101(Cr) with metal–organic polyhedral (MOP) encapsulation	Simple method. Two-bed layout. COP and SCP increase with MOP content. $SCP_{max} \approx 164$ W/kg (35% MOP content). $COP_{max} \approx 0.33$ (35% MOP content).	NA

(Continued)

Table 8.4 (Continued)

References	Year	Working pair	Adsorbent composition	Study characteristics	Heat exchanger layout
[136]	2023	MOF Water	MIL-101(Cr) with Graphene Oxide (GO)	Simple method. Two-bed layout. $SCP_{max} \approx 148$ W/kg (10% GO content). $COP_{max} \approx 0.32$ (35% GO content).	NA
[217]	2023	MOF-801 Water	MOF-801 packed into a copper foamed adsorbent bed Baseline comparison with silica gel–water	Simple method. Isotherm and diffusion/adsorption kinetic models Eq. (8.10) and Eq. (8.32). $SCP_{max} \approx 676.7$ W/kg (MOF-801). $COP_{max} \approx 0.65$ (>silica gel baseline $COP_{silica\ gel} \approx 0.69$).	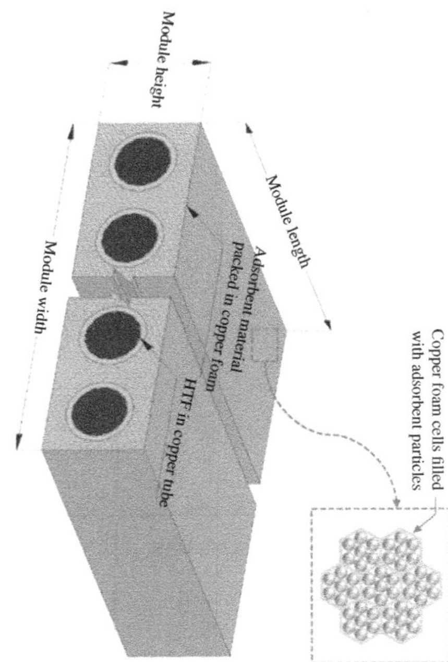

[216]	2023	Zeolite NaY Water–ethanol mixture	Zeolite NaY (Si/Al) The overall ethanol mole fraction is 0, 14, and 22 mol%.	Antifreezing enables temperatures below 0 °C. One-bed layout. SCP and COP decrease by 45% and 42%, respectively, with ethanol addition. Ethanol shows lower enthalpy of vaporization and reduces the adsorbed mass and heat transfer during evaporation.	NA
[215]	2023	Silica gel Water	Silica gel with 200–250 μm particle diameter (Geldart group B) with aluminum and carbon nanotube additives.	Silica gel mixture containing 5% carbon nanotubes shows the best mass transfer result. 5% and 15% aluminum powder reduces the sorption properties.	NA
[214]	2023	MOF Ethanol	Aluminum fumarate MIL-100(Fe) CAU-10H	COP = 0.92 and SCE = 510 kJ/kg by MIL-100(Fe).	NA
[213]	2023	Activated carbon Methanol	Activated carbon	One-bed layout. Solar collector was used. COP = 0.57 SCP = 140 W/kg	Finned condenser (144 fins/m).

(Continued)

Table 8.4 (Continued)

References	Year	Working pair	Adsorbent composition	Study characteristics	Heat exchanger layout
[212]	2023	Silica gel and aluminum Water	Triply periodic minimal surface (TPMS) configuration made of aluminum and silica gel (porosity) for the adsorber/desorber. TPMS configurations named as gyroid, diamond, IWP (I-graph-wrapped package), primitive, and lidinoid. Comparison with fins as baseline.	Two-bed layout. Higher performance for TPMS due to higher specific surface area. However, the performance is limited by the structures' porosity. SCP = 401.6 W/kg (lidinoid at a constant adsorption time) and 474.5 W/kg (lidinoid at the optimal adsorption time).	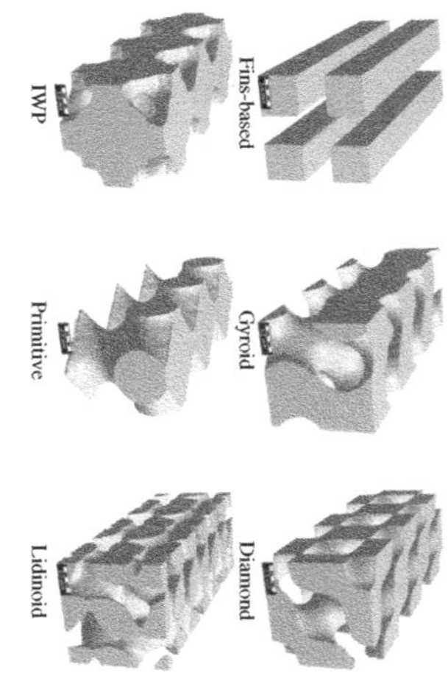
[88]	2023	Graphene oxide (GO) Water	GO from Graphitene Ltd. (0.36 porosity of 1–3 carbon atomic layers and particle sizes 0.5–20 μm) Silica gel (RD 2060 of 0.18–1 mm particle size) as baseline.	Isotherm and diffusion/adsorption kinetic models Eqs. (8.11) and (8.31). Coupled heat and mass transfer (distributed parameter model using COMSOL Multiphysics and finite element technique) and lumped parameter model (evaporator and adsorption bed).	NA

[197]	2018	High pure CO_2 gas (99.99%) Coconut shell activated carbon (CSAC)	CSAC – granular type activated carbon (IndoCarb GC E816)	Isotherm and diffusion/adsorption kinetic models Eqs. (8.7) and (8.12) models. Thermodynamics model. Two-bed layout (single-stage). Performance increases with increasing evaporator temperature but decreases with increasing adsorption/condenser temperature. SCE = 12.52 kJ/kg and COP = 0.10 (@ condenser temperature of 25 °C and evaporator temperature of 15 °C).	NA
[211]	2023	Water Covalent triazine framework (CTF), Porous organic polymer (POP), ionogel-based adsorbents. Silica gel as baseline for comparison.	Silica gel: A^{++}, A5BW, NS10, RD, RD2060 CTF: Pz-CTF6, pym-CTF500, bipy-CTF500, CTF-FL, CTF-TPC POP: PS-I, PS-II, 2Dep-POP, 3Dep-POP, MOPI-IV Ionogel: Syloid-72FP + 60 wt% Emim-Ac, Syloid-72FP + 60 wt% Emim-Oms, Syloid-AL1-FP + Emim-Ac, Syloid-AL1-FP + Emim-Oms, Syloid 72FP + 57 wt% Emim-Ac	Isotherm and diffusion/adsorption kinetic models Eq. (8.12). Clausius–Clapeyron Eq. (8.6). Thermodynamic model. One-bed configuration. COP = 0.74–0.93. Potential material: Syloid-72FP + 57 wt% Emim-Ac	NA

(Continued)

Table 8.4 (Continued)

References	Year	Working pair	Adsorbent composition	Study characteristics	Heat exchanger layout
[210]	2023	Water Activated carbon Maxsorb III (MAX)	Raw MAX, MAX treated with HCl, and activated MAX by impregnating it in $(NH_4)_2CO_3$.	Isotherm and diffusion/adsorption kinetic models Eqs. (8.12) and (8.32). Lumped parameter model. Two-bed configuration. With and without heat recovery technique. MAX $(NH_4)_2CO_3$ COP = 0.63and SCP = 373 W/kg	NA
[209]	2022	R32 refrigerant Maxsorb III activated carbon composites	Maxsorb III synthesized with the additives of: H25 graphene nanoplatelets (GNPs), 1-hexyl-3-methylimidazolium bis(trifluoromethyl sulfonyl)imide ([HMIM][Tf_2N]) ionic liquid with polyvinyl alcohol (PVA).	Isotherm and diffusion/adsorption kinetic models Eqs. (8.12) and (8.32). Lumped parameter model. MIII : H25 : IL : PVA 80 : 0 : 10 : 10 COP = 0.21 and SCP = 158.7 W/kg.	NA

[208]	2023	Water Nanoparticle (NP)-embedded silica gel (SG) composite adsorbents with $CaCl_2$	Nano-Fe, nano-Cu, and carbon nanotubes (CNTs) NP-embedded SG composite adsorbents with $CaCl_2$.	Isotherm and diffusion/adsorption kinetic models Eq. (8.32). No COP, SCP, or SCE disclosed. However, SG/CNT/$CaCl_2$ performed better than the other materials. SG/CNT/$CaCl_2$ adsorption capacity = 0.53 g/g.	NA
[82]	2022	Water Zeolites	AFI-type zeolites: FAM-Z series with AFI and CHA types. Silica gel NS10 (15 wt% Al_2O_3 and 85 wt% SiO_2) as baseline for comparison.	Modified Do-Do model. Hybrid Langmuir–Sips isotherms. Two-bed layout. Lumped parameter model. FAM-Z05 COP = 0.52, SCC = 62.82 RT/ton.	NA
[207]	2022	Water Copper foamed material	Copper foamed material coated with silico-aluminophosphate (SAPO-34) and nickel-based metal–organic framework CPO-27(Ni)	Isotherm and diffusion/adsorption kinetic models Eq. (8.32). Two-bed configuration. Transient axisymmetric 2D coupled mass, momentum, and energy conservation equations (CFD COMSOL Multiphysics). SAPO-34 COP = 0.421, SCP = 821 W/kg CPO-27(Ni) COP = 0.388, SCP = 855 W/kg	Copper foamed heat exchangers. Foam thickness: 2, 5, 10, and 15 mm, and number of tubes for each thickness SAPO-34: 266, 85, 32, and 17; CPO-27(Ni): 306, 98, 37, and 20.

(Continued)

Table 8.4 (Continued)

References	Year	Working pair	Adsorbent composition	Study characteristics	Heat exchanger layout
[182]	2022	Water Silica gel and zeolites	Silica gel: A++, A5BW, NS10, RD, and RD2060 Aluminophosphate-based zeolites (AQSOA): AQSOA-Z01, AQSOA-Z02, and AQSOA-Z05	Heat transformers. Best performance by AQSOA-Z01 zeolite 320-SCE = 370 kJ/kg.	NA
[206]	2022	Water Green synthesized transitional metal-doped aluminum fumarate metal–organic frameworks	Green synthesized transitional metal (10% Ni and 10% Co) doped aluminum fumarate metal–organic frameworks. Silica gel as baseline for comparison.	COP SMOF = 0.78 COP 10% Ni SMOF = 0.76 COP 10% Co SMOF = 0.77 SCE SMOF = 535.88 kJ/kg SCE 10% Ni SMOF = 476.33 kJ/kg SCE 10% Co SMOF = 523.97 kJ/kg	NA
[22]	2022	Water Zeolite-like aluminophosphate with SFO topology (EMM-8)	Zeolite-like aluminophosphate with SFO topology (EMM-8)	Two-bed configuration. EMM-8 COP = 0.85	NA

[205]	2022	Water Metal foam with silica gel (SG) and zeolites (AQSOA-Z02)	Aluminum and SG, copper and SG, Nickel and SG Aluminum and Z02, Copper and Z02, Nickel and Z02	Isotherm and diffusion/adsorption kinetic models Eq. (8.12) and Eq. (8.32). Transient axisymmetric 2D coupled mass, momentum, and energy conservation equations (CFD COMSOL Multiphysics). Best performance by Copper and SG. Copper and SG SCP = 291 W/kg	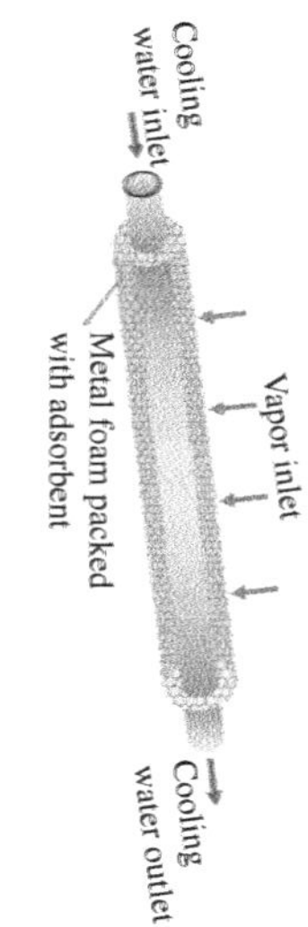
[83]	2022	Water Metal–organic frameworks (MOF)	Zirconium fumarate MOFs-801: MOF-801-RT, MOF-801-U60 $_m$, MOF-801-U90 $_m$, and MOF-801-U120 $_m$.	Isotherm and diffusion/adsorption kinetic models Eq. (8.12), Eqs. (8.14, 8.15), and Eq. (8.26). Thermodynamic model. COP depends on the surface coverage. For maximum surface coverage, MOF-801-RT, MOF-801-U90 $_m$ COP = 0.956.	NA

(Continued)

Table 8.4 (Continued)

References	Year	Working pair	Adsorbent composition	Study characteristics	Heat exchanger layout
[78]	2022	Ethanol Composites made of activated carbon with additives.	Activated carbon with MOF MIL-53Al, expanded graphite powder (EGP), silica get (SG), and polyvinylpyrrolidone (PVP).	Isotherm and diffusion/adsorption kinetic models Eq. (8.13). One-bed configuration. Composite B: Activated carbon 70 wt% with MOF MIL-53Al 10 wt%, EGP 10 wt%, SG 0 wt%, and PVP 10% COP = 0.61 and SCP = 198 W/kg	NA
[204]	2022	Water Aluminum fumarate MOF	Aluminum fumarate MOF	Isotherm and diffusion/adsorption kinetic models Eqs. (8.12) and (8.32). Two-bed configuration. 3D heat and mass transfer model using COMSOL Multiphysics based on the finite element method. Tube diameter of 10 mm, fin height of 7 mm, 20 fins, and coating thickness of 0.3 mm show the highest SCP = 585 W/kg	Heat exchanger based on a wire finned tube using coating, packing, or a combination of both. Outer tube diameters: 10, 15, and 22 mm Tube thickness: 0.5 and 0.75 mm Number of fins for each outer tube diameter: 20, 32, and 44 Fin heights: 7, 10, 15, and 20 mm Coating thickness: 0.3, 0.5, and 0.7 mm Fin spacing: 2 mm

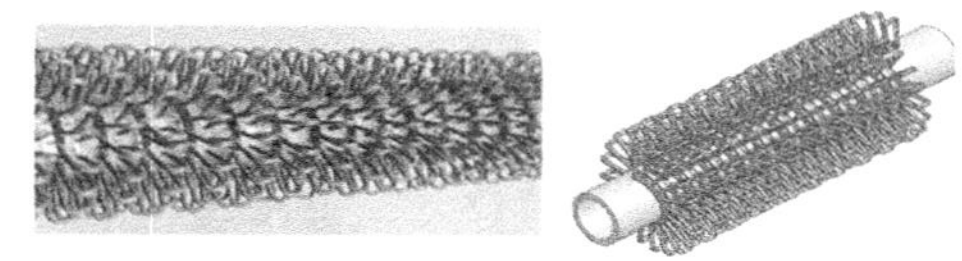

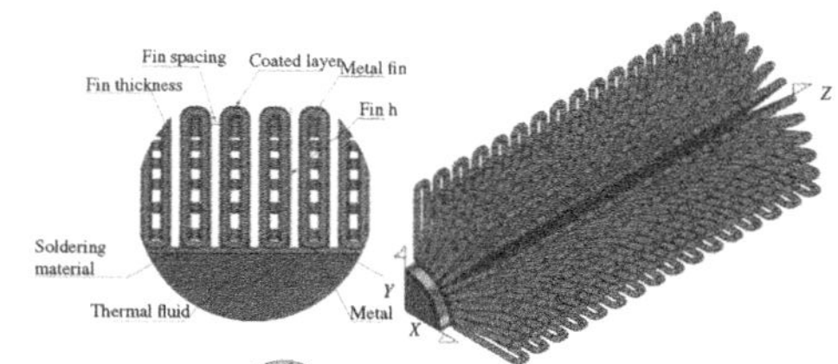

Fin spacing
Coated layer
Metal fin
Fin thickness
Fin h
Soldering material
Thermal fluid
Metal
Y
X
Z

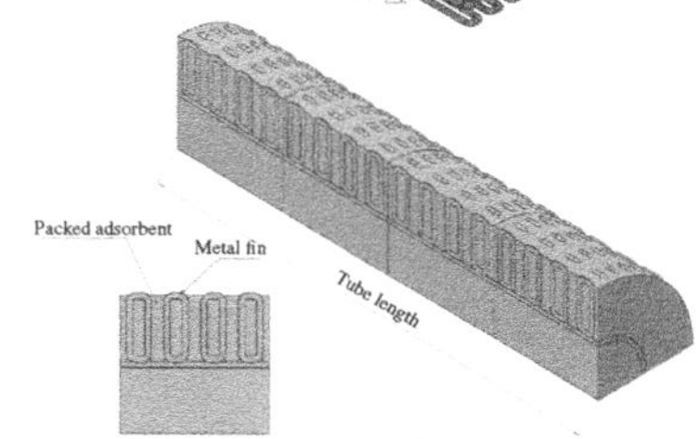

Packed adsorbent
Metal fin
Tube length

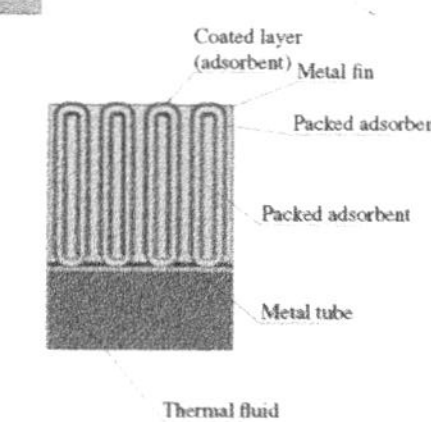

Coated layer (adsorbent)
Metal fin
Packed adsorbent
Packed adsorbent
Metal tube
Thermal fluid

(Continued)

Table 8.4 (Continued)

References	Year	Working pair	Adsorbent composition	Study characteristics	Heat exchanger layout
					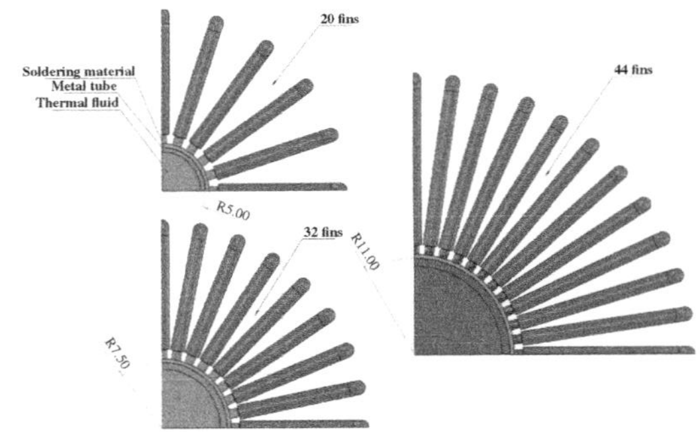 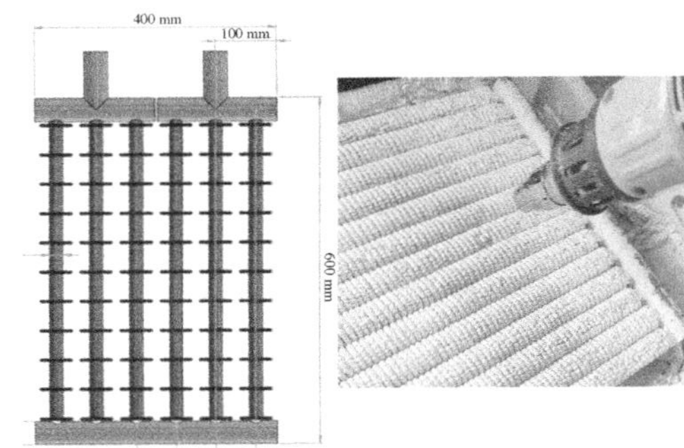

[219]	2021	Ethanol Activated carbon–graphene composite	Activated carbon–graphene nanoplatelet (GNP)-based composite with different ratios/grades of GNPs, PVA, and Maxsorb III.	Isotherm and diffusion/adsorption kinetic models Eq. (8.11) and Eq. (8.12). Isotherm and diffusion/adsorption kinetic models Eq. (8.32). Fickian diffusion model. Two-bed configuration. Thermodynamic model. Composite 1 (90 wt% Maxsorb III + 10 wt% PVA) COP = 0.76 and SCE = 621.70 kJ/kg.	NA
[200]	2018	Water Zeolite composite	Zeolite 13X/$CaCl_2$ composite. Zeolite 13X powder with average particle size 2 μm and anhydrous $CaCl_2$ powder with a concentration of 96 wt%.	Two-bed configuration. Thermodynamic model. SCP = 208.2 W/kg COP = 0.24	Coating on fins. 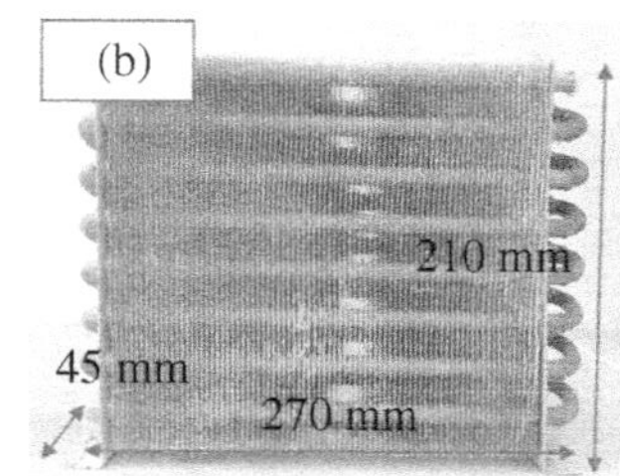

Adsorption cooling research and development have resulted in the discovery and characterization of novel porous adsorbent materials. These materials have distinct properties ideal for efficient cooling applications in adsorption systems. Recent advances in silica gel as an adsorbent for adsorption refrigeration have yielded encouraging results. Silica gel, a popular adsorbent material, has undergone numerous improvements to improve its adsorption capacity, selectivity, and effectiveness in adsorption refrigeration systems. Researchers have concentrated on improving the surface properties of silica gel by using methods such as functionalization and nanostructuring to improve its adsorption and heat transmission capacities. Efforts have also been made to optimize the pore structure of silica gel to enhance mass transfer kinetics and CC.

MOFs are one type of new porous adsorbent material. MOFs are crystalline structures composed of metal ions or clusters that have been coordinated with organic ligands. Due to their remarkable porosity and surface area, they have a high adsorption capacity and selectivity for specific refrigerants. MOF tunability stimulates the optimization of pore diameters and surface chemistries to improve adsorption efficacy and energy efficiency.

Zeolite is another type of porous adsorbent that has been widely researched for adsorption cooling applications. The microporous structures of these aluminosilicate minerals are well defined and linked. Zeolites have high adsorption capabilities and remarkable thermal stability, making them appropriate for diverse cooling applications.

Carbon-based materials, including activated carbon and carbon nanotubes, have also been investigated as innovative porous adsorbents. These materials have a high porosity, an enormous surface area, and customized surface chemistry, allowing efficient adsorption and desorption processes. Because of their capacity to integrate the benefits of diverse components, composite materials have received much attention in adsorption cooling.

Composite adsorbents are often made up of a porous matrix and an adsorbent substance that work together to improve performance. The adsorbent-impregnated composite is a popular composite material in which an adsorbent material is impregnated into a porous matrix. The porous matrix provides structural stability, while the adsorbent impregnated with the matrix improves adsorption capacity and selectivity. When compared to single-component materials, this combination improves performance. Composite materials have greater adsorption capacity, improved thermal conductivity, and enhanced stability in adsorption cooling. The porous matrix provides mechanical support and improves heat transport, while the adsorbent component has optimal adsorption qualities. The composite construction makes material composition tailored to cooling requirements, enhancing energy efficiency and system performance.

Surface modification and functionalization techniques increased adsorbent material performance in adsorption cooling systems. These methods involve modifying the surface properties of adsorbents to improve adsorption characteristics, selectivity, and heat transfer rates. Inserting functional groups onto the adsorbent surface is one surface modification method. Functional groups can improve the adsorbent–refrigerant interaction, enhancing adsorption capacity and selectivity. Chemical processes or physical deposition procedures can be used to accomplish surface functionalization approaches. Another surface modification method is to coat the adsorbent surface with thin layers. Thin films can increase mass transfer and selectivity by changing the surface chemistry or influencing diffusion rates. For example, chemical vapor or atomic layer deposition can produce uniform and controlled coatings. Surface morphological changes that create rough or hierarchical patterns can improve adsorption efficacy. These changes can increase adsorption surface area, improve pore accessibility, and improve heat transmission within the adsorbent material.

Heat exchangers play a crucial role in adsorption cooling systems by facilitating heat transfer between the working fluids and the surrounding environment. Innovative heat exchanger designs can enhance heat transfer efficiency, reduce pressure drops, and improve system performance [8, 186]. Advanced heat exchanger designs, such as compact, microchannel, and finned-tube heat exchangers, offer increased surface area, improved fluid distribution, and enhanced heat transfer characteristics. These designs optimize heat transfer, reduce thermal losses, and contribute to higher system efficiency.

8.5 Applications

Adsorption cooling systems offer versatile and sustainable cooling solutions for various applications [28]. Many commercialized adsorption chillers can be found in the literature [187], as shown in Table 8.5 from Chauhan et al. [21], and multiple patents can also be found in the study by Yong and Wang [40]. Two prominent applications are addressed: solar-driven adsorption and waste thermal-driven adsorption cooling [227, 228]. These applications demonstrate the potential of adsorption cooling technology in harnessing renewable energy sources and utilizing waste heat, contributing to energy efficiency and environmental sustainability.

Solar-driven adsorption cooling systems use solar energy as a primary heat source to drive the cooling process [17]. These systems offer a sustainable and environmentally friendly alternative to conventional cooling technologies, reducing reliance on electricity and fossil fuels [229–231].

Table 8.5 Adsorption chillers for commercial applications [21].

Manufacturer/ source	Models	Working ranges	Featured highlights	Applications	Remarks and updates
Bry-Air, India/(a)	• ADC Series:	C-FRAME: CC = 35–140 kW, 7250 kg,	• First of this kind in India,	✓ Industrial applications like food, pulp, power, beverages, paper, solar, cement industry, plastic industry, and chemical manufacturers.	Weights are shown for empty equipment. Costumers: • Adnan Menderes University, Turkey • Kolar Husky Co. Ltd., Hong Kong • IIT, Delhi, India • IIT, Jammu, India • MNIT, Jaipur, India • V S M Solar, Bangalore, India • University of Huddersfield, United Kingdom • Solution Solaire, Africa • Oorja Energy Services, India • King Abdulaziz City for Science and Technology (KACST), Saudi Arabia • Daiichi Jitsugyo, Japan • Victory Precisions Pvt. Ltd., India • Red Sea Farms Consultancy, Saudi Arabia • Zamatel Sdn. Bhd., Malaysia • EcoChill EHF, Iceland

	• C-FRAME (2.032 × 4.826 × 2.388)	D-FRAME: CC = 175–263 kW, 11,793 kg,	• Reduces CO_2 emission,	✓ In commercial areas such as buildings, malls, offices, and hotels.	• Australia Bangladesh Solar, Bangladesh Other waste heat applications: • Energy recovery • Air compressors • Aluminum • Automobile manufacturing plants • Diesel generator and gas engines • Food and beverages • Gas pipeline • General industry • Power plants • Pulp and paper Sugar
	• D-FRAME • (2.159 × 5.055 × 2.921) • E-FRAME	E–FRAME: CC = 350–527 kW, 14,969 kg, EXT E– FRAME: CC = 527–668 kW, 18,144 kg, F-FRAME: CC = 703–1160 kW, 21,319 kg,	• Reduces energy expenses, • Life expectancy > 25 years, • Ultra low electric consumption (1/10th of conventional system),	Note: Weight of chiller does not include weight of refrigerant. Size: (length(m) × width (m) × height (m))	The 11 kW model is named BryChill, and it can be stacked as modules. The adsorbent is a special silica gel BRYSORB 200. Operating weight is 525 kg. No noise and vibration.

(Continued)

Table 8.5 (Continued)

Manufacturer/ source	Models	Working ranges	Featured highlights	Applications	Remarks and updates
	• (2.489 × 5.080 × 2.921)	COP = 0.55,	• Start-up time 5–7 min only,		
	• Ext E-FRAME	Th,w,in = 52–93 °C,	• zero ozone depletion potential,		
	• (2.489 × 5.080 × 3.251)	Tchill,w,out = 3–20 °C,	• advanced microprocessor control,		
	• F-FRAME	Silica gel–water.	• wide temperature ranges allowed,		
	• (3.658 × 5.334 × 3.505)		• cut down cooling costs up to 99%,		
			• negligible maintenance,		
			• can work on process waste heat, tri-generation, solar heat, food and beverage industry, and chemical industry,		
			• green refrigerant (water).		
	• 11 kW model	CC = 11 kW,			
	(1 × 1.7 × 1.35)	COP = 0.55,			
	Weight of chiller (without refrigerant) = 450 kg	Th,w,in = 60–90 °C,			
		Tchill,w,out = 7–20 °C,			
		Silica gel–water.			

Mitsubishi Plastics, Inc., Chuo-ku, Tokyo, Japan/(b)	Cubed open-type cooling tower with energy-efficient space saving design (July 2008)	CC = 1143 kW (appx), Previous model: HT-325MQ-Ri: CC = 7.4 kW (2 units), New model: HT-325MQ-Hi: CC = 4.4 kW (2 units).	• Approximately 40% more energy efficient than previous product developed by Mitsubishi, • Requires comparatively lesser space for installation.	✓ Useful to cool the warm water of building, and factories and then cool water is used in chillers as a cooling fluid, ✓ Enhances the performance of chillers	CC ≈ 1143 kW is for the cooling tower (cooling capacity: 325 tons refrigeration, and 1 refrigeration ton = 3516.85 W).
Mitsubishi Plastics, Inc. Chuo-ku, Tokyo, Japan with Mayekawa Manufacturing Co., Ltd Koto-ku, Tokyo, Japan/(b)	Manufactured, first in the world, solar energy-powered adsorption chiller to achieve stable operation	CC = 350 kW, and 70 kW (pilot system),	• Cut downs 70% by weight of the desiccant,	✓ Building air-conditioning,	

(Continued)

Table 8.5 (Continued)

Manufacturer/ source	Models	Working ranges	Featured highlights	Applications	Remarks and updates
	(Aug 2008)	Th,w, in = 60–100 °C,	• Reduces 42% in the volume of adsorption,	✓ Food plants,	
		Tchill,w, out = 3–25 °C,	• Cut downs 44% in weight adsorption,	✓ Chemical plants, and	
		Zeolitic AQSOA™–water.	• Reduces 64% in CO_2 emissions,	✓ Beverage plants.	
			• Capable to generate 2–3 times chilled water per unit volume in comparison to silica gel-based conventional adsorption chiller.		
Mitsubishi Plastics, Inc. Chuo-ku, Tokyo, Japan/(b)	Highly efficient desiccant heating, ventilating, air-conditioning system capable of using Low-temperature waste heat	Th,w, in = 40–80 °C,	• AQSOA™-based honeycomb rotors are used,	✓ For humidification and dehumidification,	

	(July 2009)	AQSOA™–water.	• Capable to save approximately 20% of the energy in comparison to ordinary desiccant air conditioners.	✓ Semiconductor plants, food plants,
				✓ Cold/frozen storage, warehouses,
				✓ Supermarkets, hospitals, and welfare facilities.
Mitsubishi Plastics, Inc. Nagahama city, Japan/(b)	Pilot heating, ventilation, and air-conditioning system using AQSOA powered by light and heat from the sun	CC = 175 kW,	• Equipped with water-heating circuit using solar panels,	✓ Installed in main building-2 of own industry,
	(Sep 2010)	Th,w,in = 65 °C,	• Energy efficient,	✓ Reduces carbon emissions appx 40% annually.
		AQSOA™–water.	• Reduces CO_2 emissions (i.e. 1/3rd of ordinary chillers).	

(Continued)

Table 8.5 (Continued)

Manufacturer/ source	Models	Working ranges	Featured highlights	Applications	Remarks and updates
Mitsubishi Plastics, Inc., Chuo-ku, Tokyo, Japan and Union Industry Co., Ltd., Japan/(b)	Compact unit of adsorption chiller with integrated cooling tower for the U. S. market, U type chiller	CC = 10 kW,	• Cooling performance is 50% more from traditional chiller,	✓ Data centers,	
	(Jan 2012)	Th,w,in = 40–80 °C,	• On combining with a cogeneration system, electricity consumption is reduced by up to 20% compared to VCRS,	✓ Domestic houses.	
		Tchill,w, out = 10–15 °C,	• Powered by solar and industrial waste heat.		
		AQSOA™–water.			
Mitsubishi Plastics, Inc., Chuo-ku, Tokyo, Japan with Union Industry Co., Ltd., Japan/(b)	Compact design of adsorption chiller with the greatest CC per volume in the world, M type chiller	CC = 10 kW,	• AQSOA™ coated heat exchanger is used,	✓ Data centers,	

	(March 2012)	$T_{h,w,in}$ = 50–80 °C, External size: Length × Width × Height = 1.34 m × 0.8 m × 1.15 m, AQSOA™–water.	• Optimized and reduces power consumption by ➢ Up to appx. 50% (with waste thermal energy recycle system), ➢ Up to 20% (with cogeneration unit).	✓ Domestic houses.
InvenSor GmbH LTC, Germany/(b)	A small adsorption chiller (July 2013)	CC = 7–10 kW, $T_{h,w,in}$ = 60–70 °C, External size: Length × Width × Height = 1.1 m × 0.75 m × 1.37 m, light weight = 390 kg, AQSOATM–water.	• High performance AQSOA™-coated heat exchanger is used, • Negligible maintenance, • Excellent energy-saving, • Compact size and light weight, • Easy installation, • Touch panel operated.	✓ Data centers, ✓ Offices and stores, ✓ Cooling of industrial processes.

(Continued)

Table 8.5 (Continued)

Manufacturer/ source	Models	Working ranges	Featured highlights	Applications	Remarks and updates
InvenSor GmbH Berlin, Germany/(c)	Two adsorption chillers:	CC = 18 kW (HTC 18 plus),	• Patented environmentally friendly technology,	✓ HTC 18 plus is suitable for warm regions,	HTC 18 Plus • $COP_{th,max} = 0.55$ • $COP_{nominal} = 0.52$ • Length = 1.1 m × 0.75 m × 1.37 m • 420–425 kg
	HTC 18 plus (high-temperature chiller) and	CC = 10 kW (LTC 10 plus)			
	LTC 10 plus (low-temperature chiller)		• Cut downs power consumption by up to 70% compared to VCRS,	✓ LTC 10 plus is fit to use mainly in mild climatic regions,	LTC 10 Plus • $COP_{th,max} = 0.65$ • $COP_{nominal} = 0.60$ • Same dimensions of the HTC 18 Plus
			• Can operate on combined solar thermal, heat and power units, and industrial waste heat.	✓ Data centers,	
			• Capable to accomplish almost 100% of its capacity at Th,w,in = 65°C only.	✓ Offices and stores,	
				✓ Cooling of industrial processes.	

InvenSor GmbH/(d)	LTC 30 e series	CC = 10–35 kW (LTC 30 e),	• Integrated smart controller,	✓ Useful for industries,	LTC 30e Plus • $COP_{th,max} = 0.72$ • $COP_{nominal} = 0.60–0.65$ • Length = 1.56 m x 0.8 m x 1.845 m • 1200–1205 kg
	LTC 90 e series	CC = 30–105 kW (LTC 90 e)	• Up to almost 70% less power in comparison to VCRS.	✓ Free cooling function at surrounding temperatures.	
InvenSor GmbH/(e)	Plastic injection molding cooling	CC = 50 kW	• Four units of air compressors with CC = 120 kW thermal,	✓ Best practices industry cooling like tool cooling system in Bühler Motor GmbH, Monheim, Germany – 1st Project.	LTC 90e Plus • $COP_{th,max} = 0.75$ • $COP_{nominal} = 0.60–0.65$ • Length = 2.08 m × 2.54 m × 1.94 m • 4,500–4,515 kg
	5 units of InvenSor LTC 10 plus		• In operation since august 2015.	✓ Schmidt Kunststoffverarbeitung GmbH & Co. KG, Emsbüren, Germany.	
	1 unit of InvenSor LTC 90 e plus	CC = 90 kW	• Driving energy combined heat and power 2G 123 kW thermal/100 kW electricity.	✓ Busch-Jaeger Elektro GmbH, Bad Berleburg, Germany.	

(Continued)

Table 8.5 (Continued)

Manufacturer/ source	Models	Working ranges	Featured highlights	Applications	Remarks and updates
	4 units of InvenSor LTC 30 e plus and 3 units of InvenSor HTC 18 plus	CC = 174 kW	• Driving energy: combined heat and power 2G 290 kW thermal/250 kW electricity, • In operation since September 2016.		
	4 units of InvenSor LTC 30 plus	CC = 120 kW	• Driving energy: air compres-sor + combined heat and power, • In operation since October 2017.	✓ Metal processing like cold water pool in Gebrüder Eberhard GmbH & Co.KG, Nordheim, Germany.	
	11 units of InvenSor LTC 10 plus	CC = 110 kW	• Driving energy: combined heat and power Viessmann 370 kW thermal/240 kW electricity, • In operation since April 2015.	✓ Machine cooling like cooling of eroding machines in Miele & Cie. KG, Bünde, Germany.	
	3 units of InvenSor LTC 09	CC = 27 kW	• Driving source: solar energy using thermal collectors, • In operation since July 2011.	✓ Room cooling like chilled air circulation in Marburg, Germany.	

	InvenSor HTC 10	CC = 10 kW	• Driving source energy: solar energy using flat plate collectors, • In working since April 2011.	✓ Office space cooling in Turin, Italy, Tekno Energy.	
	InvenSor HTC 10	CC = 10 kW	• Driving source of energy: solar energy using thermal collectors, • In working since February 2011.	✓ Hotel room cooling in Cuiabá, Brazil.	
MAYEKAWA MFG. CO., LTD. Japan/(f)	AdRef-Noa: ADR-Z4520 Outer dimension: (5.6 m × 2.2 m × 2.35 m)	CC = 315 kW, Th,w,in = 68 °C, Tchill,w,out = 15 °C, Net weight = 11,000 kg,	• Heat source: unutilized energy (Waste heat from furnace, industry, plants, combined heat and power, etc.)	✓ Commercialized by NEDO in Japan.	• Achieving COP 10 • 70% less kgCO_2/kWh than air-cooled chiller • Study using low temperature waste heat [226]

(Continued)

Table 8.5 (Continued)

Manufacturer/ source	Models	Working ranges	Featured highlights	Applications	Remarks and updates
		Zeolite–water.			
MAYEKAWA MFG. CO., LTD. Japan/(g)	ADR-Z3515,	CC = 92 kW (ADR-Z3515),	• COP can be attained as 9.0.*1.	✓ For cooling purpose in pharmaceutical industry, and beverage industry etc.	ADR-Z3515 • 4 m × 2 m x 2.55 m • 5,800 kg
	ADR-Z3525,	CC = 184 kW (ADR-Z3525),	• Power consumption is very low.		ADR-Z3525 • 3.5 m x 3.2 m x 3 m • 10,000 kg
	ADR-Z6025	CC = 369 kW (ADR-Z6025),	• Energy source can be waste heat from factory and fuel cell, vapor condensate, hot spring, solar heat, and geothermal heat.		ADR-Z6025 • 5.7 m × 3.2 m × 5.7 m • 19,000 kg
		Th,w,in = 75 °C,	*1: Standard design conditions		
		Tchill,w,out = 9 °C,			
		Zeolite–water.			

Sources: (a). https://www.bryair.com/wp-content/uploads/2020/09/adc-technical-specs.pdf, (b). https://www.m-chemical.co.jp/en/news/mpi/201201170399.html, (c). https://www.esmagazine.com/articles/96174-chillers-invensor, (d). https://www.iea-shc.org/Data/Sites/53/media/events/meeting-09/workshop/14-kontogeorgopoulou_new-developments-and-applications-in-solar-cooling-with-invensor-adsorption-chillers.pdf, (e). https://www.iea-shc.org/Data/Sites/53/media/events/meeting-09/workshop/14-kontogeorgopoulou_new-developments-and-applications-in-solar-cooling-with-invensor-adsorption-chillers.pdf, (f). https://www.mayekawa.com.au/494/wp-content/uploads/2013/07/AdRef-Noa-Brochure-170713.pdf, (g). https://www.mayekawa.com.au/494/wp-content/uploads/2011/02/AdRef-Noa-Adsorption-Chiller1.pdf
Nomenclature: CC is cooling capacity, Th,w,in is the inlet water hot/high temperature, Tchill,w,out is the outlet chiller water temperature, $COP_{th,max}$ is the maximum theoretical COP, $COP_{nominal}$ is the nominal COP, and dimensions are shown as length × width × height and in meters.

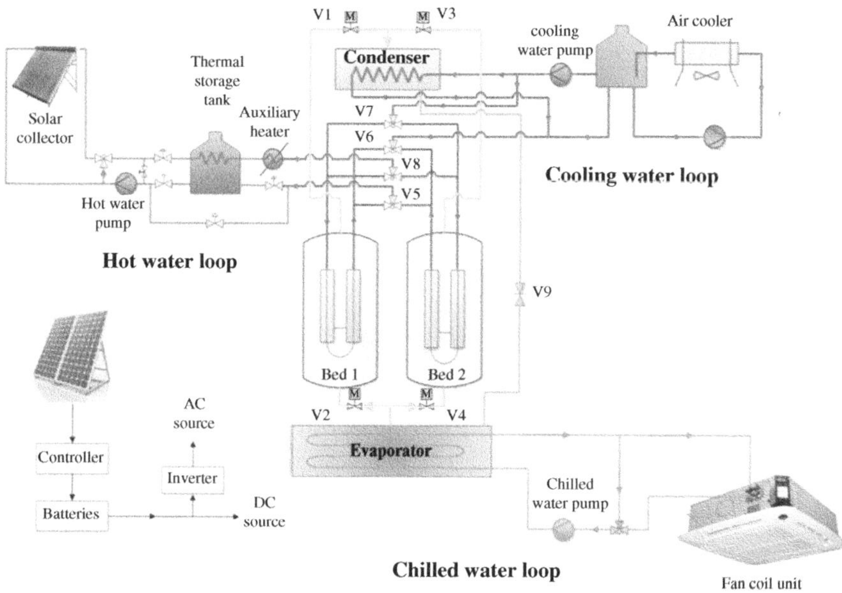

Figure 8.21 Schematic diagram of a solar adsorption system. Source: Ref. [53] / with permission of Elsevier.

In a solar-driven adsorption cooling system, solar collectors capture solar radiation and transfer the collected heat to the adsorption chiller (see Figure 8.21 and Figure 8.22). The heat is then utilized to drive the desorption process, releasing the refrigerant from the adsorbent material. The refrigerant vapor is then condensed and used for cooling purposes. The adsorption chiller operates in a closed-loop cycle, driven solely by solar energy.

Solar-driven adsorption cooling systems find applications in various sectors, including residential, commercial, and industrial buildings. They provide cooling comfort in off-grid areas, reduce electricity demand during peak load periods, and contribute to energy conservation and sustainability [26, 177, 232–244].

Waste thermal-driven adsorption cooling systems utilize waste heat as a valuable energy source for cooling applications. These systems harness waste heat from industrial processes, power generation, or other sources and convert it into generous CC [44, 245–248]. Geothermal and biomass energies can also be used as low-temperature heat sources [210, 249].

In a waste thermal-driven adsorption cooling system, waste heat streams are directed to the adsorption chiller for the desorption process. The waste heat heats the adsorbent material, causing the release of the refrigerant. The refrigerant vapor is then condensed and utilized for cooling purposes. By utilizing

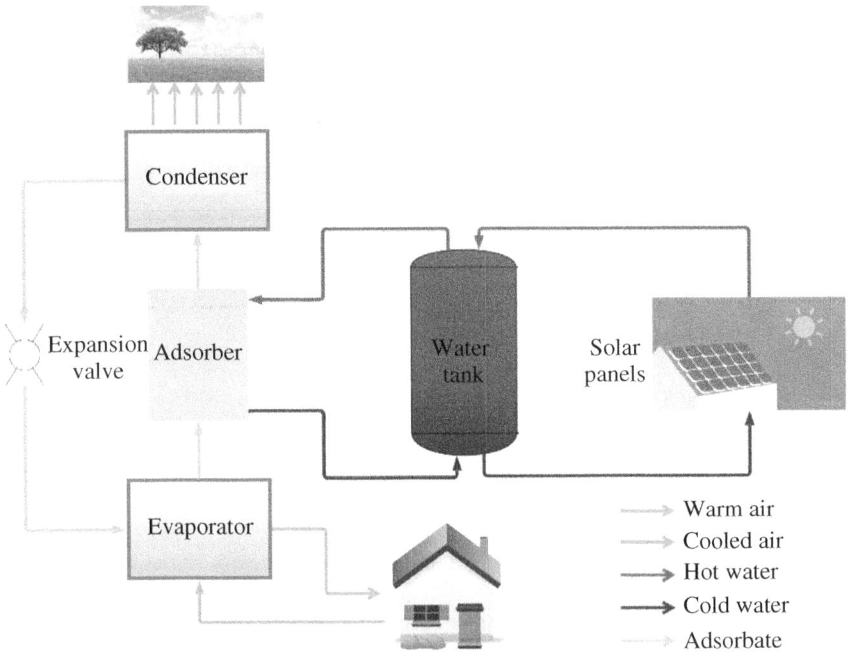

Figure 8.22 Schematic of a solar driven air-conditioning for buildings. Source: Ref. [19] / MDPI / CC BY 4.0.

waste heat, these systems offer a practical means of reducing energy waste and maximizing the energy efficiency of industrial processes. Waste thermal-driven adsorption cooling systems find applications in power generation, chemical processing, and manufacturing industries. They provide a means of utilizing waste heat, improving process efficiency, and reducing the environmental impact of waste heat discharge.

Additionally, adsorption cooling systems offer an attractive alternative for residential cooling, particularly in off-grid or remote areas. These systems can be powered by renewable energy sources such as solar or biomass, providing cooling comfort without grid-connected electricity. They have been successfully implemented in residential buildings, providing reliable and sustainable cooling solutions. Adsorption cooling finds application in commercial buildings, including offices, shopping malls, and hotels, where cooling demands are substantial. These systems can be integrated with existing heating, ventilation, and air-conditioning systems, providing efficient cooling while reducing electricity consumption and carbon emissions [26].

Adsorption cooling has also demonstrated potential in the automotive industry [250, 251], offering an energy-efficient and environmentally friendly alternative

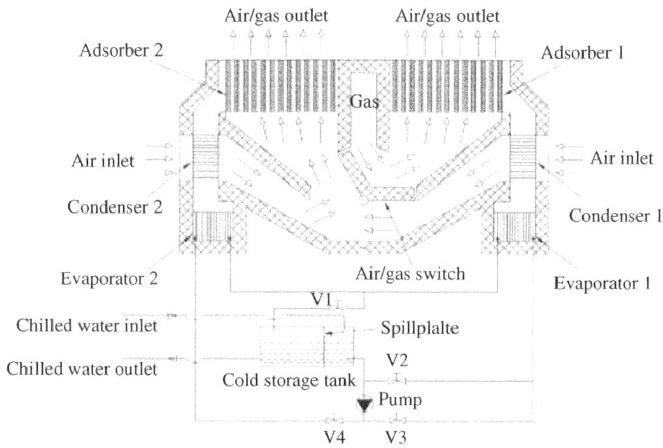

Figure 8.23 Double-bed adsorption air conditioner for continuous generation of chilled water. Source: Adapted from Refs. [248, 252].

to traditional vehicle air-conditioning systems. Adsorption cooling systems can be powered by waste heat from the vehicle's engine or integrated with hybrid or electric vehicles, reducing the reliance on the vehicle's electrical system. Adsorption cooling systems utilize vehicle waste heat [172], as seen in Figures 8.23–8.25. The findings demonstrate that this technology has the potential to enhance engine efficiency and mitigate thermal pollution emitted by engines. In many energy conversion devices, a substantial amount of energy in the exhaust stream is well spent if properly recovered or utilized. In internal combustion engines (ICEs), approximately 30–40% of the supplied energy is converted into practical mechanical work [248]. In contrast, the remaining energy is dissipated through engine cooling systems and exhaust gases, leading to increased entropy and significant environmental issues. The exhaust gas stream from ICE carries away around 30% of the combustion heat. In an adsorption system powered by exhaust heat, the adsorber is traversed by exhaust gases during the desorption phase. Subsequently, the evaporated refrigerant undergoes condensation at the condenser. Meanwhile, ambient air flows through the adsorption bed during the adsorption period, causing the refrigerant to evaporate from the evaporator and enter the adsorption bed. The adsorption cooling system in automobiles has two types of systems. One is exhaust heat-driven adsorption AC systems; the other is adsorption AC systems powered by coolant liquid. The transportation sector presents unique cooling challenges [253, 255–261], particularly in refrigerated transportation for the driver's well-being and the cooling of electronic components/equipment installed in vehicles. Adsorption cooling systems offer efficient and environmentally friendly cooling solutions, ensuring a nice temperature sensation for the driver,

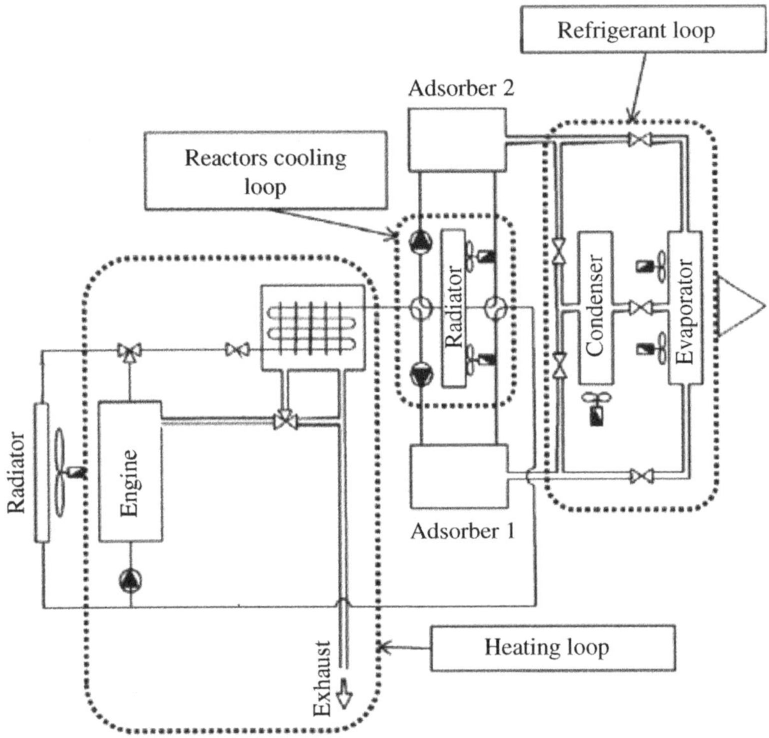

Figure 8.24 Adsorption cooling system using engine water coolant. Source: Adapted from Refs. [248, 253].

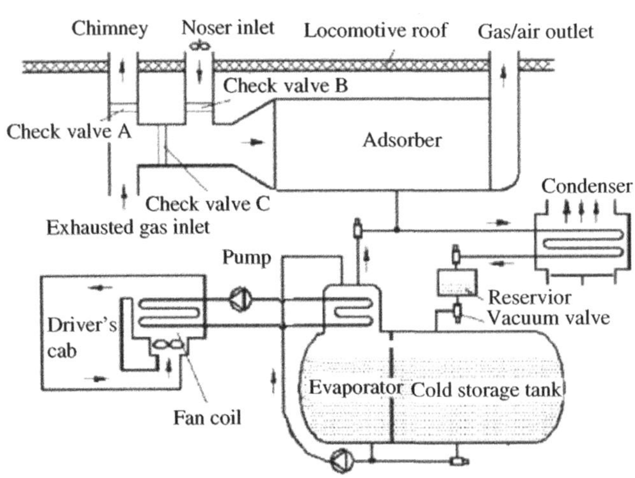

Figure 8.25 Adsorption air-conditioning worked by internal combustion engine exhaust heat. Source: Adapted from Refs. [248, 254].

the preservation of temperature-sensitive goods, and the optimal performance of electronic systems in transportation applications.

Hybridizing adsorption cooling systems with other cooling technologies, such as vapor compression or absorption systems, have been explored to enhance their performance [262, 263], such as combining adsorption and absorption. By combining the strengths of different cooling technologies, hybrid systems can achieve higher cooling capacities, improved performance coefficients, and greater operational flexibility. Another example is integrating an adsorption chiller with a vapor compression chiller [21] can improve performance during periods of high cooling demand or when the heat source temperature is insufficient for adsorption cooling alone. Integrating adsorption cooling with desalination also has significant improvements in water production and system's efficiency [264].

The food and beverage industry requires precise temperature control and preservation techniques to maintain product quality and freshness [216, 265–267]. Adsorption cooling systems offer an effective solution for cooling and refrigeration needs in this industry [268]. They provide reliable and energy-efficient cooling, reducing dependency on traditional vapor compression systems and ensuring the preservation of perishable goods. Adsorption cooling systems offer a viable solution for food storage and preservation, particularly in off-grid or remote areas where access to electricity may be limited. These systems can provide precise temperature and humidity control, ensuring the freshness and quality of perishable goods during storage and transportation.

The healthcare sector relies on precise temperature control for various applications, including medical equipment cooling (see Figure 8.26), with possible applications in laboratory testing and pharmaceutical storage [230, 269]. Adsorption cooling systems offer a reliable and energy-efficient solution, ensuring the integrity and efficacy of medical products while reducing energy consumption and environmental impact.

Figure 8.26 Adsorption cooling used in a hospital [230].

Data centers consumes a significant amount of energy for cooling purposes. Adsorption cooling systems offer an energy-efficient alternative, reducing data centers' carbon footprint and operating costs (see Table 8.5). These systems can utilize waste heat generated by data center equipment for cooling, ensuring sustainable and reliable operation. The telecommunications industry relies on cooling systems to maintain the optimum operating temperature of communication equipment and data centers [270, 271]. Adsorption cooling systems provide efficient and sustainable cooling solutions, or energy storage [272], reducing the energy consumption and environmental impact of cooling infrastructure in the telecommunications sector.

8.6 Discussion

Adsorption cooling is a promising technology for energy-efficient and sustainable cooling applications. However, as with any innovative technology, it faces numerous challenges that must be addressed before it can be widely adopted and implemented.

Obtaining a high COP is one of the most difficult challenges in adsorption cooling. Adsorption cooling systems typically have a lower COP than conventional vapor compression systems. This relatively lower COP is due to lower heat transfer rates in adsorption processes, adsorption/desorption kinetic limitations, and pressure drop within the adsorbent bed. Efforts are being made to improve the COP using advanced materials, better system designs, and optimization techniques.

Adsorbent material selection is critical for the performance and efficiency of adsorption cooling systems. The difficulties lie in identifying adsorbents with high adsorption capacity, fast kinetics, thermal stability, and refrigerant compatibility. Furthermore, adsorbent regeneration and cyclic stability are critical considerations. Ongoing research focuses on the creation of novel adsorbent materials with enhanced properties. The development of advanced adsorbent materials is still an active area of research. For example, MOFs, zeolites, and composite adsorbents hold great promise for improving the performance and efficiency of adsorption cooling systems. These materials have distinct properties, such as higher adsorption capacities, improved selectivity, and increased stability, which can lead to significant advances in adsorption cooling technology.

Heat and mass transfer efficiencies are critical for the performance of adsorption cooling systems. High heat transfer rates between the heat source and adsorbent bed and within the adsorbent materials are difficult to achieve. Improvements in heat exchanger design, fin configurations, and flow distribution can improve heat and mass transfer, increasing system efficiency.

Adsorption cooling offers a substantial opportunity for waste heat recovery. Adsorption can be driven by waste heat from various industrial processes, power generation, or other sources, improving the system's overall energy efficiency and reducing environmental impact by utilizing waste heat that would otherwise be wasted. Adsorption cooling can be integrated effectively with renewable energy sources such as solar thermal, biomass, or geothermal. This integration enables the use of sustainable energy for cooling, reducing reliance on traditional energy sources, and lowering greenhouse gas emissions. Adsorption cooling can be combined with other cooling technologies, such as vapor compression or thermoelectric cooling, to create hybrid systems and combined technologies. These hybrid systems can utilize each technology's strengths, improving system performance, efficiency, and flexibility.

While traditional adsorption cooling systems typically consist of one or more beds, there is growing interest in investigating advanced system configurations. Multistage and combined cycles with other cooling technologies are examples of these configurations. It is possible to achieve higher efficiencies, better performance, and more options for meeting various cooling needs by taking advantage of how adsorption processes differ and combining them with other cooling methods.

Heat exchanger design and optimization are critical to the performance of adsorption cooling systems. Microchannel heat exchangers, compact heat exchangers, and heat pipes are advanced heat exchanger configurations that can improve heat transfer, reduce pressure drop, and increase system efficiency. Investigating new heat exchanger designs can significantly improve the performance and compactness of adsorption cooling systems.

The possibility of integrating adsorption cooling systems with other energy systems and applications is intriguing. This includes combining adsorption cooling with thermal storage systems, utilizing waste heat, district cooling networks, and implementing decentralized energy systems. Such integration can result from synergistic effects, improved system performance, and increased energy efficiency.

Mathematical models, such as the simple, lumped parameter, and distributed parameter models, are critical in analyzing and optimizing adsorption cooling systems. These models, however, face challenges and present opportunities for further development. One challenge is the simple model's oversimplification, which ignores spatial and temporal variations, heat and mass transfer resistances, and non-uniformities within the adsorbent bed. The lumped parameter model improves on this by considering transient effects, but it still assumes uniform temperature and concentration distributions. The distributed parameter model overcomes these constraints by accounting for spatial variations, but it necessitates high computational resources. The challenges are the complexity of accurately modeling the adsorption–desorption processes, developing appropriate adsorption isotherm models, and incorporating nonideal behavior. Although opportunities are limited, opportunities exist in incorporating more advanced mathematical techniques such as finite element methods and CFD, considering nonequilibrium and dynamic behaviors, and incorporating experimental data for validation. Advances in mathematical models can lead to more accurate predictions, improved system performance, and more efficient adsorption cooling system design, especially when adsorption is combined with other cooling techniques (absorption).

Exploring new perspectives and future directions in adsorption cooling opens exciting new avenues for this technology's advancement. Adsorption cooling can continue to evolve as a sustainable and efficient cooling solution by focusing on advanced adsorbent materials, innovative system configurations, optimized heat exchanger designs, and integration with other energy systems. This integration can unlock the full potential of adsorption cooling through ongoing research and collaboration, paving the way for a greener, more sustainable future. While COP, adsorbent materials, and heat transfer are all challenges for adsorption cooling, there are significant opportunities for improvement and expansion. Adsorption cooling system advancement can lead to higher efficiency, more expansive application areas, and greater sustainability in the field of cooling technology through ongoing research and development efforts.

CO_2 captured from the air, or Direct Air Capture (DAC), shows promise in mitigating climate change by removing CO_2 from the atmosphere [273]. Integrating DAC with adsorption cooling offers several benefits, such as utilizing the captured CO_2 for cooling applications and reducing energy requirements, enhancing efficiency and economic viability, and supporting the transition to a low-carbon energy system [64, 197, 274–278]. However, challenges include the energy-intensive nature of DAC, limiting efficiency gains, and the need for large-scale deployment for significant impact. Also, finding suitable adsorbent materials for efficient cooling while maintaining high CO_2 adsorption capacity poses a technical challenge. Despite these obstacles, this integration can mitigate global warming and provide sustainable cooling.

8.7 Conclusion

This chapter has looked at adsorption cooling as a thermally driven cooling technology with a bright future regarding sustainability and energy efficiency. Insights into the working principles and configurations of adsorption cooling systems are gained through an overview of their many aspects.

The fundamental concepts of adsorption cooling have been covered, including one-bed, two-bed, multi-bed, and multistage configurations, providing a solid foundation for understanding these systems' operational principles and design considerations. Furthermore, mathematical modeling techniques for describing adsorption cooling processes, such as simple, lumped parameter, and distributed parameter models, were discussed. These modeling approaches predict system performance and optimize design parameters, with the distributed parameter model producing the best results but requiring the most computational effort. Future research should focus on mathematical modeling optimization.

Performance enhancement strategies have also been investigated, such as mass recovery, heat recovery, and combined mass-heat methods. These techniques improve the efficiency and effectiveness of adsorption cooling systems, allowing them to operate with less energy consumption and environmental impact, with the combined method showing promise for further investigation. Adsorbent materials and heat exchangers must be chosen carefully to achieve high-performance adsorption cooling systems.

Adsorption cooling applications ranging from solar-powered cooling systems to waste thermal utilization were investigated. These applications demonstrate adsorption cooling technology's adaptability and ability to meet cooling needs in various industries, including residential, commercial, and industrial settings. However, a vast opportunity is still available for further industrial usage, especially combining adsorption with other cooling techniques like absorption.

There are several intriguing possibilities for future research in adsorption cooling. Creating advanced adsorbent materials with improved adsorption properties and thermal stability holds promise for improving system performance. Adsorption cooling, integration with renewable energy sources and DAC, and advanced heat exchanger designs may improve efficiency and sustainability.

To summarize, adsorption cooling is a promising alternative to traditional cooling technologies, with numerous advantages such as lower energy consumption, lower environmental impact, and the potential use of waste heat sources. With further advancements and research, adsorption cooling systems have the potential to play a pivotal role in shaping a greener and more sustainable future.

References

1 Bahadori, M.N. (1978). Passive cooling systems in Iranian architecture. *Scientific American* 238 (2): 144–155. http://www.jstor.org.ezproxy.lib.rmit.edu.au/stable/24955643.

2 Moosavi, M. (2011). *An Analysis to Historic Roots of Climatic Design in Ancient Architecture of Central Zone of Iran*, vol. 17. International Proceedings of Economics Development & Research.

3 Mahdavinejad, M. and Javanrudi, K. (2012). Assessment of ancient fridges: a sustainable method to storage ice in hot-arid climates. *Asian Culture and History* 4 (2): https://doi.org/10.5539/ach.v4n2p133.

4 Beigli, F. and Lenci, R. (2016). Underground and semi underground passive cooling strategies in hot climate of Iran. *Journal of Environmental Sciences* 5 (3): 1–12.

5 Sørensen, B. (2018). *Renewable Energy: Four Volume Set*, 1e. Routledge.

6 Bahadori, M.N. (2018). Chapter 9 – Passive cooling systems in Iranian architecture. In: *Renewable Energy*, Vol1_87–Vol1_101. Routledge.

7 Shokoohi, K. (2023). The ancient Persian way to keep cool. *Future Planet*. www.bbc.com/future/article/20210810-the-ancient-persian-way-to-keep-cool (accessed 27 April 2024).

8 Khatibi, M., Mohammadzadeh Kowsari, M., Golparvar, B. et al. (2022). A comparative study to critically assess the designing criteria for selecting an optimal adsorption heat exchanger in cooling applications. *Applied Thermal Engineering* 215: https://doi.org/10.1016/j.applthermaleng.2022.118960.

9 Ng, K.C. and Saha, B.B. (2010). Application of adsorption technologies for energy efficiency. *Heat Transfer Engineering* 31 (11): 907–909. https://doi.org/10.1080/01457631003603816.

10 Das, R. and Saha, B.B. (2022). *Rapid Refrigeration and Water Protection*. Springer Water.

11 van Ruijven, B.J., De Cian, E., and Sue Wing, I. (2019). Amplification of future energy demand growth due to climate change. *Nature Communications* 10 (1): https://doi.org/10.1038/s41467-019-10399-3.

12 IEA (2018). *The Future of Cooling - Opportunities for Energy-Efficient Air Conditioning*. IEA https://www.iea.org/reports/the-future-of-cooling; https://www.iea.org/reports/the-future-of-cooling (accessed 14 July 2023).

13 Armstrong, M. (2020). Air conditioning biggest factor in growing electricity demand. Topics, Energy Storage, Future. https://www.statista.com/chart/14401/growing-demand-for-air-conditioning-and-energy/ (accessed 27 April 2024).

14 Alahmer, A., Ajib, S., and Wang, X. (2019). Comprehensive strategies for performance improvement of adsorption air conditioning systems: a review. *Renewable and Sustainable Energy Reviews* 99: 138–158. https://doi.org/10.1016/j.rser.2018.10.004.

15 Wang, R., Wang, L., and Wu, J. (2014). *Adsorption Refrigeration Technology*. Wiley.

16 Steven Brown, J. and Domanski, P.A. (2014). Review of alternative cooling technologies. *Applied Thermal Engineering* 64 (1–2): 252–262. https://doi.org/10.1016/j.applthermaleng.2013.12.014.

17 Ojha, M.K., Shukla, A.K., Verma, P., and Kannojiya, R. (2021). Recent progress and outlook of solar adsorption refrigeration systems. *Materials Today: Proceedings* 46: 5639–5646. https://doi.org/10.1016/j.matpr.2020.09.593.

18 Gado, M.G., Ookawara, S., Nada, S., and El-Sharkawy, I.I. (2021). Hybrid sorption-vapor compression cooling systems: a comprehensive overview. *Renewable and Sustainable Energy Reviews* 143: https://doi.org/10.1016/j.rser.2021.110912.

19 Dias, J.M.S. and Costa, V.A.F. (2022). Adsorption cooler design, dynamic modeling, and performance analyses. *Clean Technologies* 4 (4): 1152–1161. https://doi.org/10.3390/cleantechnol4040070.

20 Randazzo, T., De Cian, E., and Mistry, M.N. (2020). Air conditioning and electricity expenditure: the role of climate in temperate countries. *Economic Modelling* 90: 273–287. https://doi.org/10.1016/j.econmod.2020.05.001.

21 Chauhan, P.R., Kaushik, S.C., and Tyagi, S.K. (2022). Current status and technological advancements in adsorption refrigeration systems: a review. *Renewable and Sustainable Energy Reviews* 154: https://doi.org/10.1016/j.rser.2021.111808.

22 Liu, Z., Xu, J., Xu, M. et al. (2022). Ultralow-temperature-driven water-based sorption refrigeration enabled by low-cost zeolite-like porous aluminophosphate. *Nature Communications* 13 (1): https://doi.org/10.1038/s41467-021-27883-4.

23 Unuabonah, E.I., Omorogie, M.O., and Oladoja, N.A. (2019). Modeling in adsorption: fundamentals and applications. In: *Composite Nanoadsorbents*, 85–118. Elsevier B.V.

24 Kilic, M. and Anjrini, M. (2020). Comparative performance analysis of a combined cooling system with mechanical and adsorption cycles. *Energy Conversion and Management* 221: https://doi.org/10.1016/j.enconman.2020.113208.

25 Lewis, J., Chaer, I., and Tassou, S. (2007). *Reviews of Alternative Refrigeration Technologies*. Brunel University.

26 Sidhareddy, M., Tiwari, S., Phelan, P., and Bellos, E. (2023). Comprehensive review on adsorption cooling systems and its regeneration methods using solar, ultrasound, and microwave energy. *International Journal of Refrigeration* 146: 174–201. https://doi.org/10.1016/j.ijrefrig.2022.10.025.

27 Younes, M.M., El-Sharkawy, I.I., Kabeel, A.E., and Saha, B.B. (2017). A review on adsorbent-adsorbate pairs for cooling applications. *Applied Thermal Engineering* 114: 394–414. https://doi.org/10.1016/j.applthermaleng.2016.11.138.

28 Elsheniti, M.B., Elsamni, O.A., Al-dadah, R.K. et al. (2018). Chapter 4 – Adsorption refrigeration technologies. In: *Sustainable Air Conditioning Systems*. IntechOpen.

29 Shabir, F., Sultan, M., Miyazaki, T. et al. (2020). Recent updates on the adsorption capacities of adsorbent-adsorbate pairs for heat transformation applications. *Renewable and Sustainable Energy Reviews* 119: https://doi.org/10.1016/j.rser.2019.109630.

30 Puthumana, A.B., Paulose, P., Raveendran, R., and Shouri, D.P. (2017). Analysis of two stage solar vapour adsorption refrigeration system. *International Journal of Mechanical Engineering and Technology* 8 (4): 1–9.

31 Sharafianardakani, A. and Bahrami, M. (2012). A quasi steady state model for adsorption cooling systems: automotive applications. Presented at the ASME 2012 6th International Conference on Energy Sustainability, Parts A and B.

32 Demir, H., Mobedi, M., and Ülkü, S. (2008). A review on adsorption heat pump: problems and solutions. *Renewable and Sustainable Energy Reviews* 12 (9): 2381–2403. https://doi.org/10.1016/j.rser.2007.06.005.

33 Pons, M. and Kodama, A. (2000). Entropic analysis of adsorption open cycles for air conditioning. Part 1: first and second law analyses. *International Journal of Energy Research* 24 (3): 251–262. https://doi.org/10.1002/(sici)1099-114x(20000310)24:3<251::Aid-er578>3.0.Co;2-u.

34 Critoph, R.E. (1988). Performance limitations of adsorption cycles for solar cooling. *Solar Energy* 41 (1): 21–31. https://doi.org/10.1016/0038-092x(88)90111-9.

35 Passos, E., Meunier, F., and Gianola, J.C. (1986). Thermodynamic performance improvement of an intermittent solar-powered refrigeration cycle using adsorption of methanol on activated carbon. *Journal of Heat Recovery Systems* 6 (3): 259–264. https://doi.org/10.1016/0198-7593(86)90010-x.

36 Douss, N. and Meunier, F. (1988). Effect of operating temperatures on the coefficient of performance of active carbon-methanol systems. *Heat Recovery Systems and CHP* 8 (5): 383–392. https://doi.org/10.1016/0890-4332(88)90042-7.

37 Critoph, R.E. (1994). An ammonia carbon solar refrigerator for vaccine cooling. *Renewable Energy* 5 (1–4): 502–508. https://doi.org/10.1016/0960-1481(94)90424-3.

38 Li, M., Wang, R.Z., Xu, Y.X. et al. (2002). Experimental study on dynamic performance analysis of a flat-plate solar solid-adsorption refrigeration for ice maker. *Renewable Energy* 27 (2): 211–221. https://doi.org/10.1016/s0960-1481(01)00188-4.

39 Liu, Y. and Leong, K.C. (2005). The effect of operating conditions on the performance of zeolite/water adsorption cooling systems. *Applied Thermal Engineering* 25 (10): 1403–1418. https://doi.org/10.1016/j.applthermaleng.2004.09.013.

40 Yong, L. and Wang, R. (2007). Adsorption refrigeration: a survey of novel technologies. *Recent Patents on Engineering* 1 (1): 1–21. https://doi.org/10.2174/187221207779814725.

41 Aristov, Y.I., Sapienza, A., Ovoshchnikov, D.S. et al. (2012). Reallocation of adsorption and desorption times for optimisation of cooling cycles. *International Journal of Refrigeration* 35 (3): 525–531. https://doi.org/10.1016/j.ijrefrig.2010.07.019.

42 Rezk, H., Mohammed, R.H., Rashad, E., and Nassef, A.M. (2022). ANFIS-based accurate modeling of silica gel adsorption cooling cycle. *Sustainable Energy Technologies and Assessments* 50: 101793. https://doi.org/10.1016/j.seta.2021.101793.

43 Attan, D., Alghoul, M.A., Saha, B.B. et al. (2011). The role of activated carbon fiber in adsorption cooling cycles. *Renewable and Sustainable Energy Reviews* 15 (3): 1708–1721. https://doi.org/10.1016/j.rser.2010.10.017.

44 Sosnowski, M. (2019). Experimental and numerical analysis of multi-disc heat exchanger efficiency in adsorption chillers powered with waste heat. *Journal of Physics: Conference Series* 1398 (1): https://doi.org/10.1088/1742-6596/1398/1/012013.

45 Jribi, S., Chakraborty, A., El-Sharkawy, I.I. et al. (2008). Thermodynamic analysis of activated carbon-CO_2 based adsorption cooling cycles. *World Academy of Science, Engineering and Technology* 43: 63–66.

46 Chua, H.T., Ng, K.C., Malek, A. et al. (1999). Modeling the performance of two-bed, sillica gel-water adsorption chillers. *International Journal of Refrigeration* 22 (3): 194–204. https://doi.org/10.1016/s0140-7007(98)00063-2.

47 Alam, K.C.A., Kang, Y.T., Saha, B.B. et al. (2003). A novel approach to determine optimum switching frequency of a conventional adsorption chiller. *Energy* 28 (10): 1021–1037. https://doi.org/10.1016/s0360-5442(03)00064-1.

48 Chua, H.T., Ng, K.C., Wang, W. et al. (2004). Transient modeling of a two-bed silica gel–water adsorption chiller. *International Journal of Heat and Mass Transfer* 47 (4): 659–669. https://doi.org/10.1016/j.ijheatmasstransfer.2003.08.010.

49 Wang, X. and Chua, H.T. (2007). Two bed silica gel–water adsorption chillers: an effectual lumped parameter model. *International Journal of Refrigeration* 30 (8): 1417–1426. https://doi.org/10.1016/j.ijrefrig.2007.03.010.

50 Miyazaki, T., Akisawa, A., Saha, B.B. et al. (2009). A new cycle time allocation for enhancing the performance of two-bed adsorption chillers. *International Journal of Refrigeration* 32 (5): 846–853. https://doi.org/10.1016/j.ijrefrig.2008.12.002.

51 Saha, B.B., Chakraborty, A., Koyama, S., and Aristov, Y.I. (2009). A new generation cooling device employing $CaCl_2$-in-silica gel–water system. *International Journal of Heat and Mass Transfer* 52 (1–2): 516–524. https://doi.org/10.1016/j.ijheatmasstransfer.2008.06.018.

52 Baiju, V., Asif Sha, A., Sajid, N.K.M., and Muhammedali Shafeeque, K. (2021). Simulation and performance study of a two-bed adsorption cooling system operated with activated carbon-ethanol. *Proceedings of the Institution of Mechanical Engineers, Part C: Journal of Mechanical Engineering Science* 236 (7): 3804–3817. https://doi.org/10.1177/09544062211042040.

53 Elsheniti, M.B., Abd El-Hamid, A.T., El-Samni, O.A. et al. (2021). Experimental evaluation of a solar two-bed lab-scale adsorption cooling system. *Alexandria Engineering Journal* 60 (3): 2747–2757. https://doi.org/10.1016/j.aej.2021.01.024.

54 Yagnamurthy, S., Rakshit, D., Jain, S., and Saha, B.B. (2021). Operational envelope and performance enhancement of a two-bed adsorption cooling system. *Applied Thermal Engineering* 195: https://doi.org/10.1016/j.applthermaleng.2021.117181.

55 Baiju, V., Asif Sha, A., Shajahan, C.A., and Chindhu, V.G. (2022). Energy and exergy based assessment of a two bed solar adsorption cooling system. *International Journal of Refrigeration* 141: 90–101. https://doi.org/10.1016/j.ijrefrig.2022.06.001.

56 Elbassoussi, M.H. and Zubair, S.M. (2023). A novel non-dimensional investigation of two-bed adsorption cooling/desalination systems. *Applied Thermal Engineering* 230: https://doi.org/10.1016/j.applthermaleng.2023.120739.

57 Krishnappa, A.K., Kapilan, N., Kasthurirengan, S., and Ashwathnarayana, D.P. (2023). Experimental and simulation study of a solar assisted two bed adsorption refrigeration system using activated carbon-methanol. *International Journal of Heat and Technology* 41 (2): 293–303. https://doi.org/10.18280/ijht.410202.

58 Yin, G., Wang, Y., Li, M. et al. (2022). Experimental investigation on a two-bed adsorption refrigeration system with mass recovery. *Applied Thermal Engineering* 207: https://doi.org/10.1016/j.applthermaleng.2022.118152.

59 Pal, A., El-Sharkawy, I.I., Saha, B.B. et al. (2016). Thermodynamic analysis of adsorption cooling cycle using consolidated composite adsorbents-ethanol pairs. *ARPN Journal of Engineering and Applied Sciences* 11: 12234–12238.

60 Sanaye, S. and Jafari, H. (2023). Modeling and optimizing a two-bed adsorption cooling system with a modified mass recovery process. *Energy Conversion and Management* 277: https://doi.org/10.1016/j.enconman.2022.116649.

61 Teng, W.S., Leong, K.C., and Chakraborty, A. (2016). Revisiting adsorption cooling cycle from mathematical modelling to system development. *Renewable and Sustainable Energy Reviews* 63: 315–332. https://doi.org/10.1016/j.rser.2016.05.059.

62 Wang, X. and Chua, H.T. (2007). A comparative evaluation of two different heat-recovery schemes as applied to a two-bed adsorption chiller. *International Journal of Heat and Mass Transfer* 50 (3–4): 433–443. https://doi.org/10.1016/j.ijheatmasstransfer.2006.08.003.

63 Sato, H., Inoue, S., Tanaka, H., and Terao, T. (1995). Adsorptive type refrigeration apparatus. US Patent US5619866A Patent Appl. 530,518, 15/04/1997.

64 Jribi, S., Saha, B.B., Koyama, S., and Bentaher, H. (2014). Modeling and simulation of an activated carbon–CO_2 four bed based adsorption cooling system. *Energy Conversion and Management* 78: 985–991. https://doi.org/10.1016/j.enconman.2013.06.061.

65 Li, X.H., Hou, X.H., Zhang, X., and Yuan, Z.X. (2015). A review on development of adsorption cooling—novel beds and advanced cycles. *Energy Conversion and Management* 94: 221–232. https://doi.org/10.1016/j.enconman.2015.01.076.

66 Mitra, S., Thu, K., Saha, B.B. et al. (2017). Modeling study of two-stage, multi-bed air cooled silica gel + water adsorption cooling cum desalination system. *Applied Thermal Engineering* 114: 704–712. https://doi.org/10.1016/j.applthermaleng.2016.12.011.

67 Jribi, S., Saha, B.B., Koyama, S. et al. (2013). Study on activated carbon/HFO-1234ze(E) based adsorption cooling cycle. *Applied Thermal Engineering* 50 (2): 1570–1575. https://doi.org/10.1016/j.applthermaleng.2011.11.066.

68 Li, T.X., Wang, R.Z., and Li, H. (2014). Progress in the development of solid–gas sorption refrigeration thermodynamic cycle driven by low-grade thermal energy. *Progress in Energy and Combustion Science* 40: 1–58. https://doi.org/10.1016/j.pecs.2013.09.002.

69 Pesaran, A., Lee, H., Hwang, Y. et al. (2016). Review article: numerical simulation of adsorption heat pumps. *Energy* 100: 310–320. https://doi.org/10.1016/j.energy.2016.01.103.

70 Hassan, H.Z., Mohamad, A.A., Alyousef, Y., and Al-Ansary, H.A. (2015). A review on the equations of state for the working pairs used in adsorption cooling systems. *Renewable and Sustainable Energy Reviews* 45: 600–609. https://doi.org/10.1016/j.rser.2015.02.008.

71 Yong, L. and Sumathy, K. (2002). Review of mathematical investigation on the closed adsorption heat pump and cooling systems. *Renewable and Sustainable Energy Reviews* 6 (4): 305–338. https://doi.org/10.1016/s1364-0321(02)00010-2.

72 Langmuir, I. (2002). The adsorption of gases on plane surfaces of glass, mica and platinum. *Journal of the American Chemical Society* 40 (9): 1361–1403. https://doi.org/10.1021/ja02242a004.

73 Brennan, C.R. (2012). *Mathematical Modelling*, Mathematics Research Developments. New York: Nova Science Publishers (in English).

74 Freundlich, H. and Hatfield, H. (1926). *Colloid & Capillary Chemistry*. Methuen & Company Limited.

75 Tóth, J. (1981). A uniform interpretation of gas/solid adsorption. *Journal of Colloid and Interface Science* 79 (1): 85–95. https://doi.org/10.1016/0021-9797(81)90050-3.

76 Dubinin, M.M. and Astakhov, V.A. (1971). Description of adsorption equilibria of vapors on zeolites over wide ranges of temperature and pressure. In: *Molecular Sieve Zeolites-II*, Advances in Chemistry, 69–85. ACS Publications.

77 Dubinin, M.M. (1975). Physical adsorption of gases and vapors in micropores. *Progress in Surface and Membrane Science* 9: 1–70.

78 Asif Sha, A., Baiju, V., Rehna, R.S. et al. (2022). Performance investigations of carbon based consolidated composite adsorbents effective for adsorption cooling systems. *Applied Thermal Engineering* 217: https://doi.org/10.1016/j.applthermaleng.2022.119199.

79 Sun, B. and Chakraborty, A. (2014). Thermodynamic formalism of water uptakes on solid porous adsorbents for adsorption cooling applications. *Applied Physics Letters* 104 (20): https://doi.org/10.1063/1.4876922.

80 Do, D.D. and Do, H.D. (2000). A model for water adsorption in activated carbon. *Carbon* 38 (5): 767–773. https://doi.org/10.1016/s0008-6223(99)00159-1.

81 Neitsch, M., Heschel, W., and Suckow, M. (2001). Water vapor adsorption by activated carbon: a modification to the isotherm model of Do and Do. *Carbon* 39 (9): 1437–1438. https://doi.org/10.1016/s0008-6223(01)00077-x.

82 Woo, S.-Y., Lee, H.-S., Kim, J.-S. et al. (2022). Applicability assessment of functional adsorption zeolite materials in adsorption desalination cum cooling systems driven by low-grade heat source. *Chemical Engineering Journal* 430: https://doi.org/10.1016/j.cej.2021.131375.

83 Jahan, I., Rupam, T.H., Palash, M.L. et al. (2022). Energy efficient green synthesized MOF-801 for adsorption cooling applications. *Journal of Molecular Liquids* 345: https://doi.org/10.1016/j.molliq.2021.117760.

84 Ng, K.C., Burhan, M., Shahzad, M.W., and Ismail, A.B. (2017). A universal isotherm model to capture adsorption uptake and energy distribution of porous heterogeneous surface. *Scientific Reports* 7 (1): https://doi.org/10.1038/s41598-017-11156-6.

85 Muttakin, M., Pal, A., Rupa, M.J. et al. (2021). A critical overview of adsorption kinetics for cooling and refrigeration systems. *Advances in Colloid and Interface Science* 294: https://doi.org/10.1016/j.cis.2021.102468.

86 Suzuki, M. and Suzuki, M. (1990). *Adsorption Engineering*. Kodansha Tokyo.

87 Pollard, W.G. and Present, R.D. (1948). On gaseous self-diffusion in long capillary tubes. *Physical Review* 73 (7): 762–774. https://doi.org/10.1103/PhysRev.73.762.

88 Banda, H., Rezk, A., Elsayed, E., and Askalany, A. (2023). Experimental and computational study on utilising graphene oxide for adsorption cooling and water desalination. *Applied Thermal Engineering* 229: https://doi.org/10.1016/j.applthermaleng.2023.120631.

89 Glueckauf, E. (1955). Theory of chromatography. Part 10. Formulæ for diffusion into spheres and their application to chromatography. *Transactions of the Faraday Society* 51: 1540–1551.

90 Sircar, S. and Hufton, J.R. (2000). Why does the linear driving force model for adsorption kinetics work? *Adsorption* 6 (2): 137–147. https://doi.org/10.1023/a:1008965317983.

91 Weiss, J. (2004). Theory of chromatography. In: *Handbook of Ion Chromatography*, 13–25. Wiley.

92 Aristov, Y.I., Tokarev, M.M., Freni, A. et al. (2006). Kinetics of water adsorption on silica Fuji Davison RD. *Microporous and Mesoporous Materials* 96 (1–3): 65–71. https://doi.org/10.1016/j.micromeso.2006.06.008.

93 Aristov, Y.I., Dawoud, B., Glaznev, I.S., and Elyas, A. (2008). A new methodology of studying the dynamics of water sorption/desorption under real operating conditions of adsorption heat pumps: experiment. *International Journal of Heat and Mass Transfer* 51 (19–20): 4966–4972. https://doi.org/10.1016/j.ijheatmasstransfer.2007.10.042.

94 Sapienza, A., Santamaria, S., Frazzica, A. et al. (2014). Dynamic study of adsorbers by a new gravimetric version of the Large Temperature Jump method. *Applied Energy* 113: 1244–1251. https://doi.org/10.1016/j.apenergy.2013.09.005.

95 Mohammed, R.H., Mesalhy, O., Elsayed, M.L. et al. (2018). Revisiting the adsorption equilibrium equations of silica-gel/water for adsorption cooling applications. *International Journal of Refrigeration* 86: 40–47. https://doi.org/10.1016/j.ijrefrig.2017.10.038.

96 Patel, H. (2021). Comparison of batch and fixed bed column adsorption: a critical review. *International journal of Environmental Science and Technology* 19 (10): 10409–10426. https://doi.org/10.1007/s13762-021-03492-y.

97 Al-Ghouti, M.A. and Da'ana, D.A. (2020). Guidelines for the use and interpretation of adsorption isotherm models: a review. *Journal of Hazardous Materials* 393: https://doi.org/10.1016/j.jhazmat.2020.122383.

98 Azizian, S. and Eris, S. (2021). Adsorption isotherms and kinetics. In: *Adsorption: Fundamental Processes and Applications*, Interface Science and Technology, 445–509. Elsevier B.V.

99 Wang, J. and Guo, X. (2020). Adsorption isotherm models: classification, physical meaning, application and solving method. *Chemosphere* 258: https://doi.org/10.1016/j.chemosphere.2020.127279.

100 Quirijns, E.J., van Boxtel, A.J.B., van Loon, W.K.P., and van Straten, G. (2005). Sorption isotherms, GAB parameters and isosteric heat of sorption. *Journal of the Science of Food and Agriculture* 85 (11): 1805–1814. https://doi.org/10.1002/jsfa.2140.

101 Yagnamurthy, S., Rakshit, D., Jain, S., and Saha, B.B. (2023). Compact heat exchanger designs for difluoromethane-activated carbon composites based adsorption cooling systems. *International Communications in Heat and Mass Transfer* 140: https://doi.org/10.1016/j.icheatmasstransfer.2022.106549.

102 Salvestrini, S., Ambrosone, L., and Kopinke, F.-D. (2022). Some mistakes and misinterpretations in the analysis of thermodynamic adsorption data. *Journal of Molecular Liquids* 352: https://doi.org/10.1016/j.molliq.2022.118762.

103 Chakraborty, A., Saha, B.B., Koyama, S., and Ng, K.C. (2006). On the thermodynamic modeling of the isosteric heat of adsorption and comparison with experiments. *Applied Physics Letters* 89 (17): https://doi.org/10.1063/1.2360925.

104 Chakraborty, A., Saha, B.B., El-Sharkawy, I.I. et al. (2008). Theory and experimental validation on isosteric heat of adsorption for an adsorbent+ adsorbate system. *High Temperatures - High Pressures* 37 (2): 109.

105 Chakraborty, A., Saha, B.B., Ng, K.C. et al. (2009). Theoretical insight of physical adsorption for a single-component adsorbent + adsorbate system: I. Thermodynamic property surfaces. *Langmuir* 25 (4): 2204–2211. https://doi.org/10.1021/la803289p.

106 Rahman, K.A., Chakraborty, A., Saha, B.B., and Ng, K.C. (2012). On thermodynamics of methane + carbonaceous materials adsorption. *International Journal of Heat and Mass Transfer* 55 (4): 565–573. https://doi.org/10.1016/j.ijheatmasstransfer.2011.10.056.

107 Askalany, A.A. and Saha, B.B. (2015). Derivation of isosteric heat of adsorption for non-ideal gases. *International Journal of Heat and Mass Transfer* 89: 186–192. https://doi.org/10.1016/j.ijheatmasstransfer.2015.05.018.

108 Askalany, A.A., Saha, B., Uddin, K. et al. (2013). Adsorption isotherms and heat of adsorption of difluoromethane on activated carbons. *Journal of Chemical and Engineering Data* 58 (10): 2828–2834. https://doi.org/10.1021/je4005678.

109 Ghazy, M., Askalany, A.A., Harby, K., and Ahmed, M.S. (2016). Adsorption isotherms and kinetics of HFC-404A onto bituminous based granular activated carbon for storage and cooling applications. *Applied Thermal Engineering* 105: 639–645. https://doi.org/10.1016/j.applthermaleng.2016.03.057.

110 Askalany, A.A. and Saha, B.B. (2017). Towards an accurate estimation of the isosteric heat of adsorption – a correlation with the potential theory. *Journal of Colloid and Interface Science* 490: 59–63. https://doi.org/10.1016/j.jcis.2016.11.040.

111 Tian, Y. and Wu, J. (2017). Differential heat of adsorption and isosteres. *Langmuir* 33 (4): 996–1003. https://doi.org/10.1021/acs.langmuir.7b00004.

112 Azahar, F.H.M., Mitra, S., Yabushita, A. et al. (2018). Improved model for the isosteric heat of adsorption and impacts on the performance of heat pump cycles. *Applied Thermal Engineering* 143: 688–700. https://doi.org/10.1016/j.applthermaleng.2018.07.131.

113 Saha, B.B., Boelman, E.C., and Kashiwagi, T. (1995). Computational analysis of an advanced adsorption-refrigeration cycle. *Energy* 20 (10): 983–994. https://doi.org/10.1016/0360-5442(95)00047-k.

114 Bejan, A. (1977). The concept of irreversibility in heat exchanger design: counterflow heat exchangers for gas-to-gas applications. *Journal of Heat Transfer* 99 (3): 374–380. https://doi.org/10.1115/1.3450705.

115 Chang, S.C. and Roux, J.A. (1985). Thermodynamic analysis of a solar zeolite refrigeration system. *Journal of Solar Energy Engineering* 107 (3): 189–195. https://doi.org/10.1115/1.3267677.

116 Meunier, F. (1985). Second law analysis of a solid adsorption heat pump operating on reversible cascade cycles: application to the Zeolite-water pair. *Journal of Heat Recovery Systems* 5 (2): 133–141. https://doi.org/10.1016/0198-7593(85)90045-1.

117 Shelton, S.V., Wepfer, W.J., and Miles, D.J. (1989). Square wave analysis of the solid-vapor adsorption heat pump. *Heat Recovery Systems and CHP* 9 (3): 233–247. https://doi.org/10.1016/0890-4332(89)90007-0.

118 Cacciola, G. and Restuccia, G. (1995). Reversible adsorption heat pump: a thermodynamic model. *International Journal of Refrigeration* 18 (2): 100–106. https://doi.org/10.1016/0140-7007(94)00005-i.

119 Pons, M. (1996). Second law analysis of adsorption cycles with thermal regeneration. *Journal of Energy Resources Technology* 118 (3): 229–236. https://doi.org/10.1115/1.2793867.

120 Pons, M., Meunier, F., Cacciola, G. et al. (1999). Thermodynamic based comparison of sorption systems for cooling and heat pumping. *International Journal of Refrigeration* 22 (1): 5–17. https://doi.org/10.1016/s0140-7007(98)00048-6.

121 Gui, Y.B. and Wang, R.Z. (2001). Practical three-heat-reservoir model on heat-regenerative adsorption air-conditioning system. *Applied*

Thermal Engineering 21 (16): 1643–1656. https://doi.org/10.1016/s1359-4311(01)00034-5.

122 Thu, K., Kim, Y.-D., Myat, A. et al. (2013). Entropy generation analysis of an adsorption cooling cycle. *International Journal of Heat and Mass Transfer* 60: 143–155. https://doi.org/10.1016/j.ijheatmasstransfer.2012.12.055.

123 Sakoda, A. and Suzuki, M. (1984). Fundamental study on solar powered adsorption cooling system. *Journal of Chemical Engineering of Japan* 17 (1): 52–57. https://doi.org/10.1252/jcej.17.52.

124 Chihara, K. and Suzuki, M. (1983). Air drying by pressure swing adsorption. *Journal of Chemical Engineering of Japan* 16 (4): 293–299. https://doi.org/10.1252/jcej.16.293.

125 Karagiorgas, M. and Meunier, F. (1987). The dynamics of a solid-adsorption heat pump connected with outside heat sources of finite capacity. *Heat Recovery Systems and CHP* 7 (3): 285–299. https://doi.org/10.1016/0890-4332(87)90141-4.

126 Cho, S.-H. and Kim, J.-N. (1992). Modeling of a silica gel/water adsorption-cooling system. *Energy* 17 (9): 829–839. https://doi.org/10.1016/0360-5442(92)90101-5.

127 Cacciola, G., Hajji, A., Maggio, G., and Restuccia, G. (1993). Dynamic simulation of a recuperative adsorption heat pump. *Energy* 18 (11): 1125–1137. https://doi.org/10.1016/0360-5442(93)90085-r.

128 Sami, S.M. and Tribes, C. (1996). An improved model for predicting the dynamic behaviour of adsorption systems. *Applied Thermal Engineering* 16 (2): 149–161. https://doi.org/10.1016/1359-4311(95)00054-h.

129 Critoph, R.E. (1998). Forced convection adsorption cycles. *Applied Thermal Engineering* 18 (9–10): 799–807. https://doi.org/10.1016/s1359-4311(97)00110-5.

130 Wu, J.Y., Wang, R.Z., and Xu, Y.X. (2000). Dynamic simulation and experiments of a heat regenerative adsorption heat pump. *Energy Conversion and Management* 41 (10): 1007–1018. https://doi.org/10.1016/s0196-8904(99)00161-2.

131 Douss, N., Meunier, F.E., and Sun, L.M. (2002). Predictive model and experimental results for a two-adsorber solid adsorption heat pump. *Industrial and Engineering Chemistry Research* 27 (2): 310–316. https://doi.org/10.1021/ie00074a017.

132 Yong, L. and Sumathy, K. (2004). Modeling and simulation of a solar powered two bed adsorption air conditioning system. *Energy Conversion and Management* 45 (17): 2761–2775. https://doi.org/10.1016/j.enconman.2003.12.004.

133 Khan, M.Z.I., Saha, B.B., Alam, K.C.A. et al. (2006). Multi-bed mass recovery adsorption cycle-improving performance. *Transactions of the Japan Society of*

Refrigerating and Air Conditioning Engineers 23 (4): 399–408. https://doi.org/10.11322/tjsrae.23.399.

134 Lanzerath, F., Bau, U., Seiler, J., and Bardow, A. (2015). Optimal design of adsorption chillers based on a validated dynamic object-oriented model. *Science and Technology for the Built Environment* 21 (3): 248–257. https://doi.org/10.1080/10789669.2014.990337.

135 Xia, X., Liu, B., Zhao, B. et al. (2023). Enhanced water adsorption of MIL-101(Cr) by metal-organic polyhedral encapsulation for adsorption cooling. *Nanomaterials* 13 (7): https://doi.org/10.3390/nano13071147.

136 Xia, X. and Li, S. (2023). Improved adsorption cooling performance of MIL-101(Cr)/GO composites by tuning the water adsorption rate. *Sustainable Energy & Fuels* 7 (2): 437–447. https://doi.org/10.1039/d2se01508k.

137 Leong, K.C. and Liu, Y. (2006). System performance of a combined heat and mass recovery adsorption cooling cycle: a parametric study. *International Journal of Heat and Mass Transfer* 49 (15–16): 2703–2711. https://doi.org/10.1016/j.ijheatmasstransfer.2006.01.012.

138 Guilleminot, J.J., Meunier, F., and Pakleza, J. (1987). Heat and mass transfer in a non-isothermal fixed bed solid adsorbent reactor: a uniform pressure-non-uniform temperature case. *International Journal of Heat and Mass Transfer* 30 (8): 1595–1606. https://doi.org/10.1016/0017-9310(87)90304-8.

139 Fuller, T.A., Wepfer, W.J., Shelton, S.V., and Ellis, M.W. (1994). A two-temperature model of the regenerative solid-vapor heat pump. *Journal of Energy Resources Technology* 116 (4): 297–304. https://doi.org/10.1115/1.2906457.

140 Sun, L.M., Amar, N.B., and Meunier, F. (1995). Numerical study on coupled heat and mass transfers in an absorber with external fluid heating. *Heat Recovery Systems and CHP* 15 (1): 19–29. https://doi.org/10.1016/0890-4332(95)90034-9.

141 Zheng, W., Worek, W.M., and Nowakowski, G. (1995). Effect of design and operating parameters on the performance of two-bed sorption heat pump systems. *Journal of Energy Resources Technology* 117 (1): 67–74. https://doi.org/10.1115/1.2835323.

142 Hajji, A. and Khalloufi, S. (1996). Improving the performance of adsorption heat exchangers using a finned structure. *International Journal of Heat and Mass Transfer* 39 (8): 1677–1686. https://doi.org/10.1016/0017-9310(95)00254-5.

143 Mhimid, A. (1998). Theoretical study of heat and mass transfer in a zeolite bed during water desorption: validity of local thermal equilibrium assumption. *International Journal of Heat and Mass Transfer* 41 (19): 2967–2977. https://doi.org/10.1016/s0017-9310(98)00010-6.

144 Tatlıer, M., Tantekin-Ersolmaz, B., and Erdem-Şenatalar, A. (1999). A novel approach to enhance heat and mass transfer in adsorption heat pumps using the zeolite–water pair. *Microporous and Mesoporous Materials* 27 (1): 1–10. https://doi.org/10.1016/s1387-1811(98)00174-7.

145 Zhang, L.Z. and Wang, L. (1999). Effects of coupled heat and mass transfers in adsorbent on the performance of a waste heat adsorption cooling unit. *Applied Thermal Engineering* 19 (2): 195–215. https://doi.org/10.1016/s1359-4311(98)00023-4.

146 Zhang, L.Z. and Wang, L. (1999). Momentum and heat transfer in the adsorbent of a waste heat adsorption cooling system. *Energy* 24 (7): 605–624. https://doi.org/10.1016/s0360-5442(99)00018-3.

147 Alam, K.C.A., Saha, B.B., Kang, Y.T. et al. (2000). Heat exchanger design effect on the system performance of silica gel adsorption refrigeration systems. *International Journal of Heat and Mass Transfer* 43 (24): 4419–4431. https://doi.org/10.1016/s0017-9310(00)00072-7.

148 Zhang, L.Z. (2000). A three-dimensional non-equilibrium model for an intermittent adsorption cooling system. *Solar Energy* 69 (1): 27–35. https://doi.org/10.1016/s0038-092x(00)00010-4.

149 Chahbani, M.H., Labidi, J., and Paris, J. (2002). Effect of mass transfer kinetics on the performance of adsorptive heat pump systems. *Applied Thermal Engineering* 22 (1): 23–40. https://doi.org/10.1016/s1359-4311(01)00067-9.

150 Marletta, L., Maggio, G., Freni, A. et al. (2002). A non-uniform temperature non-uniform pressure dynamic model of heat and mass transfer in compact adsorbent beds. *International Journal of Heat and Mass Transfer* 45 (16): 3321–3330. https://doi.org/10.1016/s0017-9310(02)00045-5.

151 Miltkau, T. and Dawoud, B. (2002). Dynamic modeling of the combined heat and mass transfer during the adsorption/desorption of water vapor into/from a zeolite layer of an adsorption heat pump. *International Journal of Thermal Sciences* 41 (8): 753–762. https://doi.org/10.1016/s1290-0729(02)01369-8.

152 Restuccia, G., Freni, A., and Maggio, G. (2002). A zeolite-coated bed for air conditioning adsorption systems: parametric study of heat and mass transfer by dynamic simulation. *Applied Thermal Engineering* 22 (6): 619–630. https://doi.org/10.1016/s1359-4311(01)00114-4.

153 Li, M. and Wang, R.Z. (2003). Heat and mass transfer in a flat plate solar solid adsorption refrigeration ice maker. *Renewable Energy* 28 (4): 613–622. https://doi.org/10.1016/s0960-1481(02)00094-0.

154 Li, J., Kubota, M., Watanabe, F. et al. (2004). Optimal design of a fin-type silica gel tube module in the silica gel/water adsorption heat pump. *Journal of Chemical Engineering of Japan* 37 (4): 551–557. https://doi.org/10.1252/jcej.37.551.

155 Maggio, G., Freni, A., and Restuccia, G. (2006). A dynamic model of heat and mass transfer in a double-bed adsorption machine with internal heat recovery. *International Journal of Refrigeration* 29 (4): 589–600. https://doi.org/10.1016/j.ijrefrig.2005.10.005.

156 Dawoud, B., Vedder, U., Amer, E.H., and Dunne, S. (2007). Non-isothermal adsorption kinetics of water vapour into a consolidated zeolite layer. *International Journal of Heat and Mass Transfer* 50 (11–12): 2190–2199. https://doi.org/10.1016/j.ijheatmasstransfer.2006.10.052.

157 Demir, H., Mobedi, M., and Ülkü, S. (2009). Effects of porosity on heat and mass transfer in a granular adsorbent bed. *International Communications in Heat and Mass Transfer* 36 (4): 372–377. https://doi.org/10.1016/j.icheatmasstransfer.2009.01.008.

158 Waszkiewicz, S.D., Tierney, M.J., and Scott, H.S. (2009). Development of coated, annular fins for adsorption chillers. *Applied Thermal Engineering* 29 (11–12): 2222–2227. https://doi.org/10.1016/j.applthermaleng.2008.11.004.

159 Wu, W.-D., Zhang, H., and Sun, D.-W. (2009). Mathematical simulation and experimental study of a modified zeolite 13X–water adsorption refrigeration module. *Applied Thermal Engineering* 29 (4): 645–651. https://doi.org/10.1016/j.applthermaleng.2008.03.037.

160 Gong, L.X., Wang, R.Z., Xia, Z.Z., and Chen, C.J. (2011). Design and performance prediction of a new generation adsorption chiller using composite adsorbent. *Energy Conversion and Management* 52 (6): 2345–2350. https://doi.org/10.1016/j.enconman.2010.12.036.

161 Ilis, G.G., Mobedi, M., and Ülkü, S. (2011). A dimensionless analysis of heat and mass transport in an adsorber with thin fins; uniform pressure approach. *International Communications in Heat and Mass Transfer* 38 (6): 790–797. https://doi.org/10.1016/j.icheatmasstransfer.2011.03.001.

162 Niazmand, H., Talebian, H., and Mahdavikhah, M. (2012). Bed geometrical specifications effects on the performance of silica/water adsorption chillers. *International Journal of Refrigeration* 35 (8): 2261–2274. https://doi.org/10.1016/j.ijrefrig.2012.08.017.

163 Mahdavikhah, M. and Niazmand, H. (2013). Effects of plate finned heat exchanger parameters on the adsorption chiller performance. *Applied Thermal Engineering* 50 (1): 939–949. https://doi.org/10.1016/j.applthermaleng.2012.08.033.

164 Rezk, A., Al-Dadah, R.K., Mahmoud, S., and Elsayed, A. (2013). Effects of contact resistance and metal additives in finned-tube adsorbent beds on the performance of silica gel/water adsorption chiller. *Applied Thermal Engineering* 53 (2): 278–284. https://doi.org/10.1016/j.applthermaleng.2012.04.008.

165 Li, A., Ismail, A.B., Thu, K. et al. (2014). Performance evaluation of a zeolite–water adsorption chiller with entropy analysis of thermodynamic insight. *Applied Energy* 130: 702–711. https://doi.org/10.1016/j.apenergy.2014.01.086.

166 El Fadar, A. (2015). Thermal behavior and performance assessment of a solar adsorption cooling system with finned adsorber. *Energy* 83: 674–684. https://doi.org/10.1016/j.energy.2015.02.074.

167 Hong, S.W., Ahn, S.H., Kwon, O.K., and Chung, J.D. (2015). Optimization of a fin-tube type adsorption chiller by design of experiment. *International Journal of Refrigeration* 49: 49–56. https://doi.org/10.1016/j.ijrefrig.2014.09.022.

168 Solmuş, İ., Yamalı, C., Yıldırım, C., and Bilen, K. (2015). Transient behavior of a cylindrical adsorbent bed during the adsorption process. *Applied Energy* 142: 115–124. https://doi.org/10.1016/j.apenergy.2014.12.080.

169 Xue, B., Meng, X., Wei, X. et al. (2015). Dynamic study of steam generation from low-grade waste heat in a zeolite–water adsorption heat pump. *Applied Thermal Engineering* 88: 451–458. https://doi.org/10.1016/j.applthermaleng.2014.11.050.

170 Dias, J.M.S. and Costa, V.A.F. (2019). Which dimensional model for the analysis of a coated tube adsorber for adsorption heat pumps? *Energy* 174: 1110–1120. https://doi.org/10.1016/j.energy.2019.03.028.

171 Mohammed, R.H., Mesalhy, O., Elsayed, M.L., and Chow, L.C. (2019). Assessment of numerical models in the evaluation of adsorption cooling system performance. *International Journal of Refrigeration* 99: 166–175. https://doi.org/10.1016/j.ijrefrig.2018.12.017.

172 Yang, P.-Z. (2010). Thermodynamic analysis and performance study for adsorption refrigeration cycle driven by a fuel cell electric vehicle waste heat. *Heat Transfer—Asian Research* 39 (7): 523–538. https://doi.org/10.1002/htj.20315.

173 Alam, K.C.A., Akahira, A., Hamamoto, Y. et al. (2004). A four-bed mass recovery adsorption refrigeration cycle driven by low temperature waste/renewable heat source. *Renewable Energy* 29 (9): 1461–1475. https://doi.org/10.1016/j.renene.2004.01.011.

174 Akahira, A., Alam, K.C.A., Hamamoto, Y. et al. (2004). Mass recovery adsorption refrigeration cycle—improving cooling capacity. *International Journal of Refrigeration* 27 (3): 225–234. https://doi.org/10.1016/j.ijrefrig.2003.10.004.

175 Akahira, A., Amanul Alam, K.C., Hamamoto, Y. et al. (2005). Mass recovery four-bed adsorption refrigeration cycle with energy cascading. *Applied Thermal Engineering* 25 (11–12): 1764–1778. https://doi.org/10.1016/j.applthermaleng.2004.10.006.

176 Hu, P., Yao, J.-J., and Chen, Z.-S. (2009). Analysis for composite zeolite/foam aluminum–water mass recovery adsorption refrigeration system driven by engine exhaust heat. *Energy Conversion and Management* 50 (2): 255–261. https://doi.org/10.1016/j.enconman.2008.09.022.

177 Pan, Q.W., Liu, L., Wang, B. et al. (2023). Design and experimental study on a small-scale silica gel/water adsorption chiller with heat and mass recovery scheme for solar energy use. *Solar Energy* 252: 91–100. https://doi.org/10.1016/j.solener.2023.01.052.

178 Qu, T.F., Wang, R.Z., and Wang, W. (2001). Study on heat and mass recovery in adsorption refrigeration cycles. *Applied Thermal Engineering* 21 (4): 439–452. https://doi.org/10.1016/s1359-4311(00)00050-8.

179 Saren, S., Mitra, S., Miksik, F. et al. (2023). Impacts of the internal heat recovery scheme on the performance of an adsorption heat transformer cycle for temperature upgrade. *International Communications in Heat and Mass Transfer* 144: https://doi.org/10.1016/j.icheatmasstransfer.2023.106774.

180 El Fil, B. and Garimella, S. (2022). Heat recovery, adsorption thermal storage, and heat pumping to augment gas-fired tumble dryer efficiency. *Journal of Energy Storage* 48: https://doi.org/10.1016/j.est.2021.103949.

181 Duong, X.Q., Cao, N.V., Hong, S.W. et al. (2018). Numerical study on the combined heat and mass recovery adsorption cooling cycle. *Energy Technology* 6 (2): 296–305. https://doi.org/10.1002/ente.201700417.

182 Saren, S., Mitra, S., Miyazaki, T. et al. (2022). Adsorption heat transformer cycle using multiple adsorbent + water pairs for waste heat upgrade. *Journal of Thermal Analysis and Calorimetry* 148 (8): 3059–3071. https://doi.org/10.1007/s10973-022-11350-3.

183 Aprile, M., Freni, A., Toppi, T., and Motta, M. (2020). Modelling and performance assessment of a thermally-driven cascade adsorption cycle suitable for cooling applications. *Thermal Science and Engineering Progress* 19: https://doi.org/10.1016/j.tsep.2020.100602.

184 Koushaeian, M., Duong, X.Q., Park, M.Y. et al. (2023). Investigation of the cascade type of hybrid adsorption-vapor compression chiller. *Case Studies in Thermal Engineering* 47: https://doi.org/10.1016/j.csite.2023.103037.

185 Akisawa, A., Tamogami, A., Takeda, N. et al. (2022). Effect of adsorbent on the performance of double effect adsorption refrigeration cycle with adsorption heat recovery. *International Journal of Refrigeration* 141: 21–30. https://doi.org/10.1016/j.ijrefrig.2022.06.012.

186 Stefański, S., Mika, Ł., Sztekler, K. et al. (2019). Adsorption bed configurations for adsorption cooling application. *E3S Web of Conferences* 108: https://doi.org/10.1051/e3sconf/201910801010.

187 Chauhan, P.R., Kaushik, S.C., and Tyagi, S.K. (2022). A review on thermal performance enhancement of green cooling system using different adsorbent/refrigerant pairs. *Energy Conversion and Management: X* 14: https://doi.org/10.1016/j.ecmx.2022.100225.

188 Aristov, Y. (2014). Concept of adsorbent optimal for adsorptive cooling/heating. *Applied Thermal Engineering* 72 (2): 166–175. https://doi.org/10.1016/j.applthermaleng.2014.04.077.

189 Aristov, Y.I. (2013). Challenging offers of material science for adsorption heat transformation: a review. *Applied Thermal Engineering* 50 (2): 1610–1618. https://doi.org/10.1016/j.applthermaleng.2011.09.003.

190 Solmuş, İ., Yamalı, C., Kaftanoğlu, B. et al. (2010). Adsorption properties of a natural zeolite–water pair for use in adsorption cooling cycles. *Applied Energy* 87 (6): 2062–2067. https://doi.org/10.1016/j.apenergy.2009.11.027.

191 Gordeeva, L.G. and Aristov, Y.I. (2011). Composite sorbent of methanol "LiCl in mesoporous silica gel" for adsorption cooling: dynamic optimization. *Energy* 36 (2): 1273–1279. https://doi.org/10.1016/j.energy.2010.11.016.

192 Louajari, M., Mimet, A., and Ouammi, A. (2011). Study of the effect of finned tube adsorber on the performance of solar driven adsorption cooling machine using activated carbon–ammonia pair. *Applied Energy* 88 (3): 690–698. https://doi.org/10.1016/j.apenergy.2010.08.032.

193 Solmuş, İ., Kaftanoğlu, B., Yamalı, C., and Baker, D. (2011). Experimental investigation of a natural zeolite–water adsorption cooling unit. *Applied Energy* 88 (11): 4206–4213. https://doi.org/10.1016/j.apenergy.2011.04.057.

194 Askalany, A.A., Saha, B.B., Ahmed, M.S., and Ismail, I.M. (2013). Adsorption cooling system employing granular activated carbon–R134a pair for renewable energy applications. *International Journal of Refrigeration* 36 (3): 1037–1044. https://doi.org/10.1016/j.ijrefrig.2012.11.009.

195 Gordeeva, L., Frazzica, A., Sapienza, A. et al. (2014). Adsorption cooling utilizing the "LiBr/silica–ethanol" working pair: dynamic optimization of the adsorber/heat exchanger unit. *Energy* 75: 390–399. https://doi.org/10.1016/j.energy.2014.07.088.

196 Chan, K.C., Tso, C.Y., Wu, C., and Chao, C.Y.H. (2018). Enhancing the performance of a zeolite 13X/$CaCl_2$–water adsorption cooling system by improving adsorber design and operation sequence. *Energy and Buildings* 158: 1368–1378. https://doi.org/10.1016/j.enbuild.2017.11.040.

197 Singh, V.K., Kumar, E.A., and Saha, B.B. (2018). Adsorption isotherms, kinetics and thermodynamic simulation of CO_2-CSAC pair for cooling application. *Energy* 160: 1158–1173. https://doi.org/10.1016/j.energy.2018.07.063.

198 Solovyeva, M.V., Gordeeva, L.G., Krieger, T.A., and Aristov, Y.I. (2018). MOF-801 as a promising material for adsorption cooling: equilibrium and dynamics of water adsorption. *Energy Conversion and Management* 174: 356–363. https://doi.org/10.1016/j.enconman.2018.08.032.

199 Teo, H.W.B., Chakraborty, A., and Kayal, S. (2018). Formic acid modulated (fam) aluminium fumarate MOF for improved isotherms and kinetics with water adsorption: cooling/heat pump applications. *Microporous and Mesoporous Materials* 272: 109–116. https://doi.org/10.1016/j.micromeso.2018.06.016.

200 Zhu, L.Q., Tso, C.Y., Chan, K.C. et al. (2018). Experimental investigation on composite adsorbent–water pair for a solar-powered adsorption cooling system. *Applied Thermal Engineering* 131: 649–659. https://doi.org/10.1016/j.applthermaleng.2017.12.053.

201 Chen, H., Chen, Z., Zhang, L. et al. (2019). Toward design rules of metal–organic frameworks for adsorption cooling: effect of topology on the ethanol working capacity. *Chemistry of Materials* 31 (8): 2702–2706. https://doi.org/10.1021/acs.chemmater.9b00062.

202 Elsayed, E., Anderson, P., Al-Dadah, R. et al. (2019). MIL-101(Cr)/calcium chloride composites for enhanced adsorption cooling and water desalination. *Journal of Solid State Chemistry* 277: 123–132. https://doi.org/10.1016/j.jssc.2019.05.026.

203 Rocky, K.A., Pal, A., Rupam, T.H. et al. (2021). Recent advances of composite adsorbents for heat transformation applications. *Thermal Science and Engineering Progress* 23: https://doi.org/10.1016/j.tsep.2021.100900.

204 Albaik, I., Al-Dadah, R., Mahmoud, S. et al. (2022). Coated, packed and combined wire finned tube adsorption cooling and desalination system using metal-organic framework: numerical study. *Energy* 247: https://doi.org/10.1016/j.energy.2022.123506.

205 Li, M., Zhao, Y., Long, R. et al. (2022). Metal foam packed adsorbent bed boosting the performance of the adsorption-based desalination and cooling system. *Energy Conversion and Management* 254: https://doi.org/10.1016/j.enconman.2022.115250.

206 Rupam, T.H., Palash, M.L., Islam, M.A., and Saha, B.B. (2022). Transitional metal-doped aluminum fumarates for ultra-low heat driven adsorption cooling systems. *Energy* 238: https://doi.org/10.1016/j.energy.2021.122079.

207 Shaaban, M., Elsheniti, M.B., Rezk, A. et al. (2022). Performance investigation of adsorption cooling and desalination systems employing thermally enhanced copper foamed bed coated with SAPO-34 and CPO-27(Ni). *Applied Thermal Engineering* 205: https://doi.org/10.1016/j.applthermaleng.2022.118056.

208 Xie, W., Hua, W., Zhang, X. et al. (2023). Preparation and characteristics analysis of the new-type silica gel/$CaCl_2$ adsorbents with nanoparticles for adsorption desalination and cooling system. *Journal of Sol-Gel Science and Technology* 105 (2): 525–536. https://doi.org/10.1007/s10971-022-06010-9.

209 Yagnamurthy, S., Rakshit, D., Jain, S. et al. (2022). Adsorption of difluoromethane onto activated carbon based composites: adsorption kinetics, heat of adsorption, cooling performance and irreversibility evaluation. *Applied Thermal Engineering* 210: https://doi.org/10.1016/j.applthermaleng.2022.118359.

210 Alsaman, A.S., Ahmed, M.S., Ibrahim, E.M.M. et al. (2023). Experimental investigation of porous carbon for cooling and desalination applications. *npj Clean Water* 6 (1): https://doi.org/10.1038/s41545-022-00211-z.

211 Asfahan, H.M., Sultan, M., Farooq, M. et al. (2023). Evaluating the emerging adsorbents for performance improvement of adsorption desalination cum cooling system. *International Communications in Heat and Mass Transfer* 142: https://doi.org/10.1016/j.icheatmasstransfer.2023.106661.

212 Gado, M.G., Ookawara, S., and Hassan, H. (2023). Utilization of triply periodic minimal surfaces for performance enhancement of adsorption cooling systems: computational fluid dynamics analysis. *Energy Conversion and Management* 277: https://doi.org/10.1016/j.enconman.2023.116657.

213 Al-Maamory, N.H. and Fadel Farman, N. (2023). Performance of solar adsorption cooling system using methanol and activated carbon as a working pair. *Journal of Engineering* 29 (7): 71–85. https://doi.org/10.31026/j.eng.2023.07.05.

214 Hasan Rupam, T., Ahmed Rocky, K., Palash, M.L., and Baran Saha, B. (2023). Ethanol adsorption onto various metal organic frameworks for cooling applications. *Thermal Science and Engineering Progress* 37: https://doi.org/10.1016/j.tsep.2022.101602.

215 Kulakowska, A., Zylka, A., Krzywanski, J. et al. (2023). Influence of the adsorption bed composition on the low-pressure fluidization. *Processes* 11 (7): https://doi.org/10.3390/pr11071912.

216 Lache, M., Kappelhoff, C., Seiler, J., and Bardow, A. (2023). Water and ethanol as refrigerant mixture enabling adsorption cooling below 0 °C. *Energy Technology* 11 (3): https://doi.org/10.1002/ente.202201158.

217 Rezk, M., Elsheniti, M.B., Rezk, A., and Elsamni, O.A. (2023). Multi-objective optimisation of MOF-801 adsorbent packed into copper foamed bed for cooling and water desalination systems. *Applied Thermal Engineering* 229: https://doi.org/10.1016/j.applthermaleng.2023.120642.

218 Sapienza, A., Brancato, V., Aristov, Y., and Vasta, S. (2021). Plastic heat exchangers for adsorption cooling: thermodynamic and dynamic performance. *Applied Thermal Engineering* 188: https://doi.org/10.1016/j.applthermaleng.2021.116622.

219 Rupa, M.J., Pal, A., and Saha, B.B. (2020). Activated carbon-graphene nanoplatelets based green cooling system: adsorption kinetics, heat of adsorption, and thermodynamic performance. *Energy* 193: https://doi.org/10.1016/j.energy.2019.116774.

220 Zheng, J., Wahiduzzaman, M., Barpaga, D. et al. (2021). Porous covalent organic polymers for efficient fluorocarbon-based adsorption cooling. *Angewandte Chemie International Edition* 60 (33): 18037–18043. https://doi.org/10.1002/anie.202102337.

221 Nzila, C., Oluoch, N., Kiprop, A., Ramkat, R., and Kosgey, I. (Eds.). (2022). Advances in Phytochemistry, Textile and Renewable Energy Research for Industrial Growth: Proceedings of the International Conference of Phytochemistry, Textile and Renewable Energy for Sustainable development (ICPTRE 2020), August 12-14, Eldoret, Kenya (1st ed.). CRC Press. https://doi.org/10.1201/9781003221968.

222 Tatlier, M. and Atalay-Oral, C. (2022). Selection of a favorable zeolite for solar adsorption cooling: how straightforward is it? *Chemical Engineering Communications* 210 (8): 1247–1256. https://doi.org/10.1080/00986445.2022.2067750.

223 Cho, K.H., Borges, D.D., Lee, U.H. et al. (2020). Rational design of a robust aluminum metal-organic framework for multi-purpose water-sorption-driven heat allocations. *Nature Communications* 11 (1): 5112. https://doi.org/10.1038/s41467-020-18968-7.

224 Wang, S., Lee, J.S., Wahiduzzaman, M. et al. (2018). A robust large-pore zirconium carboxylate metal–organic framework for energy-efficient water-sorption-driven refrigeration. *Nature Energy* 3 (11): 985–993. https://doi.org/10.1038/s41560-018-0261-6.

225 Wang, L.W., Wu, J.Y., Wang, R.Z. et al. (2003). Study of the performance of activated carbon–methanol adsorption systems concerning heat and mass transfer. *Applied Thermal Engineering* 23 (13): 1605–1617. https://doi.org/10.1016/s1359-4311(03)00104-2.

226 Komatsu, F. (2014). Utilization of low-temperature heat by adsorption chiller. *Japan Journal of Food Engineering* 15 (3): 195–198. https://doi.org/10.11301/jsfe.15.195.

227 Fernandes, M.S., Brites, G.J.V.N., Costa, J.J. et al. (2014). Review and future trends of solar adsorption refrigeration systems. *Renewable and Sustainable Energy Reviews* 39: 102–123. https://doi.org/10.1016/j.rser.2014.07.081.

228 Choudhury, B., Saha, B.B., Chatterjee, P.K., and Sarkar, J.P. (2013). An overview of developments in adsorption refrigeration systems towards a sustainable way of cooling. *Applied Energy* 104: 554–567. https://doi.org/10.1016/j.apenergy.2012.11.042.

229 Jakob, U. and Mittelbach, W. (2008). Development and investigation of a compact silica gel/water adsorption chiller integrated in solar cooling systems. VII Minsk International Seminar "Heat Pipes, Heat Pumps, Refrigerators, Power Sources", Minsk, Belarus, 8–11.

230 Zhai, X.Q., Wang, R.Z., Wu, J.Y. et al. (2008). Design and performance of a solar-powered air-conditioning system in a green building. *Applied Energy* 85 (5): 297–311. https://doi.org/10.1016/j.apenergy.2007.07.016.

231 Elsheniti, M.B., Rezk, A., Shaaban, M. et al. (2021). Performance of a solar adsorption cooling and desalination system using aluminum fumarate and silica gel. *Applied Thermal Engineering* 194: https://doi.org/10.1016/j.applthermaleng.2021.117116.

232 El-Sharkawy, I.I., Saha, B.B., Koyama, S. et al. (2008). Experimental investigation on activated carbon–ethanol pair for solar powered adsorption cooling applications. *International Journal of Refrigeration* 31 (8): 1407–1413. https://doi.org/10.1016/j.ijrefrig.2008.03.012.

233 El-Sharkawy, I.I., AbdelMeguid, H., and Saha, B.B. (2014). Potential application of solar powered adsorption cooling systems in the Middle East. *Applied Energy* 126: 235–245. https://doi.org/10.1016/j.apenergy.2014.03.092.

234 Goyal, P., Baredar, P., Mittal, A., and Siddiqui, A.R. (2016). Adsorption refrigeration technology–an overview of theory and its solar energy applications. *Renewable and Sustainable Energy Reviews* 53: 1389–1410. https://doi.org/10.1016/j.rser.2015.09.027.

235 Islam, M.P. and Morimoto, T. (2016). Thermodynamic performances of a solar driven adsorption system. *Solar Energy* 139: 266–277. https://doi.org/10.1016/j.solener.2016.09.003.

236 Hoseinzadeh, S. and Azadi, R. (2017). Simulation and optimization of a solar-assisted heating and cooling system for a house in Northern of Iran. *Journal of Renewable and Sustainable Energy* 9 (4): https://doi.org/10.1063/1.5000288.

237 Sur, A., Sah, R.P., and Pandya, S. (2020). Milk storage system for remote areas using solar thermal energy and adsorption cooling. *Materials Today: Proceedings* 28: 1764–1770. https://doi.org/10.1016/j.matpr.2020.05.170.

238 Mat Wajid, N., Abidin, A.M.Z., Hakemzadeh, M. et al. (2021). Solar adsorption air conditioning system–recent advances and its potential for cooling an office building in tropical climate. *Case Studies in Thermal Engineering* 27: https://doi.org/10.1016/j.csite.2021.101275.

239 Alsaman, A.S., Hassan, A.A., Ali, E.S. et al. (2022). Hybrid solar-driven desalination/cooling systems: current situation and future trend. *Energies* 15 (21): https://doi.org/10.3390/en15218099.

240 Ramkumar, G., Arthi, B., Sundarsingh Jebaseelan, S.D. et al. (2022). Implementation of solar heat energy and adsorption cooling mechanism for milk pasteurization application. *Adsorption Science and Technology* 2022: 1–13. https://doi.org/10.1155/2022/5125931.

241 Wang, Z., Yuan, Z., Du, C. et al. (2022). Performance of solar adsorption cooling system with internal finned vacuum tube bed. *Case Studies in Thermal Engineering* 34: https://doi.org/10.1016/j.csite.2022.102063.

242 Liang, J., Zhao, W., Wang, Y. et al. (2023). Effect of cooling temperature on the performance of a solar adsorption chiller with the enhanced mass transfer. *Applied Thermal Engineering* 219: https://doi.org/10.1016/j.applthermaleng.2022.119611.

243 Wang, R.Z., Pan, Q.W., and Xu, Z.Y. (2016). Solar-powered adsorption cooling systems. In: *Advances in Solar Heating and Cooling*, 299–328. Elsevier B.V.

244 Hassan, A.A., Hassan, H., Islam, M.A., and Saha, B.B. (2023). Thermoeconomic performance assessment of a novel solar-powered high-temperature heat pump/adsorption cogeneration system. *Solar Energy* 255: 71–88. https://doi.org/10.1016/j.solener.2023.03.017.

245 Stefański, S., Nowak, W., Sztekler, K. et al. (2019). Applicability of adsorption cooling/desalination systems driven by low-temperature waste heat. *IOP Conference Series: Earth and Environmental Science* 214: https://doi.org/10.1088/1755-1315/214/1/012126.

246 Pan, Q., Peng, J., and Wang, R. (2019). Experimental study of an adsorption chiller for extra low temperature waste heat utilization. *Applied Thermal Engineering* 163: https://doi.org/10.1016/j.applthermaleng.2019.114341.

247 Chaudhari, V.H. and Desai, A.D. (2022). Experimental study of single bed prototype adsorption refrigeration unit using waste heat energy. *Materials Today: Proceedings* 49: 1799–1803. https://doi.org/10.1016/j.matpr.2021.08.022.

248 Hamdy, M., Askalany, A.A., Harby, K., and Kora, N. (2015). An overview on adsorption cooling systems powered by waste heat from internal combustion engine. *Renewable and Sustainable Energy Reviews* 51: 1223–1234. https://doi.org/10.1016/j.rser.2015.07.056.

249 Pinheiro, J.M., Salústio, S., Rocha, J. et al. (2020). Adsorption heat pumps for heating applications. *Renewable and Sustainable Energy Reviews* 119: https://doi.org/10.1016/j.rser.2019.109528.

250 Mendes, A.C.A., Pujatti, F.J.P., and Cortez, M.F.B. (2022). Conceptual design of an adsorption refrigeration system applied to vehicles. *International Journal of Refrigeration* 135: 60–74. https://doi.org/10.1016/j.ijrefrig.2021.12.022.

251 Kehinde, A.G., Ngonda, T., Raji, A., and Kanyarusoke, K. (2022). A review of different technologies for refrigerated truck. *Materials Today: Proceedings* 56: 2305–2310. https://doi.org/10.1016/j.matpr.2021.11.646.

252 Wang, D.C., Xia, Z.Z., and Wu, J.Y. (2006). Design and performance prediction of a novel zeolite–water adsorption air conditioner. *Energy*

Conversion and Management 47 (5): 590–610. https://doi.org/10.1016/j.enconman.2005.05.011.

253 Verde, M., Cortés, L., Corberán, J.M. et al. (2010). Modelling of an adsorption system driven by engine waste heat for truck cabin A/C. Performance estimation for a standard driving cycle. *Applied Thermal Engineering* 30 (13): 1511–1522. https://doi.org/10.1016/j.applthermaleng.2010.04.005.

254 Jiangzhou, S., Wang, R.Z., Lu, Y.Z. et al. (2003). Locomotive driver cabin adsorption air-conditioner. *Renewable Energy* 28 (11): 1659–1670. https://doi.org/10.1016/s0960-1481(03)00007-7.

255 Tamainot-Telto, Z., Metcalf, S.J., and Critoph, R.E. (2009). Novel compact sorption generators for car air conditioning. *International Journal of Refrigeration* 32 (4): 727–733. https://doi.org/10.1016/j.ijrefrig.2008.11.010.

256 Vasta, S., Freni, A., Sapienza, A. et al. (2012). Development and lab-test of a mobile adsorption air-conditioner. *International Journal of Refrigeration* 35 (3): 701–708. https://doi.org/10.1016/j.ijrefrig.2011.03.013.

257 Ugale, V.D. and Pitale, A.D. (2015). A review on working pair used in adsorption cooling system. *International Journal of Air-Conditioning and Refrigeration* 23 (02): https://doi.org/10.1142/s2010132515300013.

258 Sharafian, A., McCague, C., and Bahrami, M. (2015). Impact of fin spacing on temperature distribution in adsorption cooling system for vehicle A/C applications. *International Journal of Refrigeration* 51: 135–143. https://doi.org/10.1016/j.ijrefrig.2014.12.003.

259 Sharafian, A. and Bahrami, M. (2015). Critical analysis of thermodynamic cycle modeling of adsorption cooling systems for light-duty vehicle air conditioning applications. *Renewable and Sustainable Energy Reviews* 48: 857–869. https://doi.org/10.1016/j.rser.2015.04.055.

260 Golparvar, B. and Niazmand, H. (2018). Adsorption cooling systems for heavy trucks A/C applications driven by exhaust and coolant waste heats. *Applied Thermal Engineering* 135: 158–169. https://doi.org/10.1016/j.applthermaleng.2018.02.029.

261 Vasta, S. (2023). Adsorption air conditioning for automotive applications: a critical review. https://doi.org/10.20944/preprints202306.1171.v1.

262 Hassan, M., El-Sharkawy, I.I., and Harby, K. (2023). Study of an innovative combined absorption-adsorption cooling system employing the same evaporator and condenser. *Case Studies in Thermal Engineering* 42: https://doi.org/10.1016/j.csite.2022.102690.

263 Vasta, S., Palomba, V., La Rosa, D., and Mittelbach, W. (2018). Adsorption-compression cascade cycles: an experimental study. *Energy Conversion and Management* 156: 365–375. https://doi.org/10.1016/j.enconman.2017.11.061.

264 Saha, B.B., El-Sharkawy, I.I., Shahzad, M.W. et al. (2016). Fundamental and application aspects of adsorption cooling and desalination. *Applied Thermal Engineering* 97: 68–76. https://doi.org/10.1016/j.applthermaleng.2015.09.113.

265 Tassou, S.A., Lewis, J.S., Ge, Y.T. et al. (2010). A review of emerging technologies for food refrigeration applications. *Applied Thermal Engineering* 30 (4): 263–276. https://doi.org/10.1016/j.applthermaleng.2009.09.001.

266 Dakkama, H.J., Youssef, P.G., Al-Dadah, R.K., and Mahmoud, S. (2017). Adsorption ice making and water desalination system using metal organic frameworks/water pair. *Energy Conversion and Management* 142: 53–61. https://doi.org/10.1016/j.enconman.2017.03.036.

267 Alammar, A.A., Rezk, A., Alaswad, A. et al. (2022). The technical, economic, and environmental feasibility of a bioheat-driven adsorption cooling system for food cold storing: a case study of Rwanda. *Energy* 258: https://doi.org/10.1016/j.energy.2022.124560.

268 Coorey, R., Ng, D.S.H., Jayamanne, V.S. et al. (2018). The impact of cooling rate on the safety of food products as affected by food containers. *Comprehensive Reviews in Food Science and Food Safety* 17 (4): 827–840. https://doi.org/10.1111/1541-4337.12357.

269 Henning, H.-M. (2007). Solar assisted air conditioning of buildings – an overview. *Applied Thermal Engineering* 27 (10): 1734–1749. https://doi.org/10.1016/j.applthermaleng.2006.07.021.

270 Hassan, A.A., Elwardany, A.E., Ookawara, S. et al. (2020). Integrated adsorption-based multigeneration systems: a critical review and future trends. *International Journal of Refrigeration* 116: 129–145. https://doi.org/10.1016/j.ijrefrig.2020.04.001.

271 Zhang, Y., Zhao, Y., Dai, S. et al. (2022). Cooling technologies for data centres and telecommunication base stations–a comprehensive review. *Journal of Cleaner Production* 334: https://doi.org/10.1016/j.jclepro.2021.130280.

272 Zeng, Z., Zhao, B., and Wang, R. (2023). Water based adsorption thermal battery: sorption mechanisms and applications. *Energy Storage Materials* 54: 794–821. https://doi.org/10.1016/j.ensm.2022.11.024.

273 Zhu, X., Xie, W., Wu, J. et al. (2022). Recent advances in direct air capture by adsorption. *Chemical Society Reviews* 51 (15): 6574–6651. https://doi.org/10.1039/d1cs00970b.

274 Pal, A., El-Sharkawy, I.I., Saha, B.B. et al. (2016). Experimental investigation of CO_2 adsorption onto a carbon based consolidated composite adsorbent for adsorption cooling application. *Applied Thermal Engineering* 109: 304–311. https://doi.org/10.1016/j.applthermaleng.2016.08.031.

275 Pal, A., Uddin, K., Rocky, K.A. et al. (2019). CO_2 adsorption onto activated carbon–graphene composite for cooling applications. *International Journal of Refrigeration* 106: 558–569. https://doi.org/10.1016/j.ijrefrig.2019.04.022.

276 Pal, A., Rocky, K.A., and Saha, B.B. (2021). Thermodynamic analysis of promising biomass-derived activated carbons/CO_2 based adsorption cooling systems. *Journal of CO_2* Utilization 46: https://doi.org/10.1016/j.jcou.2021.101457.

277 Sghaier, W., Ben Torkia, Y., Bouzid, M., and Ben Lamine, A. (2021). CO_2 adsorption investigation by statistical physics: thermodynamic analysis for cooling cycle application. *Journal of Environmental Chemical Engineering* 9 (2): https://doi.org/10.1016/j.jece.2021.105108.

278 Gautam, Chaudhary, A., and Sahoo, S. (2023). Exploring the feasibility of next-generation CO_2-based adsorption cooling systems using different adsorbent reactor configurations. *Energy Conversion and Management* 289: https://doi.org/10.1016/j.enconman.2023.117181.

9

Building Thermal Comfort Modulation Through Phase Change Material

Rahul Kumar Sharma, Rahul Verma, Sana Fatima Ali, and Dibakar Rakshit

Department of Energy Science and Engineering, Indian Institute of Technology Delhi, New Delhi, India

9.1 Introduction

The average global temperature of the Earth is increasing accompanied by the increase in global population and human activities. In most of the countries located in the tropical zones of the Earth, the climate change has affected every sphere of life, with the weather conditions getting harsher and in summer the maximum temperature reaches up to 50 °C [1]. To make living spaces comfortable, a large amount of energy is spent over maintaining the conditions within the thermal comfort zone of the occupants. The most suitable and convenient method is to deploy heating, ventilation, and air-conditioning (HVAC) and cooling units. However, these units consume over 60% of the total energy consumption, hence increasing the load required and make buildings a less energy-efficient venture. Phase change materials (PCMs) are the materials which have high energy storage capacity while changing the phase from solid to liquid, liquid to gas, or vice versa. PCM tends to increase the thermal inertia of a system and hence maintain the system at a constant temperature condition.

A building experiences various external and internal and energy. Internal energy fluxes are generated by the equipment, human being, loads located inside the envelope, and solar radiation coming inside the envelope, while external energy fluxes are effected by prevailing ambient as well as the local weather conditions. Heat transfer processes within a building room include convection, conduction, shortwave radiation, and longwave radiation [2]. In an energy-efficient building, the thermal comfort of occupants is the main concern along with minimizing the energy consumption by employing an efficient building envelope. Thermal energy storage within the envelope plays a crucial role in achieving energy efficiency.

Advancements in Non-Conventional Cooling and Thermal Storage Strategies, First Edition.
Edited by Bidyut Baran Saha and Dibakar Rakshit.

The use of PCMs offers significant economic and environmental benefits in controlling indoor temperatures and enhancing air-conditioning (A.C.) applications [3]. The PCM stores the energy in a building envelope and thus prevents extreme air temperature fluctuations, thereby promoting thermal comfort. The energy which can be stored is a function of mass, thermal properties, and composition as well as the components of the wall envelope [4]. Incorporating PCM into the construction material can increase the overall energy storage capacity of the lightweight buildings having lower energy storage ability.

The charging phenomenon of the PCM starts when the temperature rises above the melting temperature, while the stored energy is discharged when the temperature falls below the melting temperature; thus, it can control the temperature of a building within a narrow range. The subsequent subsections of the chapter will provide an overview of studies focusing on different approaches to maintain human thermal comfort using PCM in a building envelope.

9.2 PCM as a Sustainable Alternate

9.2.1 Review of Studies on the Thermal Performance of PCM Incorporated Buildings

Sanjay et al. [5] examined numerous sustainability options, including PCMs, *green roofs, and cool roofs modeling them on the IIT Delhi Central Library under various climatic* conditions. It was determined that PCMs have little effect on cooling burdens during prime periods, such as the summer, but are effective during other months. They recommended the cold roof configuration in conjunction with OM32 PCM as the most efficient mitigation strategy. The thermal performance of PCM, the green roof, and the cool roof was not mutually exclusive, and the parametric analysis of coupled systems was identified as a research lacuna. Ali et al. [6] investigated the thermal comfort of small office buildings using predicted mean vote (PMV) and predicted percentage of dissatisfaction (PPD) indices by incorporating various types of window glazing, blinds as internal shading systems, and various thicknesses of roof and wall insulation. This study provided optimum thermal comfort by insulating the roof and walls with 40 mm thick fiberglass, scheduling internal blinds according to the sun's path, and installing low-e glass windows. The optimization study for comfort requirement by both active and passive strategies and the comfort perception research at various locations across the nation have been identified as areas requiring additional analysis. Castell et al. [7] examined the PCM with various building materials and sites. With high heat gains and minimal thermal inertia or insulation, it has been demonstrated that PCM benefits are readily apparent. However, with substantial insulation, high thermal inertia, or low solar yields, the advantages of incorporating a PCM layer are more difficult to determine. Al-yasiri [8] conducted a comprehensive evaluation of thermal comfort management and energy conservation in buildings by combining PCMs with building construction materials. Important findings include that the type, quantity, position, and encapsulation of PCM are weather dependent and the PCM layer should be placed closer to the heat source, i.e., close to the exterior layers in hot climates and close to the interior layers in cold climates. Mourid et al. [9] analyzed the influence of PCM integration in the roof of the test cell building on various room temperatures and heat fluxes. It was determined that the thickness and positioning of PCM layers substantially impact the heat flux transmitted. Integration of the PCM reduces temperature oscillations, increases ceiling thermal inertia, and shifts peak load by up to seven hours.

Prabhakar et al. [10] emphasized the advantages of a passive PCM system with regulated natural ventilation for 15 cities around the world. It has been reported that a temperature-controlled ventilation system can enhance the effectiveness of PCM by 40% in temperate climates. It was determined that the passive PCM cooling arrangement was ineffectual for hot-arid conditions. Biplab et al. [11] assessed thermal comfort in three locations (Bhubaneswar, Jodhpur, and New Delhi) with varying climatic conditions using insulation and PCM in roofs and walls. It was determined that PCM performs better than insulation at operating temperatures of at least 30 °C when it comes to maintaining interior temperatures. Liu et al. [12] numerically studied the effect of PCM thickness on the thermal modification of a PCM-filled double glazing unit in Northeast China during the winter . Therefore, PCM thicknesses of 12–30 mm and melting temperatures of 14–16 °C are suggested. Alam et al. [13] performed a parametric analysis to assess the effectiveness of an inactive PCM system with five distinct phase change temperatures of 20–2 °C for the eight cities that represent different climatic zones in Australia. The study concluded that the local weather of a place influences the choice of the PCM, its melting temperature and the energy savings achieved by a building envelope.

9.2.2 PCM Air Heat Exchanger for Building Application

The incorporation of fresh air is necessary in the building envelope to dilute the indoor air and improving the indoor air quality. However, the ambient air is available at higher temperatures, hence increasing the temperature of a building, so studies have been performed to cool the ambient air and reduce the cooling load. PCM-based heat exchangers are used to control the air temperature entering a building envelope. Research has shown that they have been effective in reducing the building cooling load. During the day, hot air is passed through the PCM-based heat exchanger to reduce its temperature, while during the nighttime, the ambient air discharges the heat stored in the PCM [14]. A numerical and experimental study conducted by Osterman et al. demonstrated that an RT22HC-based latent energy storage system for air can save 145 kWh of energy for a year [14]. For the climatic zone of China, night ventilation with a PCM system was examined experimentally and numerically, and the novel system was able to cool the building with a designed COP of 80 for the weather condition of June and July [15]. Furthermore, stacked air PCM heat exchangers are also used to reduce the heat exchanger size. A numerical study on pyramidical and cylindrical designs was performed, and different combinations were tested by offsetting and rotating the geometry. The study concluded that the cylindrical configuration with an offset was able to outperform other configurations.

9.2.3 Parameters for PCM Selection for Incorporation

Numerous factors are required to analyze before the incorporation of PCM in the building envelope to improve the charging and discharging cycle of PCM to enhance building performance. The PCM must possess certain thermal-physical, kinetic, chemical, economic, and environmental properties to be suitable for building applications. PCM transient temperature plays a crucial role in improving building envelope thermal efficiency for different climatic regions and locations of PCM settle in the building envelope. The difference in temperature between the PCM and the surrounding air accounts for heat absorption and release during daytime and nighttime. Consequently, the PCM melting temperature must fall between the season's mean maximum and minimum temperatures. The properties required to evaluate the overall PCM selection are described in Figure 9.1, while thermodynamic properties are necessary for determining PCM at specific locations. The efficacy of PCMs is dependent on the diurnal temperature variation; these materials function well when the daily temperature range is relatively large [18].

Furthermore, the optimum thickness of PCM must be determined, and it may change with the phase change temperature for specific locations. Optimally designed PCMs have sufficient thickness to withstand solar radiation and prevent

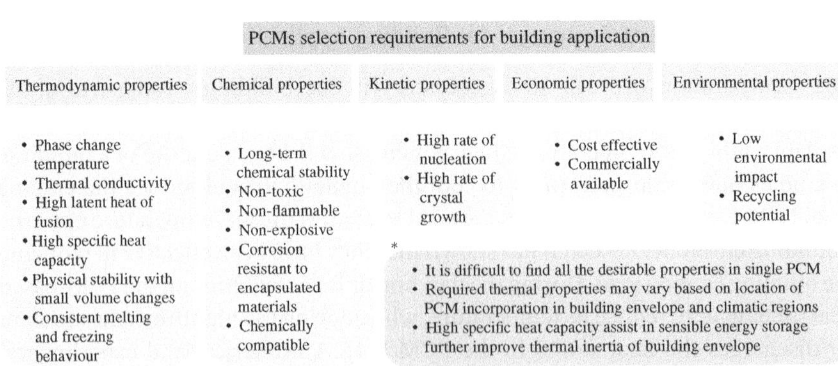

Figure 9.1 Selection of PCM properties for building applications according to [16, 17].

the increase in building indoor temperature. During night hours, when the ambient temperature falls below the PCM temperature, PCM starts releasing stored heat to the ambient. The difference between the PCM melting temperature and the nighttime minimum temperature must be sufficient for PCM to flow out stored heat, which becomes available for heat absorption and storage the following day. The high latent heat of PCM demonstrates the large absorption capacity. Therefore, it demands less thickness of the PCM material to avail the same thermal comfort.

9.2.4 PCM Identification Method for Application in Buildings

PCM is a substance that can store and discharge significant quantities of thermal energy through a phase transition, usually from solid to liquid. PCM can be used to enhance thermal comfort and energy efficiency in buildings. The primary method is to determine the precise application and function of PCM within the building. This may involve thermal energy storage, regulating the temperature, or energy conservation. Several PCM cooling methods, such as solar cooling systems with PCM, PCM in A.C. systems, PCM in building envelopes, etc., are highlighted in the review research [19]. Furthermore, building envelope cooling through PCM is classified into active, passive, and ventilated facade principle-based techniques. We assess the thermal requirements of the building, including peak cooling or heating demands, diurnal temperature variation, and energy demand patterns. This data will assist in identifying the necessary PCM properties, such as the melting point temperature and the latent heat. This may entail encapsulating PCM inside building materials, including gypsum board or concrete, or utilizing PCM modules or containers distinct from the building materials. We consider installation simplicity, required maintenance, and compatibility with present building systems [19–21].

9.2.5 PCM Incorporation Techniques

Numerous methods exist for integrating PCMs into structures. The choice of technique is determined by variables such as the building's design, the PCM application, and the desired energy efficiency. Figure 9.2 illustrates a few typical methods of PCM incorporation in building envelopes. The development can be majorly divided into direct incorporation, immersion, stabilization, and encapsulation [22]. The direct incorporation method integrates PCM into building material directly to provide temperature regulation and energy storage capabilities. While during immersion, the building components and materials are submerged in the liquid PCMs. The main downsides of these technologies, notwithstanding their affordability and viability, include PCM leakage,

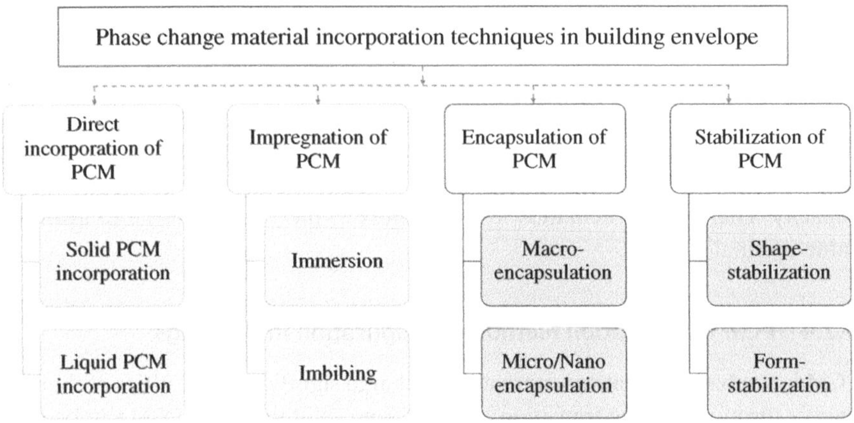

Figure 9.2 Phase change material integration methods in building.

incompatibility, and flammability with building materials. PCM can be contained in containers or capsules composed of materials such as metal or plastic. Then, these capsules could be included or implanted inside construction materials. PCM leakage and interaction with other construction materials are prevented via encapsulation. Encapsulated PCM containers can be installed on floors, ceilings, or walls to strategically store and release heat energy [23, 24]. The process of micro/nanoencapsulation includes encasing microscopic PCM droplets in a protective covering. Finally, PCM stabilization entails their incorporation into a porous structure that supports them, like high-density polyurethane [22, 25, 26].

9.2.6 Modified PCM Air Heat Exchanger for Retrofitting to Air Conditioner

Thermal comfort A.C. has become necessary due to the rising global temperatures. In 2019, it was the second warmest year ever recorded, marking the culmination of the hottest decade on record [1]. Currently, approximately 1.6 billion A.C. units are in use worldwide, with the United States and China accounting for over half of this total. However, in India, owning an air conditioner is still considered a luxury as the residential penetration rate is only 7–9% [2, 3]. Nevertheless, the demand for cooling is rapidly growing in India, driven by economic and population growth. This anticipated surge in residential A.C. units could strain the country's electrical grid. Globally, air conditioners consume over 2000 TWh of electricity annually, which is more than twice the total electricity consumption of Africa. Cooling accounts for more than a fifth of electricity usage in buildings, with space cooling alone responsible for 50% or more of residential peak electricity demand (OECD/IEA, 2018).

PCM incorporation in passive means has been used mainly to reduce the cooling load by increasing the thermal inertia of building or reduce the temperature of the incoming air. However, one more approach is to use the PCM-incorporated heat exchanger with an A.C. system to reduce the recirculation temperature or the fresh air temperature. Chaiyat [27] used a PCM-packed bed storage with an RT20 heat exchanger to cool the air coming out of the room before recirculating it back to the A.C. unit. The modified system was able to achieve an energy saving of 9% while having a payback period of 4.12 years. Sharma et al. [28] used a PCM embedded pin fin heat exchanger to offset the excessive load in an HVAC system by cooling the ambient air during peak summer. Furthermore, the study was conducted for three different climatic zones, and the system was able to offset a peak load of 4% for composite climate using a dodecane-based heat exchanger. Hence, PCM can be used in different configurations for making the HVAC systems more energy efficient.

9.3 Energy Transfer Evaluation Across PCM

9.3.1 Temperature Variation Across PCM Integrated Envelope

The current study assesses the building envelope's PCM thickness and incorporation position. According to Pomianowski and Osterman [20, 21], the research was conducted using a two-story building model for peak summer and winter conditions with office occupancy and equipment generation load. The reference building model in Figure 9.3 has only been updated to include PCM in the roof. Properties of Bio PCM29, with thicknesses ranging from 10 to 60 mm, were picked from the DesignBuilder collection to evaluate the effects on temperature changes. Additionally, the outer surface of the roof and the area below 12.5 mm of plaster were chosen as two locations to incorporate the PCM in the building roof. The

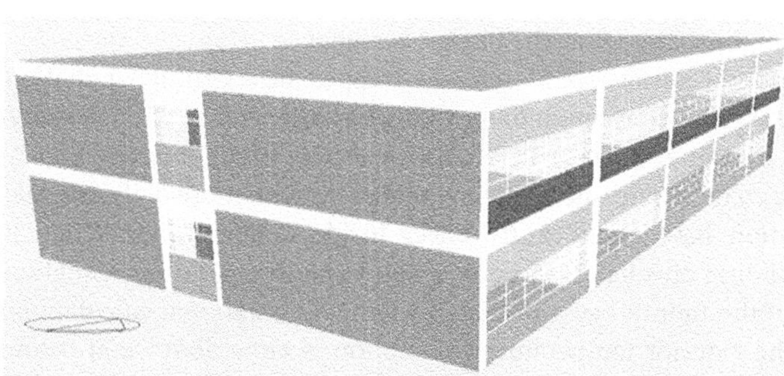

Figure 9.3 Reference building rendered view model in DesignBuilder.

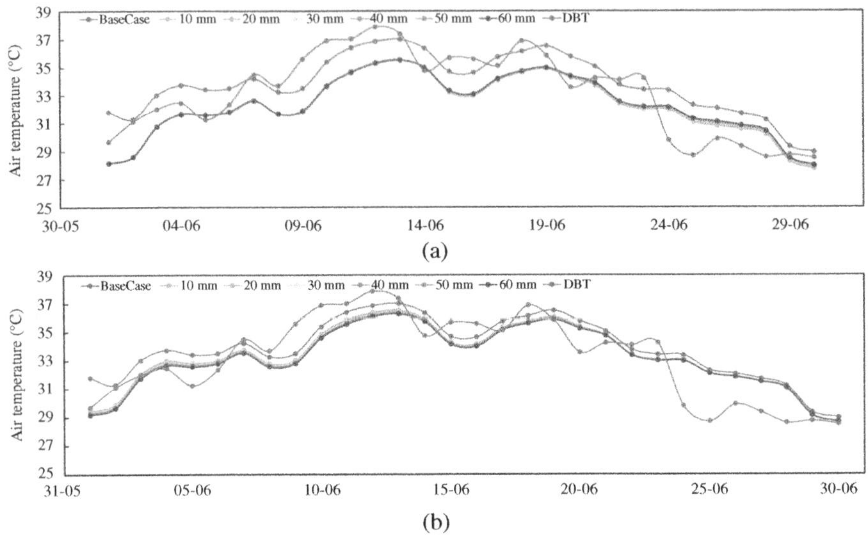

Figure 9.4 Daily temperature variation for the June month: (a) PCM at the outer surface and (b) PCM below 12.5 mm plaster from the outer surface.

investigation took into account the presumption that PCM is encapsulated and has its thermal properties preserved for both cases, below 12.5 mm of plaster and the outside surface.

The study's findings are shown in Figure 9.4 for a typical summer month. Figures 9.4a and b demonstrate how the thickness increases, while the indoor air temperature decreases slightly. Complete solidification does not occur during this time because the average outside temperature is higher most of the time than the PCM melting point. Furthermore, compared to PCM below the 12.5 mm plaster surface, PCM at the outer surface shows a significant decrease in the daily interior temperature. Because during the night, the thermal barrier for releasing the stored heat from the PCM below the plaster is higher.

In contrast, the outer surface PCM releases trapped heat through radiation exchange from the sky and convection mode of heat transfer due to wind velocity. Figure 9.5 shows the outer surface PCM case and also highlights the hourly inside temperature change. The findings show that areas with more daily external temperature variation have significantly lower inside temperatures. When it comes to PCM building envelopes, there is a noticeable temperature reduction at the beginning of the month because that is when computational calculations start and where the exterior temperature fluctuation is large above and below the PCM melting temperature. These conditions enhance PCM melting and solidification amounts. In addition, a typical winter month or cold region indoor

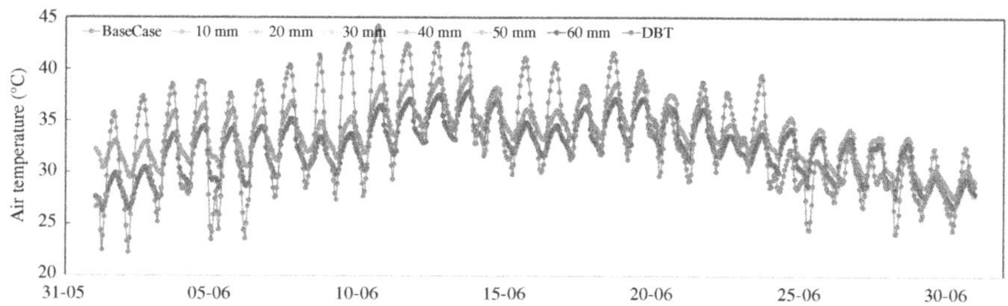

Figure 9.5 Hourly temperature variation for typical summer month.

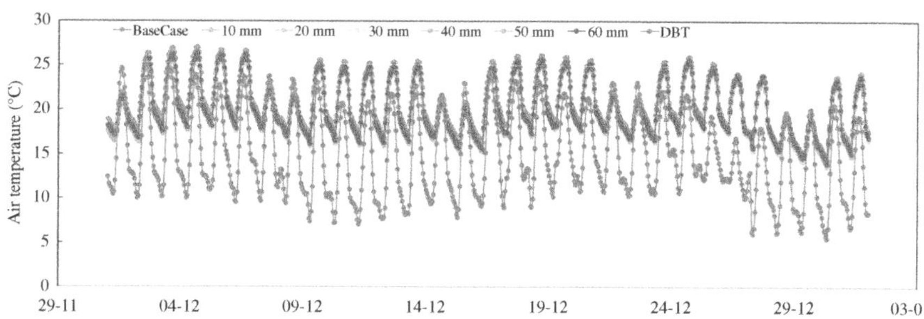

Figure 9.6 Hourly temperature variation for a typical winter month.

hourly temperature variation is shown in Figure 9.6. PCM located on the exterior surface of the roof or near the exterior surface promotes daytime heat storage when radiation is available and during off-sunshine hours releases heat to an outdoor environment. Hence, in the winter or cold climatic region, there is a loss in building performance because both phenomena restrict heat penetration inside the building. In winter or cold climatic regions, building performance might improve by selecting lower melting temperature PCM and locating the inner side of the roof.

9.3.2 Assessment of Heat Transfer

The finite difference algorithm (CondFD) is used to analyze the reduction in energy consumption in a building while integrating PCM in building wall conduction. The algorithm implements the liquid- and solid-state PCM-specific heat capacity. The time step values, space discretization constant, and thermal diffusivity were considered as 12, 3, and 6.70×10^{-8}, respectively. To accurately account for PCM, an implicit finite difference scheme with 1-D heat transfer conduction and a 16-point enthalpy–temperature curve is used [13, 29–31].

The equation below calculates the changes in thermal characteristics caused by phase transitions [32, 33],

$$\frac{\rho C_P \Delta x (T_{i,n} - T_{i,o})}{\Delta t} = \frac{K_w (T_{i+1,n} - T_{i,n})}{\Delta x} + \frac{K_e (T_{i-1,n} - T_{i,n})}{\Delta x} \tag{9.1}$$

$$K_w = \frac{K_{i+1,n} + K_{i,n}}{2} \tag{9.2}$$

$$K_e = \frac{K_{i-1,n} + K_{i,n}}{2} \tag{9.3}$$

In each iteration, the specific heat capacity of PCM is updated based on the temperature–enthalpy input data,

$$C_P = \frac{h_{i,n} - h_{i,o}}{T_{i,n} - T_{i,o}} \tag{9.4}$$

According to the following equation, the ConFD algorithm's finite difference layer thickness is highly dependent on space discretization constant (c), time step, and thermal diffusivity (α),

$$\Delta x = \sqrt{c\alpha\Delta t} \tag{9.5}$$

In this investigation, solar air temperature combines radiation intensity and ambient temperature as ordinary climatic parameters. A building element's heat transmission is an intricate phenomenon. When solar radiation falls on a surface, the surface's temperature increases beyond the ambient temperature, a phenomenon commonly referred to as the sol-air effect. Sol-air temperature" refers to the outside air temperature in the absence of all radiation changes, which causes the same rate of heat transfer into the surface of the building envelope as the sum of incident solar radiation, radiant energy exchange with the sky, and convective heat transfer with the outside air, as illustrated in Figure 9.7 and

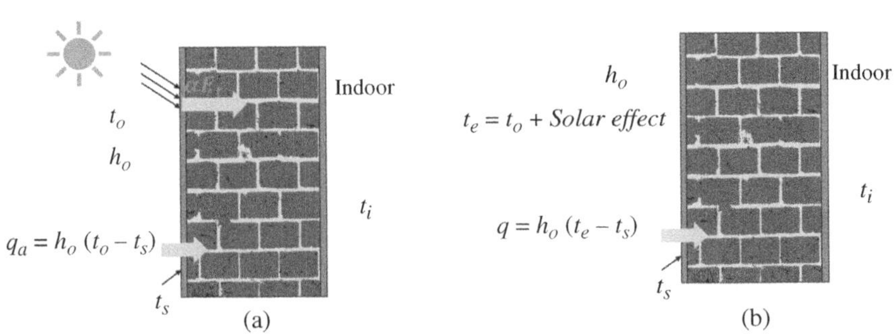

Figure 9.7 Illustration of solar air temperature: (a) Actual case and (b) idealized case (no sun).

calculated through Eq. (9.6) [32],

$$t_e = t_o + \frac{\alpha E_t}{h_o} - \frac{\varepsilon \Delta R}{h_o} \tag{9.6}$$

where α = surface absorption for solar radiation, E_t = total solar radiation incident on the surface (W/m^2), and h_o = outer surface coefficient of heat transfer by long-wave radiation and convection W/(m^2 K). Furthermore, t_s = surface temperature (°C), ε = hemispherical emittance of surface, ΔR = difference between longwave radiation incident on the surface from sky and surroundings and radiation emitted by a blackbody at outdoor air temperature (W/m^2), t_e = sol-air temperature (°C), and t_0 = outdoor air temperature (°C).

Figure 9.8 shows how much heat is gained via the roof of the building with and without PCM. The statistics show that the heat gain through the roof ceiling decreased more significantly. There is no additional improvement in building performance, as shown in the temperature analysis section, so the ceiling heat gain scenario with findings up to 20 mm of PCM incorporation is presented for comparison. As illustrated in Figure 9.8, the drop in ceiling heat gain reduced significantly when PCM was installed below the plaster surface for the same thickness of PCM.

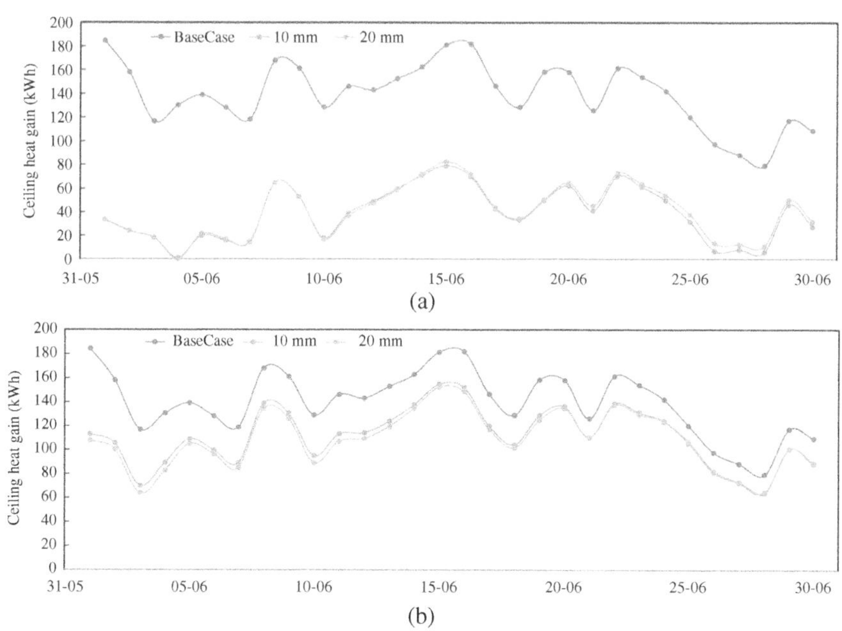

Figure 9.8 Heat gain through ceiling of the building for a typical summer month: (a) PCM at the outer surface and (b) PCM below 12.5 mm plaster from the outer surface.

9.4 Thermal Comfort Performance of Building with PCM

9.4.1 Thermal Comfort Assessment Model

Thermal comfort refers to the occupants' expression of contentment with their thermal environment in a building [34]. To ensure the well-being, physical health, and productivity of occupants, it is imperative for them to be in a thermally comfortable environment. The perception of thermal comfort can differ among individuals and is affected by various environment-related factors, which include relative humidity, air temperature, mean radiant temperature, and air velocity, as well as physiological factors, such as the rate of metabolism and clothing insulation [34].

Literature contains several indices that assist in evaluating the thermal comfort status of occupants, viz., Tropical Summer Index (TSI), operative temperature (O.T.), effective temperature (E.T.), corrected effective temperature (CET), etc., which have been devised considering the influencing factors [2]. ASHRAE Standard 55 recommends two extensively used and acknowledged indices for thermal comfort assessment, which has been formulated by P. O. Fanger: the PPD and the PMV [34].

The balancing of the heat transfer equations accounting for the thermal load on a person forms the basis of PMV, which measures the mean thermal experience of building occupants on a 7-point scale ranging from +3, representing a hotter environment, to −3.0, representing a colder environment [35, 36]. As per ASHRAE Standard 55, for the desired thermal environment, the PMV value should fall within the band from −0.5 to +0.5 [34]. Nevertheless, the actual satisfaction of people with their thermal environment is determined by PPD, which is estimated utilizing the correlation between the people's satisfaction and the obtained PMV value, thus quantifying the percentage of people dissatisfied with their thermal conditions. It is recommended by ASHRAE Standard 55 that the PPD value should not surpass 20%, implying that for a thermal environment to be deemed acceptable, a minimum of 80% of the building inhabitants should be contended with the prevailing thermal conditions [34]. Besides the environmental and physiological factors, the occupants' behavioral actions also influence thermal comfort, such as changes in location, clothing, and activity.

Although depending upon the static models that examine the deterministic cause-and-effect relationship for thermal comfort, the indices described before ignore the adaptability of the occupants [37]. Furthermore, these indices do not account for the psychological aspects of the people, which are vital for controlling and modifying thermal comfort [35]. As a result, researchers have now made a transition toward models on adaptive comfort, which work on the tenet that if a modification in the occupants' thermal environment leads to discomfort, they respond in ways that aim to restore their comfort levels [37].

The theory of adaptive thermal comfort states that there is a correlation between occupants' connection to the external environment and their ability of adaptation to a broader spectrum of comfort conditions beyond conventional notions of comfort in an indoor environment [38]. Numerous researchers have explored this adaptive approach to assess comfort in naturally ventilated buildings by establishing a linear relationship between the inhabitant's comfort temperature and the external temperature. Based on the findings of these studies on the adaptive comfort, a model for naturally ventilated buildings was also adopted by ASHRAE Standard 55 [39].

Due to the lack of recognition of climate and workplace considerations of Indian buildings, the utilization of the model proposed in ASHRAE Standard 55 for assessing the comfort conditions in India could result in excessively high thermal comfort expectations, which could further lead to unsustainable demand for energy consumption without commensurate improvement in actual comfort levels [40]. Thus, the Indian Model for Adaptive Comfort (IMAC) was established to provide design as well as operational guidance for different types of buildings. The development of this model involved conducting year-long survey studies in office buildings across different climatic zones in India. These studies aimed to establish correlations that determine the temperatures at which the inhabitants would perceive a state of neutrality in relation to their thermal environment. Additionally, the model identifies a temperature range within which 90% of occupants feel satisfied with their thermal conditions. By providing a broader comfortable temperature range, the IMAC model contributes to reduced energy consumption for space conditioning [38]. The following linear correlations were developed as part of the IMAC model to estimate the indoor O.T. [41]:

For naturally ventilated buildings:

$$T_{i,op} = (0.54 \times \overline{T_o}) + 12.83 \tag{9.7}$$

$$90\%\ acceptability\ range = T_{i,op} \pm 2.38 \tag{9.8}$$

For mixed-mode buildings:

$$T_{i,op} = (0.28 \times \overline{T_o}) + 17.87 \tag{9.9}$$

$$90\%\ acceptability\ range = T_{i,op} \pm 3.46 \tag{9.10}$$

For air-conditioned buildings:

$$T_{i,op} = (0.078 \times \overline{T_o}) + 23.25 \tag{9.11}$$

$$90\%\ acceptability\ range = T_{i,op} \pm 1.5 \tag{9.12}$$

Here,

$T_{i,op}$: Indoor O.T. eliciting a neutral perception from occupants toward their thermal environment (°C),

$\overline{T_o}$: 30-day outdoor running mean air temperature (°C), and

90 % *acceptability range*: Range of temperature within which 90% of the people experience satisfaction with their thermal environment (°C).

9.4.2 Impact of PCM Incorporated within the Building on Thermal Comfort

In this study, a two-story building model has been selected to investigate the effects of incorporating PCM in a building on the thermal comfort. The main focus has been to explore how the integration of PCM, as a passive measure, enhances thermal comfort. Additionally, the study examines how the location and thickness of the PCM influence thermal comfort. The construction details and building type have previously been specified.

To evaluate the impact of PCM incorporation, as well as variations in its location and thickness, simulations have been conducted using the DesignBuilder software, covering the peak summer and winter months. Two specific configurations of PCM locations have been assessed in the study: Configuration I, involving the application of PCM on the external surface of the roof, and Configuration II, incorporating a layer of PCM below the 12.5 mm plaster layer. The thickness of the PCM has been varied between 10 and 60 mm.

The assessment of thermal comfort conditions in relation to PCM incorporation, its location, and thickness has been carried out using the IMAC model. Weather data for the location of the case study building has been utilized to estimate the 30-day running mean temperatures for the peak summer and winter months.

Equations [7, 8] were utilized to ascertain the indoor O.T. necessary for occupants to experience thermal satisfaction. Subsequently, these calculated indoor O.T. were compared with the indoor O.T. achieved by incorporating BioPCM 29 at different thicknesses and locations. This evaluation aimed to assess the performance of BioPCM 29 in enhancing thermal comfort within the building.

Equations [7, 8] have been employed for determining the indoor O.T. required for inhabitants to experience thermal satisfaction. Subsequently, the calculated indoor O.T. have then been compared with the indoor O.T. accomplished by incorporating PCM with different thicknesses and at various locations to evaluate the performance of BioPCM 29. Detailed discussions and analysis of the corresponding plots and results have been presented in the subsequent section.

9.4.3 Observations and Findings

9.4.3.1 Configuration I of PCM Application

To conduct the analysis, temperature variation plots for a typical week during the peak summer and winter months have been presented here. The simulation results for the case study building reveal that in the base case (without PCM application), the indoor O.T. ranges from 30 to 36 °C during the typical summer week, as illustrated in Figure 9.9a. The yellow shaded area in the figure represents the comfort band established according to the IMAC model, which ensures that 90% of occupants are thermally satisfied. The red line at the top edge of the yellow-shaded area marks the highest permissible indoor O.T., while the blue line at the bottom edge indicates the lowest permissible indoor O.T. Positioned between these two extremes, the black line represents the ideal indoor O.T. for comfort for the given conditions, as per Figure 9.9. The figure noticeably demonstrates that during most of the period, the peak indoor O.T. in the base case exceeds the comfort band. The temperatures within the comfort band primarily occur during nighttime or early morning when the ambient temperature is naturally lower. This indicates a prevailing hotter environment than what is considered comfortable for occupants in the base case, as evident from the figure. Consequently, there is an evident need for passive strategies, such as incorporating PCM within buildings, to effectively narrow the temperature range and create a more comfortable environment for occupants.

In configuration I, where the PCM layer has been applied on the roof's external surface, the observed temperature range is lower compared to the base case, with temperatures approximately ranging between 27.5 and 34 °C for each of the considered PCM thicknesses. This is due to the interaction between solar radiation and the PCM layer. During the day, when solar radiation directly falls on the PCM layer, its temperature begins to rise, resulting in sensible heating. However, once the PCM reaches its phase change temperature, it starts melting, absorbing the incoming heat and limiting the heat flow to the indoor space. As a result, the indoor temperature is maintained lower than the ambient temperature. This process continues until the PCM undergoes complete melting, at which point its sensible heating accelerates the temperature rise more rapidly. However, even in this case, the peak temperature remains lower compared to the scenario without PCM application due to the heat absorbed by the PCM layer itself.

With the sun setting and the external air temperature decreasing, the PCM retains a partial or fully melted state, with its temperature exceeding both the indoor and external air temperatures. Consequently, there is a thermal energy flow from PCM toward the indoor as well as outside environments. However, due to the larger temperature gradient between the temperature of PCM and the external air temperature, the majority of thermal energy flow occurs in the

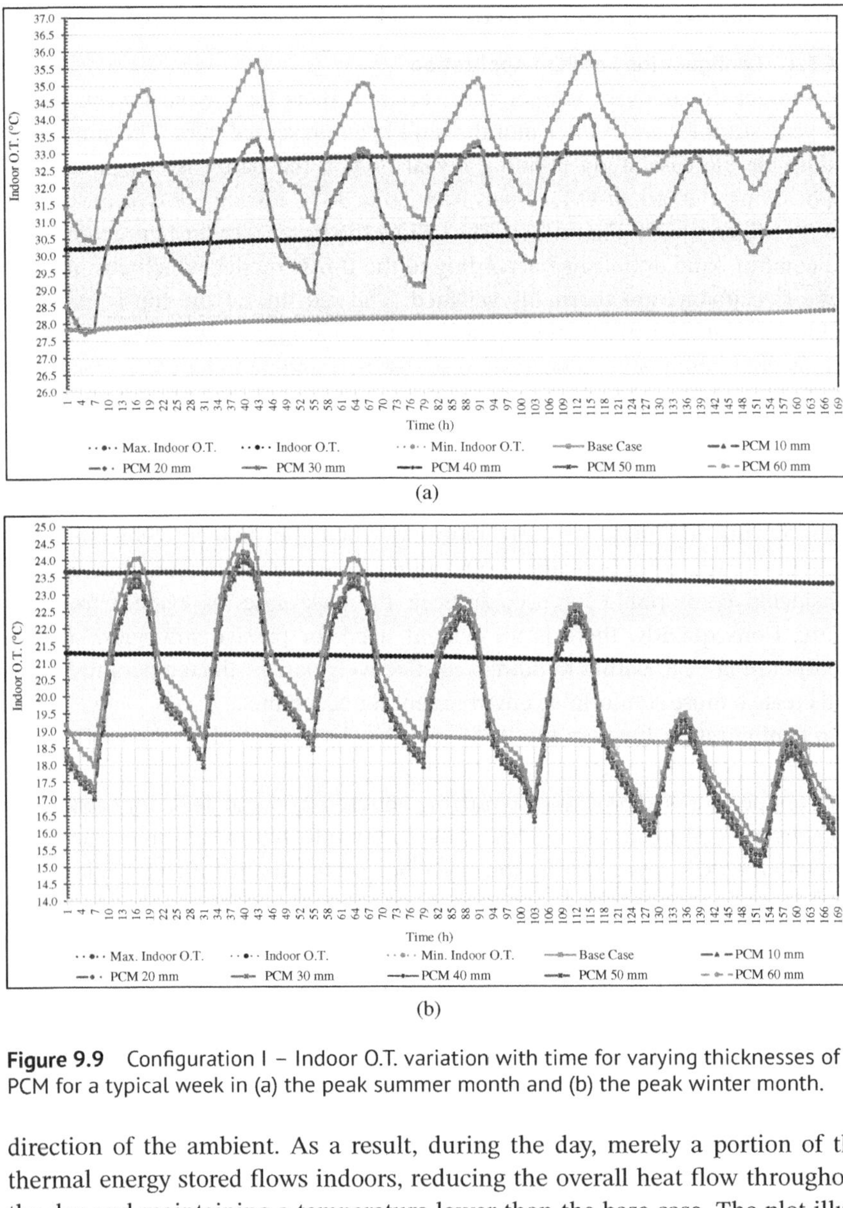

Figure 9.9 Configuration I – Indoor O.T. variation with time for varying thicknesses of PCM for a typical week in (a) the peak summer month and (b) the peak winter month.

direction of the ambient. As a result, during the day, merely a portion of the thermal energy stored flows indoors, reducing the overall heat flow throughout the day and maintaining a temperature lower than the base case. The plot illustrates that with the application of the PCM layer, the indoor O.T. predominantly remain within the comfort band throughout the study period.

Figure 9.9a also reveals that increasing the thickness of the PCM layer from 10 to 60 mm results in only a slight decrease in the indoor O.T. This marginal reduction can be attributed to the incomplete charging–discharging cycle of the PCM. The average ambient temperature remains higher than the melting point of the PCM for most of the time. Consequently, the charging–discharging process for each PCM layer thickness is not fully completed, leading to a minimal impact on temperature reduction within the building. Therefore, considering that the increase in PCM thickness only provides a marginal improvement, it cannot outweigh the associated input cost and energy required for increasing the thickness of the PCM layer.

In contrast, during the winters, the indoor O.T. in the base case ranges approximately from 15.5 to 24.5 °C, as illustrated in Figure 9.9b. It is evident that at the beginning of the week, when the ambient temperatures are relatively higher, the indoor O.T. mostly falls within the comfort band established by the IMAC model. However, when the ambient temperatures become very low, the indoor O.T. drop below the comfort band, indicating a colder indoor environment.

Upon applying PCM layers of different thicknesses in configuration I, it is observed that the indoor O.T. are not significantly improved. The temperature ranges in these cases range from approximately 14 to 24 °C. In fact, the performance of PCM application is marginally poorer than the base case. This can be attributed to the promotion of heat storage by the PCM during the daytime, which inhibits heat transfer into the building, resulting in even lower temperatures. Conversely, during nighttime, when the external temperatures are lower, the PCM releases heat to the external environment, further lowering the indoor O.T. At the beginning of the week, when ambient temperatures are relatively higher, the differences in indoor O.T. between the base case and the cases with PCM layers are comparatively greater, primarily because of the daytime heat storage phenomena. However, even during this period, the indoor O.T. with PCM application remain within the comfort band established by the IMAC model. In contrast, for cases with extremely low ambient temperatures, the temperature differences between the base case and the cases with PCM application are again more pronounced due to the thermal energy transfer between the PCM layer and the external environment. Similar to the base case, the indoor O.T. in these cases fall below the minimum comfortable indoor O.T. Another contributing factor to the poor performance of PCM layer application in configuration I in colder environments is the mismatch between the PCM's melting temperature and the specific environment, as well as insufficient temperature differentials, resulting in incomplete charging and discharging of the PCMs and consequently leading to inferior performance.

Figure 9.9b also illustrates that, in configuration I for the winter month, the thickness of the PCM layer does not exert a significant influence on enhancing the indoor O.T., as previously explained. Although a slight improvement in temperatures can be observed with an increase in the PCM layer thickness, it is essential to consider the trade-off between this temperature increase and the associated costs and energy requirements for increasing the thickness of the PCM layer.

9.4.3.2 Configuration II of PCM application

To examine the influence of the PCM location within the building construction, the PCM layer has then been applied according to configuration II, while maintaining the same base case as in the previous scenario. Figure 9.10 displays the variation in indoor O.T. over time after incorporating PCM for a typical week during the peak summer and winter months.

The plot for a typical week in the peak summer month (Figure 9.10a) illustrates that, similar to configuration I, the application of PCM in configuration II also enhances the indoor O.T. compared to the base case. This improvement is attributed to the PCM layer acting as a heat sink during the daytime and releasing a significant portion of heat to the ambient environment during non-sunshine hours, as explained earlier. However, the performance of PCM in configuration II is observed to be less effective compared to configuration I. This is primarily due to the limited heat absorption caused by the shielding of the PCM layer from direct solar radiation. Consequently, the PCM absorbs less heat compared to the scenario where it is directly exposed to solar radiation. As a result, more time is required to reach its melting point, leading to a prolonged period of higher temperatures before the PCM becomes fully activated. Furthermore, during non-sunshine hours, the insulating effect of the plaster layer impedes heat dissipation from the PCM. This further hinders the efficient charging and discharging process of the PCM, limiting its ability to effectively cool the indoor spaces. Therefore, although the indoor O.T. achieved with PCM application in configuration II tend to fall within the thermal comfort band compared to the base case, they still remain on the higher side compared to the performance of PCM applied in configuration I. The temperature range for configuration II PCM is found to be between 29 and 35 °C.

Yet again, Figure 9.10a demonstrates that increasing the thickness of the PCM layer from 10 to 60 mm results in only a minimal decrease in the indoor O.T., similar to what was observed in configuration I. This outcome can be attributed to the same factors mentioned earlier. It again emphasizes that if there is only a marginal improvement in temperature regulation with an increase in PCM layer thickness, it is not justifiable to invest in the additional input cost and energy required for increasing the PCM thickness. Therefore, such an option should not be pursued.

Furthermore, in the peak winter month, as illustrated in Figure 9.10b, the indoor O.T. achieved through the application of PCM in configuration II are found to be nearly identical to those of the base case, regardless of the PCM thicknesses considered. This similarity can be attributed to the limited heat penetration into the PCM layer, as previously mentioned, and the thermal barrier effect of the

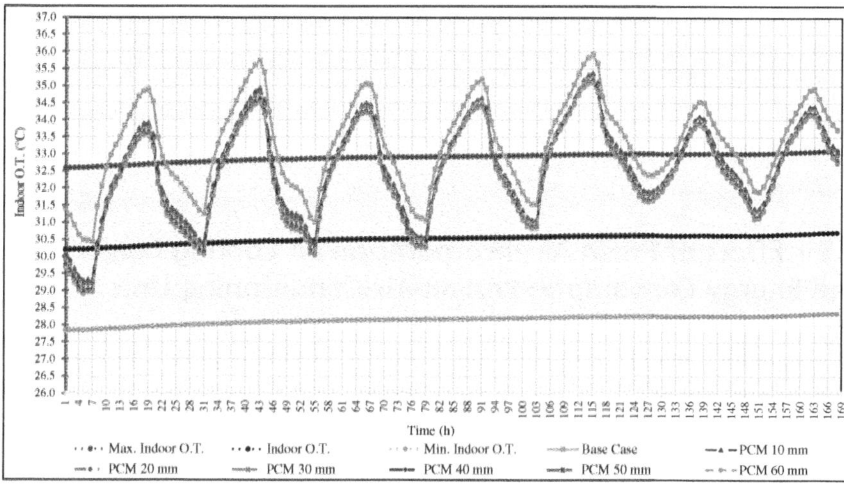

(a)

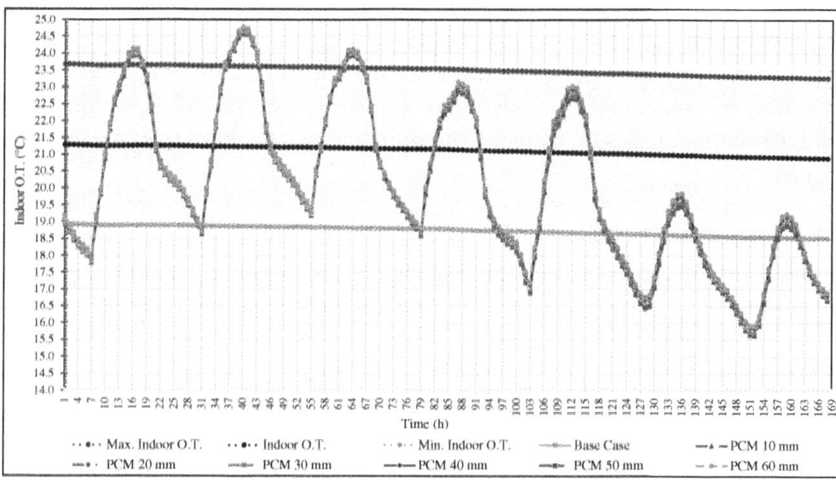

(b)

Figure 9.10 Configuration II – Indoor operative temperature variation with time for varying thicknesses of PCM for a typical week in (a) the peak summer month and (b) the peak winter month.

plaster layer. Consequently, the charging and discharging of the PCM become inefficient, resulting in the PCM layer functioning primarily as an insulation layer. As stated earlier, the thicknesses of the PCM layers also have a minimal impact on the variations in indoor O.T. in this case as well.

The study highlights the importance of carefully considering the building location and PCM properties to ensure year-round applicability in a specific place. Nevertheless, despite selecting an appropriate PCM, additional methods of space conditioning remain essential to ensure that the indoor temperature falls within the comfortable range. Moreover, when selecting a specific PCM from a range of options, it is imperative to consider factors such as cost-effectiveness and accessibility.

9.5 Effect of Fresh Air Incorporation on Cooling Load and Energy Consumption on an Air-Conditioning Unit

The degrading indoor air quality and recurring airborne diseases have affected humans and will affect the energy consumption of A.C. units. A.C. societies, as well as health societies over the world, have recommended incorporating ambient air [42, 43]. The ambient air is used to dilute the indoor air, hence decreasing the level of contaminants and improving the air quality. On the contrary to an A.C. unit which reuses the room air and supply to the room after cooling it, the modified system will need more energy. Figure 9.11 represents the layout of a system with fresh air incorporation. The increase in energy consumption is due to the availability of ambient air at much higher temperatures, so the energy required to cool the mixture of air, i.e., ambient and room air, will be more than one with no fresh air.

In a conventional HVAC system, as represented by Figure 9.11, a part of ambient air is mixed with a fraction of recirculated air at f. Then the mixed air is sent to the A.C. unit, where the air mixture is cooled to the supply condition s and then sent to the cooling space, where it absorbs the heat from the room and is sent

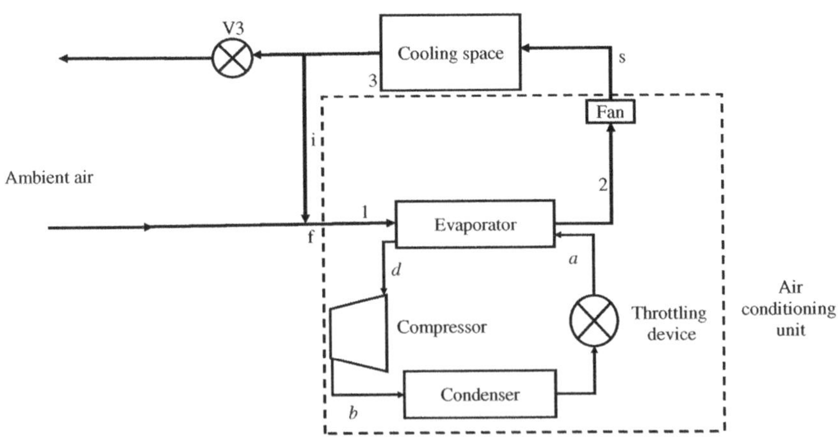

Figure 9.11 Schematic layout of a conventional HVAC system.

out of the system. One of the techniques to reduce the A.C. load is reducing the temperature of the ambient air before mixing the fresh air and recirculated air from the room. Hence, a PCM heat exchanger is numerically designed to reduce the temperature of ambient air and, thereby, the energy consumption of the whole HVAC unit.

9.6 Cooling Load Assessment and Energy Savings for a Room with Retrofitted Air-Conditioning Unit for New Delhi

Figure 9.11 depicts the conventional A.C. system incorporating fresh air and is based on a vapor compression and refrigeration (VCR) system. The inside condition of the cooling space is maintained according to the ISHRAE standards, which are a relative humidity of 50% and a dry bulb temperature of 27 °C [42, 44, 45]. The recirculated air and fresh air mix adiabatically at point f [45].

The sensible heat to be removed for thermal comfort (Q_s) can be calculated as

$$Q_s = \dot{m}_a(1.005 + 1.88\omega)(T_{mix} - T_s) \tag{9.13}$$

Here, $\dot{m}_a$ represents the mass flow rate of air, given as

$$\dot{m}_a = \dot{m}_{ai} + \dot{m}_{ao} \tag{9.14}$$

For 2 t of refrigeration with 50% recirculation air, a mass flow rate of 0.42 kg/s is required. The temperature (T_{mix}) and enthalpy (h_{mix}) of the mixed air can be obtained using the following equations:

$$h_{mix} = \frac{\dot{m}_{ai}h_i + \dot{m}_{ao}h_{out}}{\dot{m}_{ai} + \dot{m}_{ao}} \tag{9.15}$$

$$T_{mix} = \frac{\dot{m}_{ai}T_i + \dot{m}_{ao}T_o}{\dot{m}_{ai} + \dot{m}_{ao}} \tag{9.16}$$

The chiller load ($Q_{chiller}$) is a function of mixed and supply air temperatures, which is assessed as

$$Q_{chiller} = \dot{m}_a(h_1 - h_s) \tag{9.17}$$

The refrigerant mass flow rate for the desired effect is determined by

$$\dot{m}_{ref} = \frac{Q_{chiller}}{(h_a - h_d)} \tag{9.18}$$

Since the compressor is the primary energy-consuming component in a VCR system, the energy consumed by the compressor is assumed to be the energy

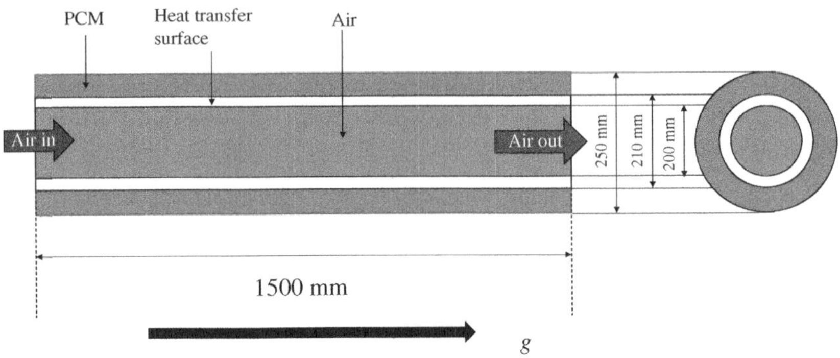

Figure 9.12 Schematic of the tube heat exchanger configuration with PCM in the outer ring.

consumed by the system. The work required by the compressor can be calculated using the following equation:

$$W_{comp} = \dot{m}_{ref}(h_{bs} - h_d) \tag{9.19}$$

The efficiency of the compressor depends on the operating pressures [46]. Therefore, the work required is calculated accordingly,

$$W_{ac} = {}^{W_{comp}}\!/_{\eta_{comp}} \tag{9.20}$$

9.6.1 Model for PCM Air Heat Exchanger

The study involves the design of a shell-and-tube latent heat energy storage system (LHESS) to cool fresh air provided to an HVAC system. The LHESS has a dimension of 1500 mm in length, and, air is directed through the inner pipe with a radius of 100 mm, while the PCM is located in the outer ring with a width of 20 mm, as depicted in Figure 9.12. The specific PCM chosen for this investigation is calcium chloride hexahydrate, which melts at 303 K.

Table 9.1 presents the thermophysical properties of PCM, air, and aluminum. After passing through the exchanger, the air is combined with some recirculated air and routed toward the A.C. To maintain a working environment temperature of 300 K, the A.C employs a vapor compression refrigeration cycle (VCRC) with a compression ratio of 5.

9.6.2 Assessing the Influence of Heat Exchanger on the Energy Savings of the Air-Conditioning System

To analyze the transient heat transfer and fluid flow within the heat exchanger, a three-dimensional axisymmetric numerical model is employed. The model incorporates the phase change behavior of the PCM using an enthalpy porosity

Table 9.1 Thermophysical properties of PCM, air and aluminum.

Properties	$CaCl_2 \cdot 6H_2O$ [47]	AlumAluminum	Air
k (W/m/K)	0.546	202.4	0.242
ρ (kg/m^3)	1538	2719	1.225
C_p (kJ/kg K)	2.230	8.710	1.006
μ (kg/m s)	0.1	—	1.789×10^{-5}
Latent heat (kJ/kg)	170	—	—
T_m (K)	303	—	—
β (K)	0.0005	—	—

approach on a fixed grid. Several assumptions are made during the development of the computational model:

(a) The phase transformation takes place at 303 K.
(b) The effects of contraction and expansion during the melting processes and solidification are neglected.
(c) The thermophysical properties of the PCM, air, and pipe material are considered isotropic and temperature invariant, except for the density of the PCM, which varies along the direction of gravity (z-direction).

The continuity, momentum, and energy conservation equations are formulated in vector form as follows [48]:

Continuity equation:

$$\frac{\partial(\rho_i)}{\partial t} + \nabla.(\rho_i V_i) = 0 \tag{9.21}$$

Momentum equation:

$$\frac{\partial(\rho_i V_i)}{\partial t} + V_i \nabla.(\rho_i V_i) = -\nabla P + \nabla^2[\mu V_i] + AV_i + S_z \tag{9.22}$$

Energy equation:

$$\frac{\partial(\rho_i c_{p,i} V_i)}{\partial t} + V_i \nabla.(\rho_i c_{p,i} V_i) = \nabla^2[k_i T_i] + S_{TP} \tag{9.23}$$

The momentum equation for PCM includes a source term that incorporates the Boussinesq approximation in the z-direction, assuming no variation in thermophysical properties except for the density of the fluid. This term can be expressed as

$$S = -A\frac{(1-\beta)^2}{\delta^3 + b} \tag{9.24}$$

Here, A is the mushy zone parameter and β represents the thermal expansion constant of the PCM.

Initial and boundary conditions:

The initial inlet air temperature condition is expressed as

$$T(r, \theta, z, 0) = 315\ \text{K} \tag{9.25}$$

To solve the mass, momentum, and energy conservation equations, appropriate boundary conditions are applied and stated as follows:

(a) At the wall of the tube, no slip condition, i.e., $u = 0$, for air and PCM.
(b) At the inlet for air, i.e., at $z = 0$, temperature = 315 K and velocity = $u_{in} = 6$ m/s
(c) At the outlet for air, $p = 0$.
(d) For PCM:
  a. At sidewall of shell, $\left.\frac{\partial T}{\partial n}\right|_{shell\ sidewall} = 0.$
  b. At inner wall of shell, $\left.\frac{\partial T}{\partial r}\right|_{r=\frac{D_i}{2}} = 0.$
(e) Symmetric boundary condition at $\theta = -30°$ and $\theta = 30°$, i.e., $\left.\frac{\partial \phi}{\partial n}\right|_{sym} = 0$ and $\phi \cdot \vec{n} = 0$.

n is the unit vector, and Φ represents the velocity or temperature.

The Finite Volume Method (FVM) is utilized to solve the governing Eqs. (9.21), (9.22), and (9.23) iteratively, and the SIMPLE scheme is employed for pressure–velocity coupling. The power law is used to solve the spatial discretization of momentum and energy equations. Ansys Fluent V2021, a commercial software, is used with appropriate initial and boundary conditions with convergence conditions of 10^{-3}, 10^{-3}, and 10^{-6} for the continuity, momentum, and energy equations, respectively.

The developed model is validated by comparing with experimental data from the work of Agyenim et al. [49] with erythritol (Tm = 390 K) for a horizontal concentric tube heat exchanger. The temperature variation with time for erythritol is displayed in Figure 9.13 and compared with the numerical findings from the model, indicating that the current predictions and experimental data are in good agreement.

An analysis of grid independence is conducted to determine the optimal grid size for the LHESS. Three different grid sizes are examined, with the coarse grid consisting of 354,806 cells, while the fine and finer grids consist of 685,350 and 1,211,560 cells, respectively. Figure 9.14 represents the output air temperature over time in the computational domain for different mesh sizes. When the number of cells is increased from 354,806 (coarse grid) to 685,350 (fine grid), the maximum variation in the output air temperature is 0.16 °C. This variation further decreases to 0.014 °C when the number of cells is increased from 685,350 to 1,211,560 (finer grid). Based on this analysis, the fine grid can be deemed suitable for further investigation. To ensure accuracy, a time step of 0.1seconds

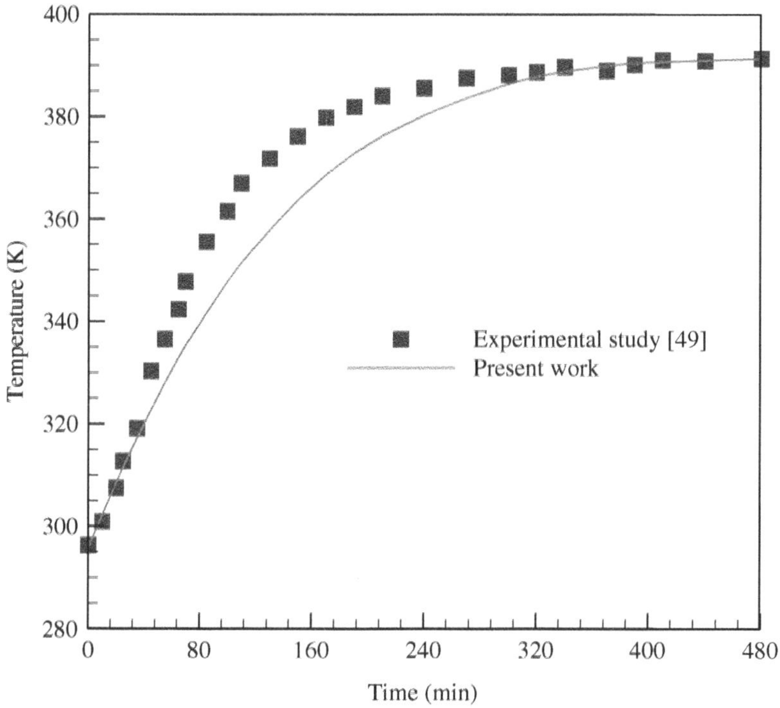

Figure 9.13 Model validation for PCM temperature using the experimental study of Agyenim et al. [49]. Source: Adapted from Ref. [49].

is chosen since diminishing differences in temperature distribution within the LHESS are observed as the time step is reduced.

9.6.3 Observations and Findings

Figure 9.15 illustrates the changes in the melt fraction of the PCM over time. In the case where the PCM is located in the annulus, melting begins at 186 seconds of operation, and after 3600 seconds, approximately 93.5% of the PCM has melted. The melting process is a result of the energy absorbed from the air flowing through the tube, with an inlet temperature of 315 K. During the initial 1200 seconds of operation, the rate of melting is relatively slow. However, a sudden increase in the melting rate occurs thereafter. This abrupt change is attributed to the enhanced heat transfer within the melted PCM region, facilitated by convection currents. Prior to the 1200-second mark, only the PCM adjacent to the wall has melted, limiting the available domain for convection currents to effectively transfer heat.

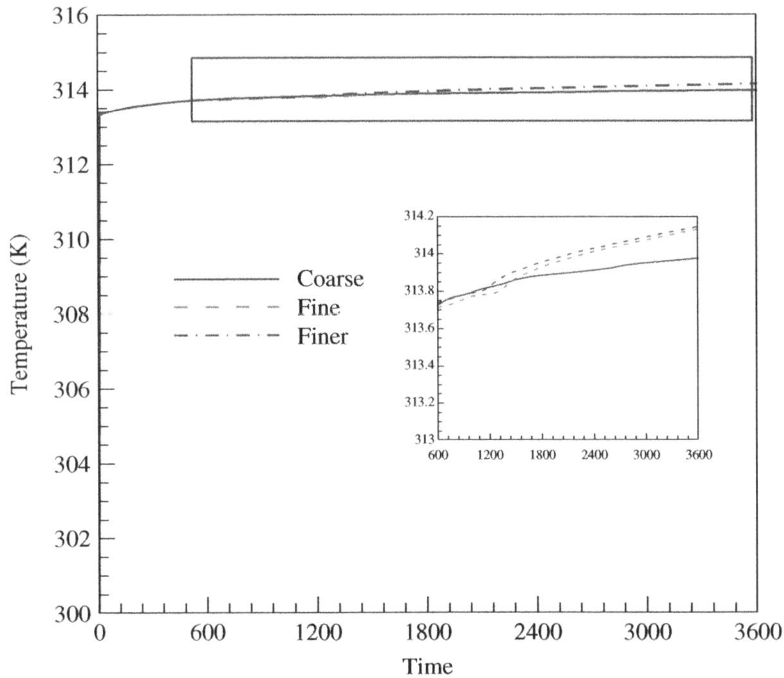

Figure 9.14 Grid independence study.

Figure 9.16 displays the contour plot of the melt fraction for PCM in the LHESS. Initially, during the early stages of operation, conduction plays a more significant role in heat transfer compared to convection. However, as the PCM begins to melt, the impact of heat transfer due to convection becomes dominant as well as surpasses the heat transmission due to conduction over time. As the PCM melts, it experiences upward buoyancy due to temperature differences, and heat transfer occurs through natural convection. The contour plot shows that the melt fraction of PCM is 54.5% after 1800 seconds of operation which increases to 93.3% after 3600 seconds.

9.6.3.1 Air Outlet Temperature

The temperature distribution of the air is visualized in Figure 9.17, specifically at a cross-section positioned at the outlet. The contours indicate that the air temperature encounters a substantial drop near the surface in contact with the PCM. This temperature drop can reach up to 5 K. The average reduction in the temperature after 3600 seconds, 1800 seconds, and 600 seconds of operation at the outlet is found to be 1.13 K, 1.31 K, and 1.63 K lower than the inlet air temperature, respectively.

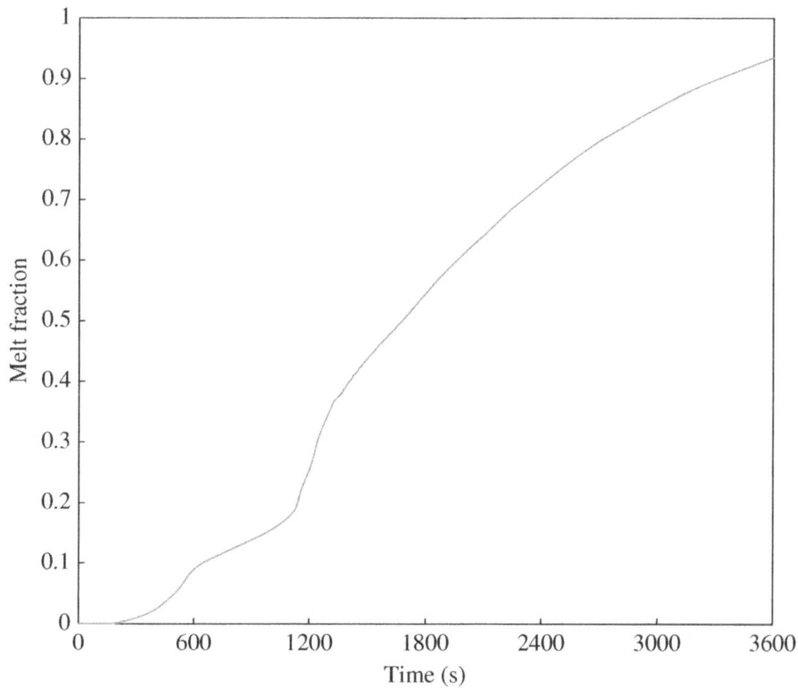

Figure 9.15 Variation of melt fraction of PCM with time.

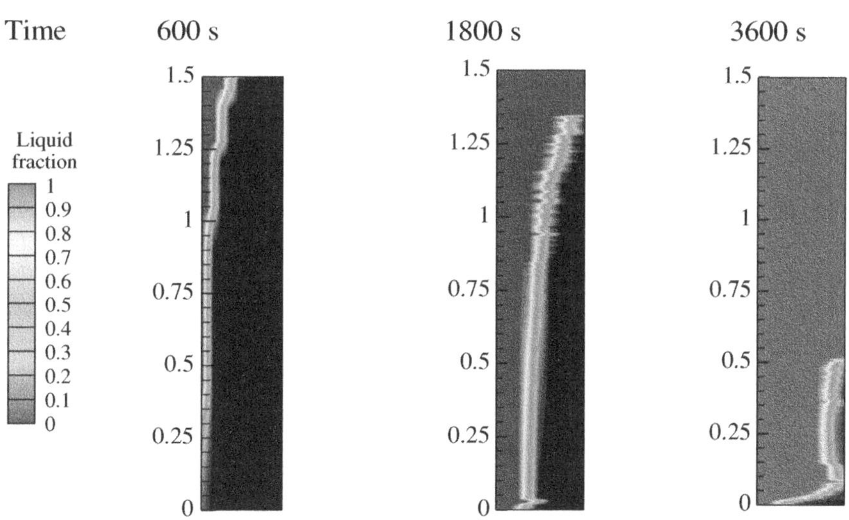

Figure 9.16 Distribution of melted PCM in the heat exchanger at various time instants.

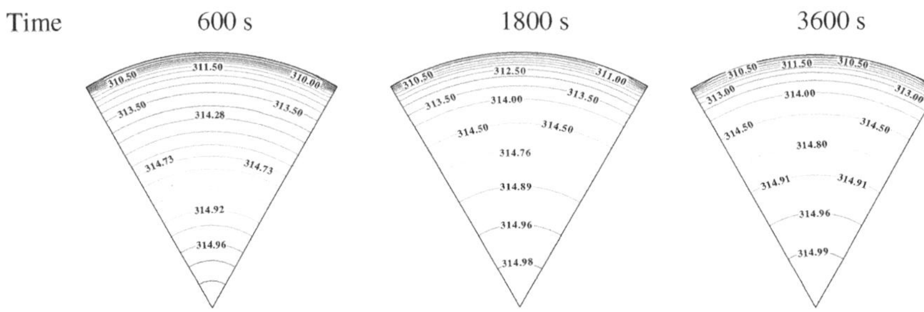

Figure 9.17 Air temperature distribution at the outlet of the heat exchanger at different time intervals.

The change in the temperature of air obtained at the outlet of the heat exchanger over is depicted in Figure 9.18. Initially, the incoming air at the inlet is assumed to flow at a temperature of 315 K. It transfers its thermal energy to the PCM and exits the heat exchanger at a lower temperature from the opposite end. Since a significant amount of thermal energy is stored in the form of latent heat, lesser variations in the air outlet temperature are observed. The average air outlet temperature is observed to increase with time and tends to move toward the air inlet temperature. Furthermore, it can be observed that the temperature difference is greater in the initial phase of operation and tends to reduce toward the end of operation. It is mainly due to the high-temperature gradient between the air and PCM in the initial stage due to higher heat transfer. However, with time, the temperature of the PCM increases. The gradient reduces, and, hence, the heat transfer rate also reduces, leading to a lower drop in the air temperature at the outlet.

9.6.3.2 Energy Saving

The current study focuses on the utilization of a heat exchanger to cool the ambient air, which is then supplied to an HVAC unit. In vapor compression refrigeration systems (VCRS), the energy consumed by the compressor accounts for a substantial fragment of the total energy consumption. To simplify the analysis, the total energy consumed by the HVAC unit is equivalent to the energy consumed by the compressor. Figure 9.18 illustrates the energy savings by the retrofitted system over time. During the first 100 seconds, energy savings are not considered as the maximum heat is stored in the form of sensible heat in the solid PCM.

Figure 9.16 also suggests that energy savings are dependent on the temperature at the outlet of the heat exchanger. A greater temperature drop results in higher energy savings. The energy savings remain relatively constant when the heat exchanger utilizes PCM in the annulus, exhibiting minimal fluctuations. Initially, the energy savings amount to 3.57%. However, over the course of 3600 seconds, the energy savings gradually decrease to 2.75%.

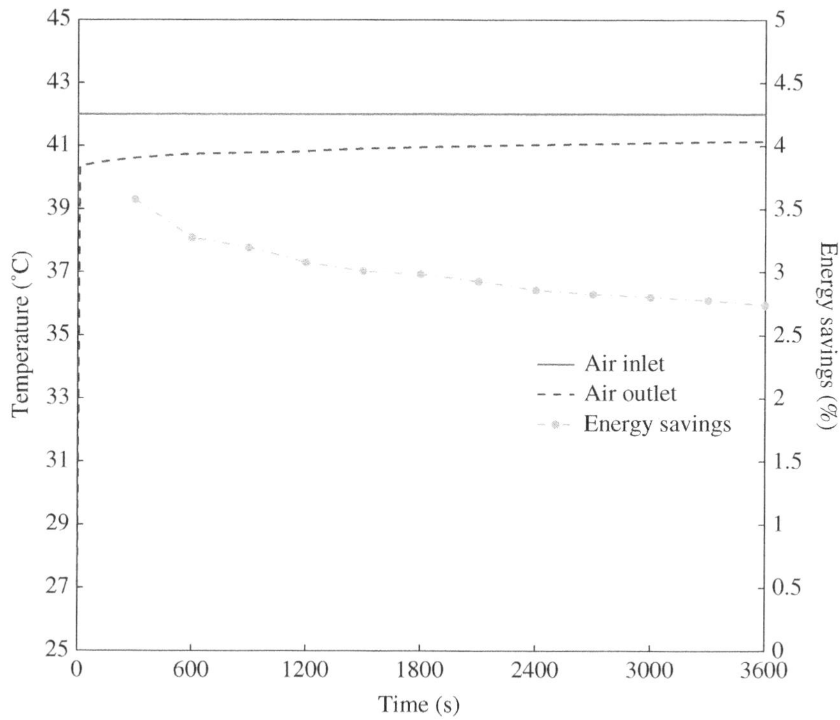

Figure 9.18 Variation of air temperature at the outlet of the heat exchanger and energy saving by the retrofitted HVAC system.

9.7 Conclusions and Future Recommendations

The thermal performance of a PCM-incorporated roof is established with the long-term goal of developing one as a passive system for building applications. Followed by the analysis of an HVAC system retrofitted with the designed LHESS. In summary, the current research demonstrating the potential of PCM in building applications can be useful in the conceptual stage of the climate-responsive building design. It can be a reference for building designers, architects, and researchers in India working on energy-efficient buildings. The following are the key components of the current investigations:

- The position of the PCM layer in the building's roof for the reduction of solar heat gain is examined. According to the results, the PCM layer placed on the exterior of the roof surface performs better than the exterior plaster surface beneath. In hot climates or in summer near its melting point, the BioPCM29 melting range provides lower room temperatures, resulting in less solar heat gain by the building.

- The location of the PCM and the local weather greatly impact how well the integrated PCM functions on the roof. According to a current study, an outer location is preferred for summer or hot climates but is unsuitable for cold or winter. PCM can recommend a varied placement position for the different regions or seasons.
- The ideal PCM thickness for attaining substantial building performance is around 10 mm as adding PCM adds needless expenditure without providing much of a benefit in moderating indoor temperatures.
- The thermal comfort conditions of occupants are notably affected by the incorporation of PCM in buildings. To investigate this impact, a simulation study was conducted using BioPCM 29 for both peak summer and winter months, representing warmer and cooler environments, respectively. Under the given conditions, the thickness of the PCM had only a minimal effect on enhancing indoor O.T. Therefore, investing in PCMs with higher thicknesses may not be a sustainable option.
- On the other hand, the location of PCM placement had a profound impact on its performance. In regions with warmer climates, applying PCM on the external surface of the building was found to be more effective, while in regions with colder climates, sandwiching PCM between other construction layers was preferred. Consequently, the proper placement of PCM is crucial and should be carried out with careful consideration.
- In addition to direct incorporation into the building, PCM-based LHESS is an approach to reduce the energy consumption of a building. An A.C. system is retrofitted with PCM-based LHESS. The utilization of PCM in the outer ring leads to a notable reduction in the outlet air temperature, averaging a decrease of 1.13 K over a 3600-second operational period. When this heat exchanger is integrated into a VCR-based AC system, it yields significant energy savings, with maximum savings of 3.57% and an average of 3% over the same 3600-second duration. This innovative technology offers an efficient solution to address the increasing energy consumption associated with A.C. systems.
- For regions with a composite climate, an optimization study should be conducted to maximize the effectiveness of PCM incorporation based on its specific placement and thickness. Additionally, it was noted that although PCM alone may not fully sustain the desired comfort temperature, it does facilitate the attainment of comfortable conditions with substantially reduced energy consumption. This is evident from the narrower discrepancy between the indoor O.T. and the comfortable temperature range observed in buildings incorporating PCM compared to conventional buildings without PCM.

References

1 Zahir, M.H., Irshad, K., Shafiullah, M. et al. (2023). Challenges of the application of PCMs to achieve zero energy buildings under hot weather conditions: a review. *Journal of Energy Storage* 64: 107156. https://doi.org/10.1016/J.EST.2023.107156.

2 Auzeby, M., Wei, S., Underwood, C. et al. (2017). Using phase change materials to reduce overheating issues in U.K. residential buildings. *Energy Procedia* 105: 4072–4077. https://doi.org/10.1016/J.EGYPRO.2017.03.861.

3 Kuznik, F., Virgone, J., and Johannes, K. (2011). In-situ study of thermal comfort enhancement in a renovated building equipped with phase change material wallboard. *Renewable Energy* 36: 1458–1462. https://doi.org/10.1016/J.RENENE.2010.11.008.

4 Kuznik, F., David, D., Johannes, K., and Roux, J.J. (2011). A review on phase change materials integrated in building walls. *Renewable and Sustainable Energy Reviews* 15: 379–391. https://doi.org/10.1016/J.RSER.2010.08.019.

5 Sanjay, A., Saikia, P., and Rakshit, D. (2020). Thermal energy performance of an academic building with sustainable probing and optimization with evolutionary algorithm. *Thermal Science and Engineering Progress* 17: 100374. https://doi.org/10.1016/j.tsep.2019.100374.

6 Ali, S.F., Sharma, L., Rakshit, D., and Bhattacharjee, B. (2020). Influence of passive design parameters on thermal comfort of an office space in a building in Delhi. *Journal of Architectural Engineering* 26: https://doi.org/10.1061/(asce)ae.1943-5568.0000406.

7 Castell, A. and Farid, M.M. (2014). Experimental validation of a methodology to assess PCM effectiveness in cooling building envelopes passively. *Energy and Buildings* 81: 59–71. https://doi.org/10.1016/j.enbuild.2014.06.011.

8 Al-Yasiri, Q. (2021). Incorporation of phase change materials into building envelope for thermal comfort and energy saving: a comprehensive analysis. *Journal of Building Engineering* 36: https://doi.org/10.1016/j.jobe.2020.102122.

9 Mourid, A. and El Alami, M. (2017). Thermal behavior of a building provided with phase-change materials on the roof and exposed to solar radiation. 139: 1–10. https://doi.org/10.1115/1.4037905.

10 Prabhakar, M., Saffari, M., De Gracia, A., and Cabeza, L.F. (2020). Improving the energy efficiency of passive PCM system using controlled natural ventilation. *Energy and Buildings* 228: 110483. https://doi.org/10.1016/j.enbuild.2020.110483.

11 Biplab, K. and Rakshit, D. (2017). Comparative assessment of thermal comfort with insulation and phase change materials utilizations in

building roofs and walls. *Advanced Materials Proceedings* 2: 393–397. https://doi.org/10.5185/amp.2017/609.

12 Liu, C., Wu, Y., Li, D. et al. (2017). Effect of PCM thickness and melting temperature on thermal performance of double glazing units. *Journal of Building Engineering* 11: 87–95. https://doi.org/10.1016/j.jobe.2017.04.005.

13 Alam, M., Jamil, H., Sanjayan, J., and Wilson, J. (2014). Energy saving potential of phase change materials in major Australian cities. *Energy and Buildings* 78: 192–201. https://doi.org/10.1016/j.enbuild.2014.04.027.

14 Osterman, E., Butala, V., and Stritih, U. (2015). PCM thermal storage system for 'free' heating and cooling of buildings. *Energy and Buildings* 106: 125–133. https://doi.org/10.1016/J.ENBUILD.2015.04.012.

15 Yanbing, K., Yi, J., and Yinping, Z. (2003). Modeling and experimental study on an innovative passive cooling system—NVP system. *Energy and Buildings* 35: 417–425. https://doi.org/10.1016/S0378-7788(02)00141-X.

16 Memon, S.A. (2014). Phase change materials integrated in building walls: a state of the art review. *Renewable and Sustainable Energy Reviews* 31: 870–906. https://doi.org/10.1016/j.rser.2013.12.042.

17 Liu, L., Hammami, N., Trovalet, L. et al. (2022). Description of phase change materials (PCMs) used in buildings under various climates: a review. *Journal of Energy Storage* 56. https://doi.org/10.1016/j.est.2022.105760.

18 Hassan, A.M., Lee, H., and Oh, S. (2016). Challenges of passive cooling techniques in buildings: a critical review for identifying the resilient technique. *Jurnal Teknologi* 78: 149–162. https://doi.org/10.11113/jt.v78.5748.

19 Souayfane, F., Fardoun, F., and Biwole, P.H. (2016). Phase change materials (PCM) for cooling applications in buildings: a review. *Energy and Buildings* 129: 396–431. https://doi.org/10.1016/j.enbuild.2016.04.006.

20 Pomianowski, M., Heiselberg, P., and Zhang, Y. (2013). Review of thermal energy storage technologies based on PCM application in buildings. *Energy and Buildings* 67: 56–69. https://doi.org/10.1016/j.enbuild.2013.08.006.

21 Osterman, E., Tyagi, V.V., Butala, V. et al. (2012). Review of PCM based cooling technologies for buildings. *Energy and Buildings* 49: 37–49. https://doi.org/10.1016/j.enbuild.2012.03.022.

22 Al-Absi, Z.A., Isa, M.H.M., and Ismail, M. (2020). Phase change materials (PCMs) and their optimum position in building walls. *Sustainability* 12. https://doi.org/10.3390/su12041294.

23 Shi, X., Memon, S.A., Tang, W. et al. (2014). Experimental assessment of position of macro encapsulated phase change material in concrete walls on indoor temperatures and humidity levels. *Energy and Buildings* 71: 80–87. https://doi.org/10.1016/j.enbuild.2013.12.001.

24 Memon, S.A., Cui, H.Z., Zhang, H., and Xing, F. (2015). Utilization of macro encapsulated phase change materials for the development of thermal energy storage and structural lightweight aggregate concrete. *Applied Energy* 139: 43–55. https://doi.org/10.1016/j.apenergy.2014.11.022.

25 Mukherjee, A., Deepmala, Srivastava, P., and Sandhu, J.K. (2021). Application of smart materials in civil engineering: a review. *Materials Today Proceedings* https://doi.org/10.1016/j.matpr.2021.03.304.

26 Sawadogo, M., Duquesne, M., Belarbi, R. et al. (2021). Review on the integration of phase change materials in building envelopes for passive latent heat storage. *Applied Sciences* 11. https://doi.org/10.3390/app11199305.

27 Chaiyat, N. (2015). Energy and economic analysis of a building air-conditioner with a phase change material (PCM). *Energy Conversion and Management* 94: 150–158. https://doi.org/10.1016/J.ENCONMAN.2015.01.068.

28 Kumar Sharma, R., Yagnamurthy, S., and Rakshit, D. (2022). Energy analysis of a phase change material embedded heat exchanger for air conditioning load reduction in different Indian climatic zones. *Sustainable Energy Technologies and Assessments* 53: 102776. https://doi.org/10.1016/J.SETA.2022.102776.

29 Sovetova, M., Memon, S.A., and Kim, J. (2019). Thermal performance and energy efficiency of building integrated with PCMs in hot desert climate region. *Solar Energy* 189: 357–371. https://doi.org/10.1016/j.solener.2019.07.067.

30 Soares, N., Costa, J.J., Gaspar, A.R., and Santos, P. (2013). Review of passive PCM latent heat thermal energy storage systems towards buildings' energy efficiency. *Energy and Buildings* 59: 82–103. https://doi.org/10.1016/j.enbuild.2012.12.042.

31 Ramakrishnan, S., Wang, X., Sanjayan, J., and Wilson, J. (2017). Thermal performance of buildings integrated with phase change materials to reduce heat stress risks during extreme heatwave events. *Applied Energy* 194: 410–421. https://doi.org/10.1016/j.apenergy.2016.04.084.

32 Edition SI. *Fundamentals*. 2017.

33 Occupant Thermal Comfort: Engineering Reference — EnergyPlus 9.4. https://bigladdersoftware.com/epx/docs/9-4/engineering-reference/occupant-thermal-comfort.html#fanger-comfort-model (accessed 22 March 2022).

34 ASHRAE (2010). *ANSI/ASHRAE Standard 55-2010, Thermal Environmental Conditions for Human Occupancy*, 1ste. Atlanta: American Society of Heating, Refrigerating and Air-conditioning Engineers, Inc. https://doi.org/ISSN 1041-2336.

35 Albatayneh, A., Alterman, D., Page, A., and Moghtaderi, B. (2019). The significance of the adaptive thermal comfort limits on the air-conditioning loads in a temperate climate. *Sustainability* 11: 328. https://doi.org/10.3390/su11020328.

36 Ali, S.F. and Rakshit, D. (2020). *Utilising Passive Design Strategies for Analysing Thermal Comfort Levels inside an Office Room Using PMV-PPD Models. Solar Energy: Systems, Challenges, and Opportunities. Energy, En*, 35–57. Singapore: Springer, https://doi.org/10.1007/978-981-15-0675-8_4.

37 Nicol, J.F. and Humphreys, M.A. (2002). Adaptive thermal comfort and sustainable thermal standards for buildings. *Energy and Buildings* 34: 563–572. https://doi.org/10.1016/S0378-7788(02)00006-3.

38 Manu, S., Shukla, Y., Rawal, R. et al. (2014). *An Introduction to the India Model for Adaptive (Thermal) Comfort IMAC 2014*. Ahmedabad, Gujarat: CEPT University.

39 Santamouris, M. (2006). Adaptive Thermal Comfort and Ventilation, Ventilation Information Paper n° 12.

40 Manu, S., Shukla, Y., Rawal, R. et al. (2016). Field studies of thermal comfort across multiple climate zones for the subcontinent: India Model for Adaptive Comfort (IMAC). *Building and Environment* 98: 55–70. https://doi.org/10.1016/j.buildenv.2015.12.019.

41 Bureau of Energy Efficiency (2017). *Energy Conservation Building Code 2017*. New Delhi: Bureau of Energy Efficiency.

42 Kapur, V., Agarwal, B., Baliga, G. et al. (2020). *ISHRAE COVID-19 Guidance Document for Air Conditioning and Ventilation*, 15. Indian Society of Heating, Refrigerating & Air Conditioning Engineers.

43 Washington State Department of Health (2020). Ventilation and Air Quality for Reducing Transmission of COVID-19;19:1–4.

44 Indian Society of Heating Refrigerating & Air Conditioning Engineers (ISHRAE), "Draft of Indoor Environmental Quality Standard," 2019. [Online]. Available: https://ishrae.in/Content/Download/ISHRAE_IEQ_Feb_26_2019_public_draft.pdf.

45 Refrigeration and Air Conditioning. https://www.expresslibrary.mheducation.com/pdfreader/refrigeration-air-conditioning50163303 (accessed 18 May 2021).

46 Barthwal, M., Dhar, A., and Powar, S. (2021). The techno-economic and environmental analysis of genetic algorithm (GA) optimized cold thermal energy storage (CTES) for air-conditioning applications. *Applied Energy* 283: 116253. https://doi.org/10.1016/J.APENERGY.2020.116253.

47 Das, D., Sharma, R.K., Saikia, P., and Rakshit, D. (2021). An integrated entropy-based multi-attribute decision-making model for phase change material selection and passive thermal management. *Decision Analytics Journal* 1: 100011. https://doi.org/10.1016/J.DAJOUR.2021.100011.

48 Kumar, A. and Saha, S.K. (2018). Latent heat thermal storage with variable porosity metal matrix: a numerical study. *Renewable Energy* 125: 962–973. https://doi.org/10.1016/J.RENENE.2018.03.030.

49 Agyenim, F., Eames, P., and Smyth, M. (2009). A comparison of heat transfer enhancement in a medium temperature thermal energy storage heat exchanger using fins. *Solar Energy* 83: 1509–1520. https://doi.org/10.1016/J.SOLENER.2009.04.007.

Index

Note: Page numbers in *italics* indicate figures and **bold** page numbers indicate tables in the text

Advancements in Non-Conventional Cooling and Thermal Storage Strategies, First Edition.
Edited by Bidyut Baran Saha and Dibakar Rakshit.

b

c

u

Printed and bound by CPI Group (UK) Ltd, Croydon, CR0 4YY

17/09/2026

14964309-0003